Chatter and Machine Tools

Brian Stone

Chatter and Machine Tools

Springer

Brian Stone
University of Western Australia
Perth, WA
Australia

Additional material to this book can be downloaded from http://extras.springer.com

ISBN 978-3-319-05235-9 ISBN 978-3-319-05236-6 (eBook)
DOI 10.1007/978-3-319-05236-6
Springer Cham Heidelberg New York Dordrecht London

Library of Congress Control Number: 2014940061

Printed on acid-free paper

Springer is part of Springer Science+Business Media (www.springer.com)

To Beryl my wife of 50 years
'Who can find a virtuous and capable wife?
She is more precious than rubies.'
Proverbs 31v10 (NLT)

Preface

The Genesis of the Book

So it was arranged, Prof. H. Sato and Dr. T. Senba were coming from Japan to visit me in Western Australia. Dr. Senba had been before but I never thought to see Prof. Sato in Perth. They were arriving late on September 2, 2007 and though staying in a hotel had agreed to stay one night at my home. This was an unexpected pleasure. There was something important in the air. As it turned out Prof. Sato had come to request that I should agree to write a book with him on vibration and machine tools. Writing a book was not something that I had contemplated but how could I refuse as he had come so far?

It was one of those providences that arranged for me to first meet and talk with Prof. Sato. We had both attended the CIRP Annual Assembly in Bruge, Belgium in 1982. I knew him by sight but there we were sitting opposite each other in a train on the way to the airport. We got on well and had much in common. Over the years we met at CIRP conferences and then when I moved to Australia we met more often. The request was for each of us to write one half of a book on vibration and machine tools. Professor Sato was concerned that much experience was in danger of being lost and wished to record it for posterity. In the event the book became too large and so my part was submitted on its own and resulted in the book before you. Professor Sato's book is yet to be published.

The Objective

It is very important to stress that this book is about preventing chatter from occurring in machining processes. It is not about forced vibration that arises from cyclic excitation but about an unstable oscillation that grows with time and that is most unwanted. The objective is to stop chatter from starting. This means that the vibration is not just reduced in amplitude but eliminated completely, it never starts. Others have had an apparent goal of modelling the machining process more and more accurately and even modelling what happens after chatter starts. This approach does not often result in major advances in chatter avoidance as the models begin to be as complex as the real process and do not immediately suggest major innovations. The alternative approach adopted in this book is to use simpler models that, while noting their limitations, give an overview and do allow for more frequent innovations.

So why another book on machine tool vibration, and in particular chatter? The author combines the following attributes:

- He was a young researcher when the foundational advances relating to machine tool vibration were made and has a particular insight into those early days.
- He applied his research expertise to industry problems and found novel solutions.
- He is familiar with and able to be critical of the progress or otherwise that has been made since then.
- He has many years of experience dealing with real vibration problems and has been careful to listen to experienced workshop practitioners. There is an immense amount of practical knowledge that is not apparently known to many current research students.
- He is aware of solutions to machine tool vibration that appear to have been long forgotten.
- He has made contributions to reducing machine tool chatter that cover turning, milling, drilling and grinding. This range of applications is not common in most texts on machine tool chatter.
- He has written many animation programs that can help with the comprehension of what to many is a complex and difficult topic.

The book is concerned with the suppression of chatter, that is, preventing it from starting. The objective is to present sufficiently complex models of chatter that allow an understanding to be gained of how a particular solution works. As already noted it is possible to develop models that are so complex that solutions cannot be understood or even imagined. The test of the approach adopted is whether the solutions have been found to work in practice. As this is the case, the simplifying assumptions have been justified as a means of finding solutions.

It is hoped that those who read this book will join in the goal of stopping chatter in machining operations with the result of improved surface finish, longer tool and grinding wheel life and significant reductions in noise.

Acknowledgments

The author is indebted to many people. He has been blessed with many outstanding, diligent and productive students. He has also been helped on his way by many interactions with fellow researchers.

Dr. Rodney Entwistle was a part-time doctoral student with me while being the Head of Mechanical Engineering at Curtin University of Technology in Western Australia. The chapter on grinding in this book has much that comes with his permission from his Ph.D. thesis. He also read through the first draft of the book and it is much improved because of his comments. However any remaining deficiencies in the book are not his responsibility but mine.

The role of Prof. Sato in the genesis of the book has already been mentioned. However, I am much indebted to the interactions I have had with other colleagues in Japan. Western Australia has little manufacturing and no manufacturer of machine tools. Many of my visits to Japan have been organised by Prof. Y. Ito and Dr. T. Senba. Those visits have been most helpful in introducing me to many machine tool manufacturers in Japan.

We all have to start somewhere and so I am greatly in debt to Prof. C. Andrew who accepted me as one of his doctoral students. As the research was sponsored by MTIRA it was of practical application. The Director of MTIRA, Mr. A. E. DeBarr convinced me, after my Ph.D., to continue in the same area of work and so I had the opportunity to gain a great deal of practical experience with the member companies of MTIRA. Also on staff at MTIRA were many experienced professionals. One of these, Mr. Harold Barrow, was my sounding board for any new idea. He would immediately be able to say if it would be practically possible. Access to such experience saved a great deal of time that could have been spent re-inventing what was already known. Also, it saved me from displaying my ignorance of what it was really like on the shop floor.

Some of the work described in this book was conducted while at MTIRA and was published in confidential reports to the member companies. As MTIRA no longer exists I wish to thank Mr. Andy de Vicq for his permission to publish work on the characteristics of bearings and his opinion that the work on bi-helix cutters might be published without affecting the interests of the member companies.

I have had many doctoral students and some of these have worked on projects concerned with chatter. Dr. J. S. Sexton was the first and investigated the possibility of using varying speed to prevent chatter. Later, as a postdoc he worked on flexible grinding wheels in collaboration with DeBeers Diamonds. Dr. T. L. H. Walford made groundbreaking measurements of the stiffness and damping in rolling element bearings under working conditions. This work was extended by two undergraduate students, Richard Lambert and now Prof. A. Pollard. Dr. R. J. Baylis was involved with the early work on chatter in grinding, particularly growth rates. Dr. Simon Drew enabled the forces resulting from torsional vibration in grinding to be investigated after Dr. R. D. Entwistle had shown that torsional vibration should not be ignored.

The measurement of grinding forces under oscillating conditions would not have been possible without the collaboration of the National University of Singapore. Professor Andrew Nee encouraged this collaboration with Dr. M. A. Mannan and his Master's students K. L. Ong and R. A. Qureshi.

My last study leave was spent in the UK looking at chatter in centreless grinding. This was with my friend of many years Dr. T. R. A. Pearce, who was my colleague at the University of Bristol. Centreless grinding was a mystery to me until this collaboration.

Finally but not least, my thanks are due to Dr. N. W. Scott. Nathan was a doctoral student with me but not in the area of chatter. However, he was brilliant at writing Java Applets for use in teaching vibration to students. Many of these Applets provided shells that allowed new programs to be written. All of the programs referenced from this book could not have been written without him. Nathan turned ideas into reality.

The Publishers

My thanks are due to Anthony Doyle and Garrett Ziółek of Springer UK and also Christine Velarde formerly with Springer. Compared with my previous experience of publishers I cannot thank them enough for their help that was always given extremely promptly. The same has applied to the printers and I am particularly grateful to Ravi Vengadachalam and Kavitha of Scientific Publishing Services.

Contents

Notations

General Vibration

A	Area of cross-section
c	Viscous damping coefficient
c_c	Critical viscous damping coefficient
E	Young's modulus
F	Force
G	Shear modulus
h	Hysteretic damping coefficient
I	Second moment of area
i	Complex variable. $\mathrm{i} = \sqrt{-1}$
k	Spring stiffness
m	Mass of rigid body
t	Time
x	Displacement
X	Amplitude of sinusoidal motion
α	Hysteretic damping ratio OR angle on force vector diagram
α_{pq}	Receptance $\alpha_{pq} = X_p/F_q$
β_{pq}	Receptance of system B. $\beta_{pq} = X_{bp}/F_{bq}$
γ_{pq}	Receptance of system C. $\gamma_{pq} = X_{cp}/F_{cq}$
δ	The logarithmic decrement
η	Hysteretic damping ratio
ρ	Density
ϕ	Phase angle
ξ	Viscous damping ratio. $\xi = c/2\sqrt{km}$
ω	Frequency
ω_n	Natural frequency $= \sqrt{k/m}$

Metal Cutting and Grinding

b	Width of cut
b_{lim}	Width of cut at stability boundary

c_{w}	Viscous damping coefficient for torsional vibration
d	Depth of cut
F_{r}	Regenerative force
F_{n}	Non-regenerative force
G	Grinding ratio = Volume removed from work/Volume removed from wheel
$G(\omega)$	Chatter receptance of machine tool
$G_{\mathrm{I}}(\omega)$	Imaginary part of the chatter receptance
$G_{\mathrm{R}}(\omega)$	Real part of the chatter receptance
I_{w}	Inertia of torsional mode
k_{c}	Contact stiffness in grinding
k_{w}	Stiffness of torsional mode
m_1, m_2	Slope of helical teeth
R	Cutting force coefficient. Cutting force/chip area
R_{w}	Radius of work
R_{g}	Radius of grinding wheel
z	Number of teeth on a cutter
z_{c}	Average number of teeth in cut
σ	Exponential growth rate
τ	Time for one revolution OR the time between teeth
τ_{w}	Time for one revolution of work
τ_{g}	Time for one revolution of grinding wheel
ξ_{w}	Viscous damping ratio of torsional mode
ω_{n1}	Natural frequency of first mode
ω_{n2}	Natural frequency of second mode
ω_{nw}	Natural frequency of torsional mode
Ω	Rotational speed
Ω_{w}	Rotational speed of work
Ω_{g}	Rotational speed grinding wheel

Chapter 1
Basic Chatter Theory

1.1 Introduction

It is informative to present the development of chatter theory in its historical context. Many lessons may be profitably learned from this approach. In particular,

1. Experience and experimental work are crucial.
2. Since chatter is an instability, the amplitude of vibration increases from very small amplitudes. If chatter is to be avoided, then it is necessary to stop these very small amplitudes increasing and a chatter model that assumes very small amplitudes is appropriate. The fact that large amplitudes subsequently develop and are limited by nonlinearities is a secondary matter and not relevant if chatter can be prevented from starting at all.
3. The use of high-powered computers for dynamic simulations does not necessarily result in solutions to chatter problems. The early advances were achieved without such computers.

In many of the first (1950s) publications on chatter, the earliest reference was to Arnold [1], who in 1945 presented a report at a meeting of the Institution of Mechanical Engineers in London. The work reported was for a subcommittee of the Institution that was concerned with carbide turning tools. The state of knowledge at that time is illustrated by the following quote from the introduction to the report.

> The cutting of metals is frequently accompanied by vibration, the action of which may at times become extremely violent. Few types of machine tool can claim immunity from this phenomenon, yet scientific literature appears to offer little guidance on the mechanism by which it is produced. Commonly known as "chatter", this type of vibration is often of a very complex nature and its existence is almost invariably unwelcome; not only does the vibration retard production and lead to unnecessary deterioration of machine components, but in many cases the ensuing noise causes acute discomfort to those situated in the near vicinity. This type of vibration has become of increasing importance as a result of the increase in cutting speeds made possible by the introduction and development of carbide tools, and in this respect the Committee feel justified in attempting to elucidate the problem. As the subject covers a wide field, and is one of extreme complexity, this report will be confined almost exclusively to the vibration of carbide tools when cutting rigid steel components.

B. Stone, *Chatter and Machine Tools*, DOI: 10.1007/978-3-319-05236-6_1,

The reference to the lack of guidance from the scientific literature is significant as this indicates how little was known about chatter by the scientific community. Arnold's summary of previous research reads as follows—note new reference numbers are added for this book,

> As little published research is devoted specifically to the above phenomenon a survey of previous work is necessarily limited. The early investigators of metal cutting, principally Taylor [2] and Smith [3] were not unaware of tool vibration, but the effect they studied seems unconnected with that encountered at the high speeds of modern production. Taylor suggested that the vibration was due to the variable force created by the periodic shearing action occurring as the metal was removed. Smith measured the periodic force at extremely slow cutting speeds and found that the variation decreased and became almost negligible at cutting speeds in excess of 15 ft. per min. Some years before, Brooks [4] published photographs of slow-speed cutting in which the intermittent shearing action gave place to steady flow over the tool when large top-rake angles were used.
>
> Following this early work little attention appears to have been paid to the subject until recent years. It is true that many short articles have appeared from time to time, describing how vibration was eliminated in some specific cutting process, but, valuable as these may be, they do not contribute greatly to our knowledge of the subject. The work by Doi (Japan) [5] ascribes the vibration to resonance due to fluctuations of the cutting action synchronizing with a natural frequency of the lathe. The same author [6] deals with vibration experienced when turning long bars, but records the experimental results without attempting to explain their significance.
>
> A comprehensive work entitled "Manual on Cutting of Metals" published by the American Society of Mechanical Engineers [7] devotes two pages to "chatter", of which it states "its severity undoubtedly is determined by the degree of resonance between periodic variations in the cutting force, and the natural frequencies of the work, and of the work- and tool-supporting structures". More recently Schmidt [8] has published a paper summarizing the present state of knowledge on the dynamics of machine tools as deduced from German publications. The section devoted to vibration is somewhat vague, and merely suggests the possibility of forced and self-induced vibration as a result of periodic shearing of the chips.
>
> The general impression gained from the above publications is that tool vibration is due to a recurrent variation of cutting force due to shearing of the chip. Reference is also made to resonant vibration but no satisfactory explanation it offered to account for its existence. Statements of this kind are very misleading, and it is hoped that the present report will give a more satisfactory interpretation of the problem.

It seems that Arnold was not aware of the vast amount of experience that could have been found on the shop floor from experienced machinists. In fact, Arnold devised an experimental set-up to produce chatter that was not typical of the then current practice. This is confirmed from the very informative record of the discussion that followed the presentation of his report. To quote,

> Dr. G. SCHLESINGER, M.I.Mech.E., said that the author had obviously intentionally chosen a tool which would vibrate either by the cutting action or by forced vibration or by a combination of both, the remarkable tool shape shown in Fig. 2a, p. 263, besides the ordinary knife tool of Fig. 2c, had therefore been designed. *Although quite unusual* (*underline* added), it was suitable, of course, for the object of the investigation, in conjunction with an apparatus which was exceedingly fine and sensitive (as was shown by the various diagrams).

He also stated (remembering that at that time, Schlesinger was an expert in acceptance tests for machine tools),

> The author stated that the measured vibrations of his sample tool were created by the cutting action with or without forced vibration. The tests were, of course, an interesting investigation of intentional tool vibration in the cutting of steel, but he wondered whether the conditions necessary to obtain a surface free of vibration marks when machining steel on a lathe in the ordinary way were fulfilled. The acceptance tests for machine tools made this condition compulsory, and if the author's research work could give the rules for correct design and workmanship of machine tools it would be a remarkable step forward. The author had all the apparatus required and the Appendices I–V gave the theoretical calculation of the different kinds of tool vibration and their theoretical influence on the machine parts only. He would, however, like to know if the working conditions of the tests corresponded to those which were demanded from a good lathe.

And later he is reported as saying,

> He believed that it was not permissible to draw conclusions from a theoretical cutting test, with unusual overhang and not very suitable shape of tool, or from forced vibration, from a practical test on a lathe, unless the possible deficiencies of the lathe were taken into consideration, but this point had not been mentioned in the paper.

Further comments from the discussion that are of interest include.

> Mr. O. V. S. BULLEID, M.I.Mech.E. (Vice-President), said…… He could not help feeling that some of the older turners would be rather horrified by the acceptance of chatter, and he was reminded of the old practice of bedding tools by lapping on leather or lead, which removed damping and furthermore, he thought, illustrated the fact that it was not necessarily the lathe or the tool that was responsible, but probably both…………… His main criticism of the paper was that it did not tell him (he was not highly trained in the subtleties of the subject) what to do to prevent chatter. He thought the paper should state in very clear and precise language what should be done to avoid it.

Here, we have reference to the importance of the experience of practising turners and that what is really required is to be told how to avoid chatter. This latter point was expanded by,

> Dr. R. F. J. Weil (Stafford) As was well known, tool vibration was the enemy of surface finish, and when engineers knew better the ways and nature of that enemy they would be able sooner to defeat it.

Here, we have the suggestion of one approach to defeating chatter, by gaining a better understanding (the ways and nature) of chatter. There followed two comments that showed how important experience was and that chatter had been avoided in practice.

> Mr. A. E. PARNACOTT (London) said he had been very interested, when at Liege, to see a number of lathes in which the turning tool was in the opposite position to the normal. Some remarkably good results were obtained, and he would like to know whether the author thought he was right in his interpretation. When a force was put on to a piece rotating on a film of oil, the thickness of the film of oil was varied by the pressure that was applied. Some machining difficulties could be minimized by impressing a far bigger load on the bearing than that due to the cut. He had obtained surprising results by not only

loading the bearing so as to have a very thin oil film but also adding resistance to the feed of the tool into the work.......

Here, from experience is the suggestion that the machine characteristics were important and could be changed to advantage. The idea of changing the cutting force direction (*turning tool was in the opposite position to the normal*) was also taken up by,

Mr. ALFRED HARRISON, A.I.Mech.E., commenting on Mr. Pamacott's remarks, said he had found that a great improvement could be achieved by down-cutting, either by putting the tool on the opposite side of the slide or by reversing the direction of the lathe and inverting the tool.......

As we shall find, these practical observations of real situations were significant. In fact, a much earlier publication from a practitioner had anticipated many of the more significant characteristics of chatter and ways of avoiding it. The family firm of Holtzapffel (1843–1897) produced beautiful ornamental turning lathes and attachments for their wealthy customers; many of the individually numbered machines exist today. Instruction to his clients was furnished in the five-volume work *Turning and Mechanical Manipulation. Intended as a Work of General Reference and Practical Instruction, on the Lathe, and the Various Mechanical Pursuits Followed by Amateurs.* The following quotation from volume four of the work, entitled *Hand or Simple Turning: Principles and Practice* [9], is startling.

...The flat and other finishing tools cut very readily upon brass and similar alloys, but their use presents a difficulty incidental to these materials, which are readily set in vibration by the process of turning. The vibration causes the work and the tool to "chatter" upon each other, and to leave the cylinder covered by numerous irregular, but parallel, fine lines or striae, also called "utters" by the brass turners; the latter term probably arising from the sound emitted by the work when set in vibration against the tool. The formation of the striae is still more frequent in turning the brass surface, and when they are once established upon this or the cylinder, they incite the work to increased vibration, and deepen and spread under the continued use of the turning tool; they can only be obliterated by returning to the use of the roughing tool, and recommencing the entire process.

To avoid this latter inconvenience, which would probably also unduly diminish the size of the work, the formation of striae is carefully guarded against from the commencement. A rather narrow flat tool is chosen, and this is first employed to turn the cylinder smooth, only a small portion at a time. The tool is held with considerable firmness, with its shaft presented to the work at a small horizontal angle, so that in making the separate cuts only about half the width of the cutting edge penetrates the work. So soon as the line of the work becomes fairly straight, the separate cuts are connected by traversing the tool, during which time it is more liable to "chatter"; this may be avoided in some cases, by retaining the edge during its traverse still in the same position, one corner penetrating and the other just free of the work, when either the cutting or the disengaged corner may precede, as the tool travels along the cylinder. Many brass turners for the same purpose prefer to tilt the flat and finishing tools upon one corner, and not at one constant angle as in turning hardwood, but continually moving the cutting edge up and down upon the corner on which it rests; so as to constantly vary its angular elevation or tilt, during the entire traverse. The impulses to vibration are then given to the work at slightly, but constantly, varying angles, when their effects cross, and to some extent neutralize each other.

The finishing shavings, which are thinner, may be removed with a wider tool, held also in the horizontal manner; the fingers of both hands close about the end of the tool and upon

> the rest, the degree of tilt being constantly varied as before. The diminished thickness of the cut sets up less vibration, while the fingers, being more or less interposed between the tool and the surface of the rest, act as a spring or cushion, and largely absorb or check the vibrations by associating them with the elastic frame of the operator.

Entwistle [10] seems to have been the first in recent times to have found this reference and he comments, while referring to chatter as the foe,

> Here, buried in the writings directed at amateurs, is the identification of many of the characteristics of our foe. He is unpleasantly audible and renders our surface finish degraded; increasing the width of the cut is like opening a door to welcome the tyrant in; given the chance he will feed upon himself and grow ever more awesome.
>
> But, the master craftsman has learnt how to foil many of the foe's intentions. Select a narrow tool (get below the limiting width of cut); hold the tool firmly (stiffen the structure and maximise damping); constantly vary the angle of tilt (impart a varying period to the waves on the workpiece); thus causing their effect to neutralise each other when they cross (time delayed regeneration is the cause!); the final thin cut sets up less vibration (the feed rate plays a role); allow the fingers to act as a cushion to absorb the vibrations into the elastic frame of the operator (provide damping or large flexibility).
>
> So, it seems, the weapons of war effective against our foe were known more than a century ago and probably much longer. Only in more recent times have the weapons been analysed and explained but the deployment of the weapons remains essentially unchanged; they have simply been mechanised. Bi-helix milling cutters now vary the pitch of the waves automatically. Damping inserts in boring bars introduce local damping. Workpiece speed variation foils the regeneration.

This book is directed at describing many of these solutions and how they improve chatter performance. To this end, the model used to represent any particular solution will be kept as simple as possible in order to illustrate the benefits of that particular approach. It is for Ph.D students to generate more complex models that may include secondary effects and/or multiple effects. We now return to the historical context of the development of chatter theory and consider Arnold, or as it was categorised by Tobias and Fishwick, Type B chatter.

1.2 Arnold or Type B Chatter

It is apparent that Arnold obtained chatter by using a lathe tool with an excessive overhang. He varied the frequency of chatter by changing the overhang, and much of the theory he presents relates to the bending vibration of cantilever beams as shown in Fig. 1.1. This diagram confirms that for Arnold, chatter did not involve any regeneration as the surface coming into cut has no surface waves. When there were surface markings, Arnold designated this as forced vibration.

A simple model of this situation is shown in Fig. 1.2.

If the cutting force is not constant but may vary with speed in the manner shown in Fig. 1.2b, then the force reduces with increasing surface speed. To illustrate the mechanism of Arnold chatter, a simple and approximate model of the cutting force component in the tangential direction may be assumed to be of the form,

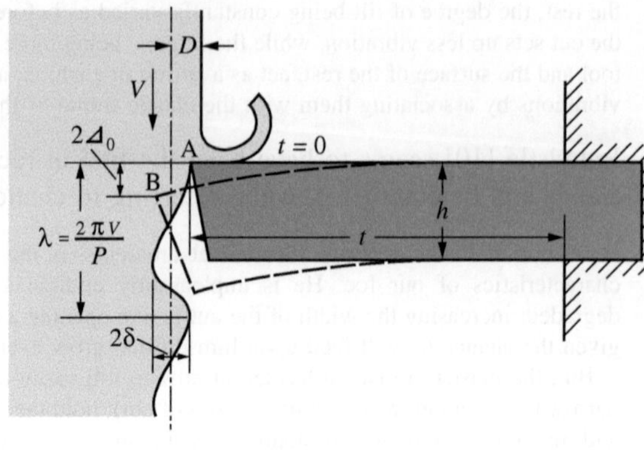

Fig. 1.1 Cantilever-type vibration of a lathe tool (after Arnold [1])

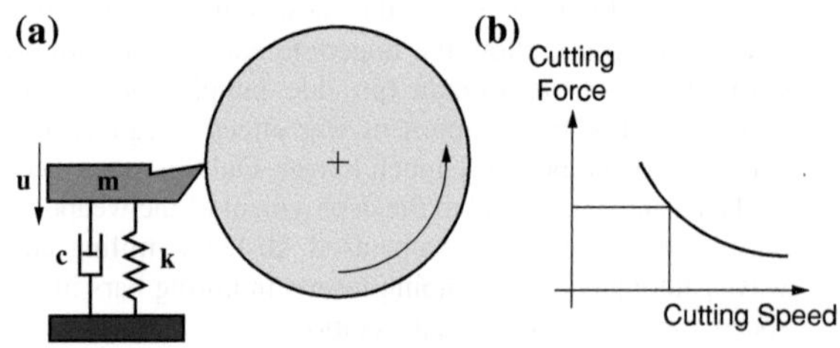

Fig. 1.2 Model of Arnold, Type B chatter

$F = bd(R - \alpha u)$, where b is the width of cut, d the depth of cut (hence, bd is the area of the undeformed chip cross section), u is the instantaneous surface speed of the work relative to the tool, and R_o and α are constants depending on numerous factors such as work material, condition and geometry of the cutting edge.

If the tool moves up and down, there is, for realistic feeds, depths of cut and small initial amplitudes, a negligible change in depth of cut so that if $x(t)$ represents the displacement of the tool in the direction of the force, the tangential cutting force is given by

$$F = bd\left(R - \alpha\left[R_{\mathrm{w}}\Omega - \frac{\mathrm{d}x(t)}{\mathrm{d}t}\right]\right) \tag{1.1}$$

where R_{w} is the work radius and Ω is the rotational speed of the work (assumed constant).

For the one degree of freedom (see Appendix A for a summary of basic vibration theory), the equation of motion is

$$m\frac{\mathrm{d}^2x(t)}{\mathrm{d}t^2} + c\frac{\mathrm{d}x(t)}{\mathrm{d}t} + kx(t) = F = bd\left(R - \alpha\left[R_{\mathrm{w}}\Omega - \frac{\mathrm{d}x(t)}{\mathrm{d}t}\right]\right) \tag{1.2}$$

The constant force term $bd(R - \alpha R_{\mathrm{w}}\Omega)$ is ignored as it causes no ongoing oscillation and the equation of motion becomes,

$$m\frac{d^2x(t)}{dt^2} + c\frac{dx(t)}{dt} + kx(t) = bd\alpha\frac{dx(t)}{dt}$$

And rearranging

$$m\frac{d^2x(t)}{dt^2} + (c - bd\alpha)\frac{dx(t)}{dt} + kx(t) = 0 \quad (1.3)$$

This is unstable when the coefficient of the velocity term becomes negative (i.e. equivalent to negative damping). Thus, unstable vibration occurs when

$$(c - bd\alpha) < 0 \quad \text{or} \quad bd\alpha > 0 \quad (1.4)$$

In this case, the model suggests that the onset of chatter depends on both the width and depth of the cut, the original damping in the tool (c) and the factor α in the assumed force equation.

This now raises a most important question that will be asked again and again as we proceed. "Is the model correct?" Entwistle [10] has the following quote after the cover page of his thesis, "Models are to be used, but not to be believed—Henri Theil". For those interested in practical solutions to chatter, the response can be, "If the model allows us to make changes that stop chatter then the model has served a useful purpose". For Arnold chatter, reducing the width of cut and/or the depth of cut is found to be helpful. Also, simply reducing the overhang of the tool will be found to increase the effective value of c because the machine will also vibrate and not just the tool. The damping in machine tools is generally an order of magnitude greater than that in a cantilever bar "fixed" at one end. With a small overhang of the tool, the machine as well as the tool will vibrate and more energy will be dissipated effectively increasing c. We will return to this idea of energy dissipation when methods of stopping chatter using structural modifications are considered in Chap. 4.

1.3 Regenerative Chatter

Regenerative chatter arises during turning when a small oscillation of the cutting tool results in a wave being left on the work surface. One revolution later, this surface has to be removed, and if the width of cut is large enough, a wave of greater amplitude will be left on the surface. Each succeeding revolution results in a wave of even greater amplitude, resulting in a vibration of increasing amplitude. Hahn [11] was one of the first, if not the first, to use the term regenerative chatter. He presented a model that included feedback from the surface wave left by vibration of the cutting tool. This form of chatter is the more common and thus the most serious. It is appropriate (even essential) to experience this form of chatter in order to appreciate how bad it is. Program 1.1 presents a short movie of such

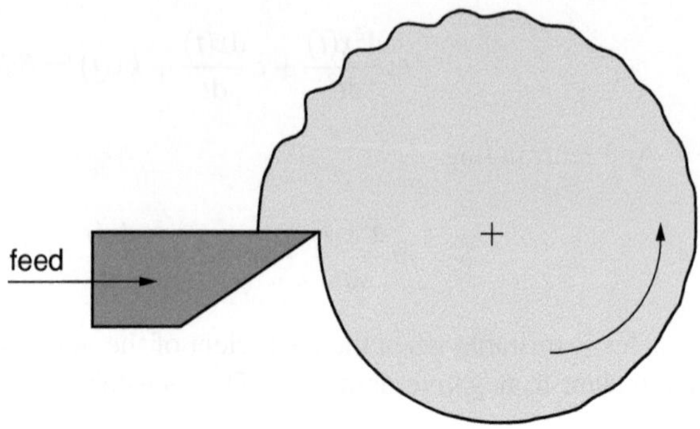

Fig. 1.3 Initial stages of regenerative chatter during facing while turning

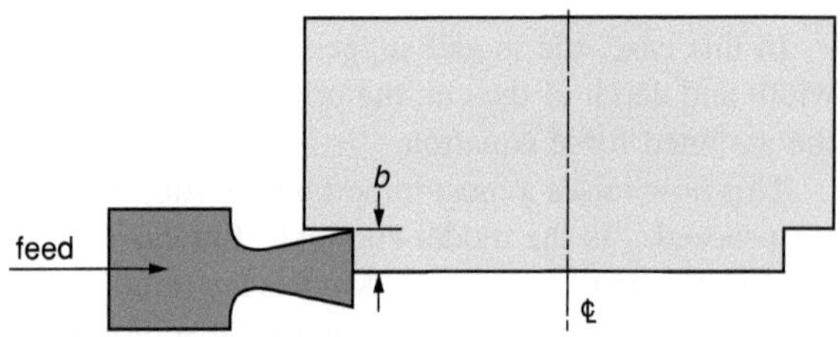

Fig. 1.4 Facing operation, while turning, showing the width of the cut (*b*)

chatter with the accompanying noise and consequent chatter marks. A second Program 1.2 is a simulation (which is based on many assumptions) that illustrates how regenerative chatter builds up.

The illustration in Fig. 1.3 is of a facing operation in turning as shown in Fig. 1.4. The tool moves in a radial direction, and the feed/rev determines the mean chip thickness.

There are several approaches to the analysis of chatter. Some methods give a better understanding of the effect and allow solutions to be discovered. Other methods are largely mathematical and do not normally allow solutions to chatter to be easily discovered. In this chapter, a mathematical method and a graphical method will be developed for the simplest model of regenerative chatter. Models will be developed for the cutting force, the machine response and finally chatter. As noted previously, models are approximations and the assumptions used in any model have to be critically examined.

It appears that Tlusty [12] was the first (1954) to produce a model of regenerative chatter that allowed machine tools with improved chatter performance to be produced. He founded and directed a machine tool research institute, known as VUOSO, in Prague, Czechoslovakia. He had the great advantage of access to numerous machine tools and was not limited to testing a single machine. (In debates at conferences, he would often ask those who disagreed with him, "How many machines have you tested?"). His model of regenerative chatter included the machine response and a simple cutting force model, and this is considered in the next section.

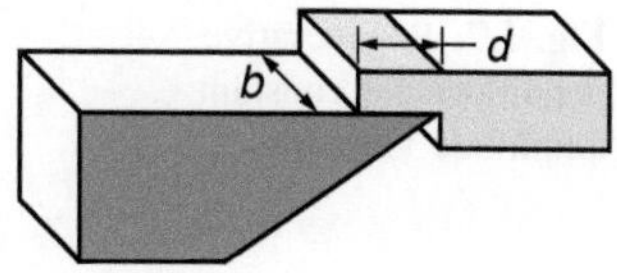

Fig. 1.5 Undeformed chip cross section

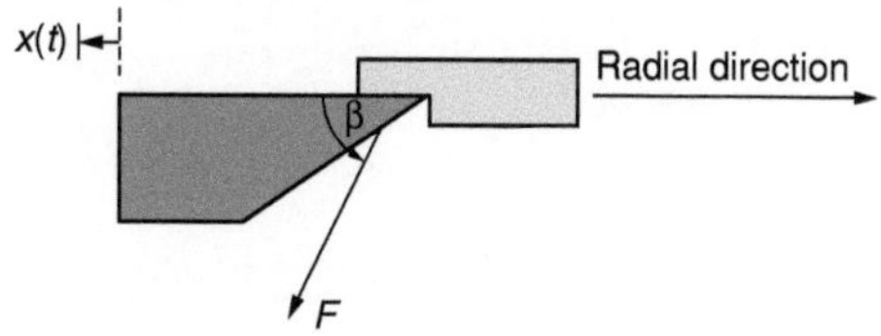

Fig. 1.6 Assumed direction of cutting force

1.3.1 Cutting Force Model

The simplest model for the cutting force assumes the force (*F*) is proportional to the undeformed chip cross-sectional area (see Fig. 1.5) which is equal to the width of cut (*b*) multiplied by the chip thickness (*d*) so that

$$F = Rbd \tag{1.5}$$

where R is conventionally called the cutting force coefficient and bd is the undeformed area of the chip.

It is assumed that this force does not depend on the cutting speed and so only vibration in the chip thickness direction is important. If the chip thickness (*d*) varies, the cutting force will vary as defined by Eq. 1.5, independent of the rate of change of the chip thickness. These assumptions will be re-examined when other models of regenerative chatter are presented.

There is a final set of assumptions to be made, which relate to the direction of the cutting force. The cutting force is assumed to act in a fixed direction at a constant angle to the radial direction as shown in Fig. 1.6. It is assumed that this direction is independent of cutting speed and does not change as the chip thickness changes. There is of course an equal and opposite force on the work that is not shown in Fig. 1.6.

1.3.2 Instantaneous Oscillating Force

In the early days of chatter research, computational facilities were limited and so time-domain simulations were not possible. The approach adopted was to consider conditions at a hypothetical, stability boundary condition. It was assumed that the vibration was sinusoidal with constant amplitude. This is not achieved in practice, but this condition proves to be accurate in determining the onset of chatter. Thus,

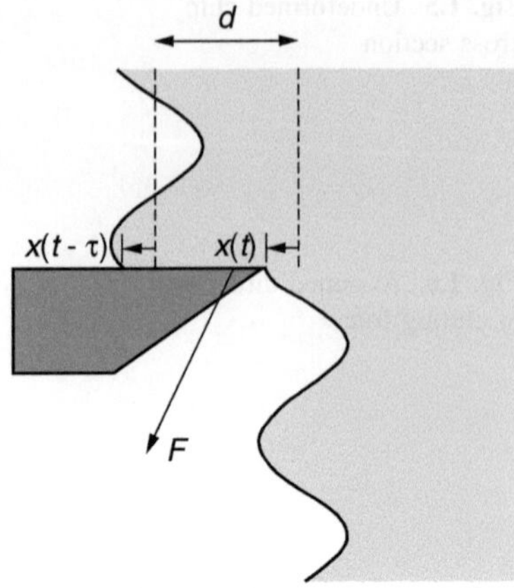

Fig. 1.7 Regenerative conditions for constant amplitude vibration

the stability boundary conditions were/are assumed to be as shown in Fig. 1.7. Note that $x(t)$ is defined as positive if the tool moves out from the work and Fig. 1.7 shows with dotted lines the mean chip thickness that would result if there was no vibration.

It was/is common to consider the resultant force to have three components. A constant mean force, that depends on the feed per revolution. A non-regenerative component caused by the current motion of the tool and a regenerative component resulting from the surface wave left on the previous revolution.

Consider first the mean force. If the feed per revolution is h, then the mean force will be given by

$$F_{\text{mean}} = Rbh \tag{1.6}$$

Next, consider the non-regenerative force. To do this, assume there is no wave on the surface being cut (see Fig. 1.8 where the tool is vibrating sinusoidally). For a tool displacement $x(t)$, the oscillating force is given, using Eq. 1.6, by

$$F_{\text{non-regenerative}} = -Rbx(t) \tag{1.7}$$

Note the negative sign, since as $x(t)$ increases, the tool moves out of cut (see Fig. 1.7), and hence, the force reduces. For the sinusoidal motion used in the illustration, the oscillating force is anti-phase with the tool motion. Also note that the cut surface has a surface wave defined by $x(t)$.

Finally, consider the regenerative force. To do this, assume there is no vibration of the tool but that a sinusoidal wave has been left on the surface one revolution earlier. This surface wave is thus defined by $x(t-\tau)$, where τ is the time taken for one revolution of the work. The regenerative component of the force on the tool is given by

$$F_{\text{regenerative}} = Rbx(t - \tau) \tag{1.8}$$

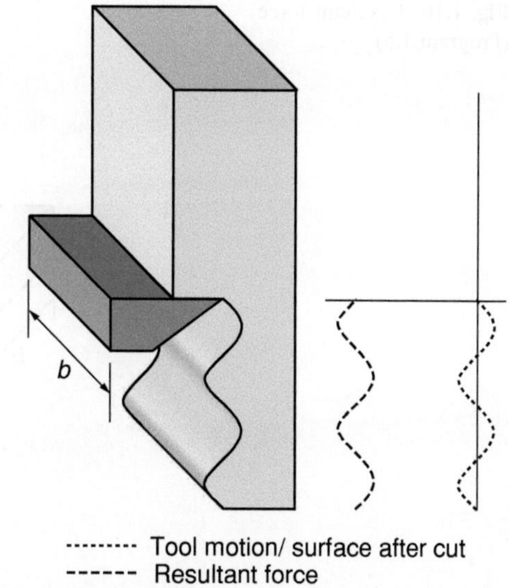

Fig. 1.8 Non-regenerative force (Program 1.3)

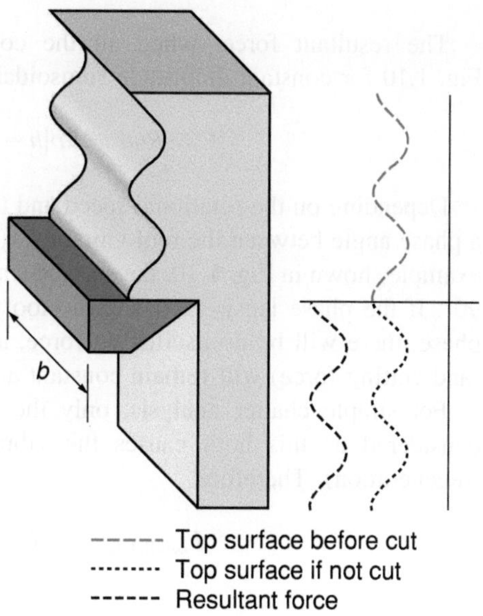

Fig. 1.9 Regenerative force (Program 1.4)

Note that as a positive value of $x(t-\tau)$ increases the chip thickness, the oscillating force increases (see Fig. 1.9). For the sinusoidal motion used in the illustration, the oscillating force is in phase with the surface wave.

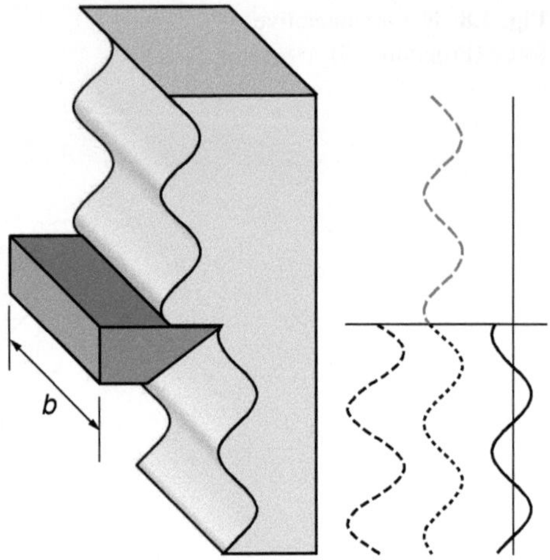

Fig. 1.10 Resultant force (Program 1.5)

The resultant force, when all the components are included, is shown in Fig. 1.10 for constant amplitude sinusoidal motion and is given by

$$F = Rbd = Rb[h - x(t) + x(t - \tau)] \tag{1.9}$$

Depending on the rotational speed and the frequency of vibration, there will be a phase angle between the tool vibration $x(t)$ and the surface wave $x(t-\tau)$. For the example shown in Fig. 1.10, the surface wave $x(t-\tau)$ lags the tool vibration $x(t)$ by 90°. If the phase lag is zero, i.e. the tool vibration and the surface wave are in phase, there will be no oscillating force, as $x(t) = x(t-\tau)$ and the chip thickness (and cutting force) will remain constant at h.

For simple chatter analysis, only the oscillating component of the force is considered as this both causes the vibration and is caused by the vibration (regeneration). Therefore,

$$F_{\text{oscillating}} = -Rb[x(t) - x(t - \tau)] \tag{1.10}$$

1.3.3 Machine Response

The machine response that is of interest is the relative deflection, $x(t)$, between the tool and work in the chip thickness direction due to the equal and opposite forces, $-Rb[x(t) - x(t - \tau)]$, in the cutting force direction. This response, $G(\omega)$, is often called the chatter receptance of the machine and is a function of frequency ω. Its in-phase and out-of-phase parts can be represented as real and imaginary components $G_R(\omega)$ and $G_I(\omega)$ so that

$$G(\omega) = G_R(\omega) + iG_I(\omega) \tag{1.11}$$

If the assumed steady vibration is to continue, then the oscillating cutting force must act on the structure to produce the assumed vibration. Thus, by definition, the response is given by

$$x(t) = G(\omega)F_{oscillating} = G(\omega)\{Rb[-x(t) + x(t-\tau)]\} \tag{1.12}$$

And rearranging and substituting for $G(\omega)$ from (1.11)

$$\frac{x(t)}{x(t-\tau)} = \frac{G_R(\omega) + iG_I(\omega)}{\frac{1}{Rb} + G_R(\omega) + iG_I(\omega)} \tag{1.13}$$

Tlusty [12] now notes that at the boundary of stability, the magnitude of $x(t)$ and $x(t-\tau)$ will be the same so that

$$\frac{|x(t)|}{|x(t-\tau)|} = \left|\frac{G_R(\omega) + iG_I(\omega)}{\frac{1}{Rb} + G_R(\omega) + iG_I(\omega)}\right| = \frac{\sqrt{G_R^2(\omega) + G_I^2(\omega)}}{\sqrt{\left(\frac{1}{Rb} + G_R(\omega)\right)^2 + G_I^2(\omega)}} = 1 \tag{1.14}$$

Squaring both sides and rearranging

$$G_R^2(\omega) = \left(\frac{1}{Rb} + G_R(\omega)\right)^2 = \left(\frac{1}{Rb}\right)^2 + 2\left(\frac{1}{Rb}\right)G_R(\omega) + G_R^2(\omega)$$

so that

$$\frac{1}{Rb} = -2G_R(\omega)$$

and hence, the width of cut at the stability boundary is given by

$$b = \frac{-1}{2RG_R(\omega)} \tag{1.15}$$

Further, when cutting, the width of cut is positive and so the minimum value of b (usually defined as b_{lim}) is determined by the maximum negative value of $G_R(\omega)$—commonly termed the maximum negative in-phase component of the chatter receptance, $G_{R,max}(\omega)$.

It should be noted that the chatter receptance has not been assumed to contain only a single mode of vibration. In fact, it is possible to measure $G_R(\omega)$ for real machines, and if the assumptions made in the derivation are valid, a method for finding the unconditional width of stable cut b_{lim} is thus available. It is appropriate to restate those assumptions.

- The cutting force is proportional to the undeformed chip cross section.
- The cutting force does not depend on the surface speed.
- The oscillating cutting force acts in a fixed direction.
- The boundary of stability can be determined using linear vibration models.
- The phase between the vibration $x(t)$ and $x(t-\tau)$ one revolution previously can be ignored.

With these assumptions in mind, there are some important conclusions to be drawn with regard to avoiding chatter.

- There is a width of cut below which chatter will never occur, regardless of rotational speed. This is called the unconditional width of cut, b_{lim}.
- Machines, with improved chatter performance, can be produced by reducing $G_{R,\max}(\omega)$.

Nothing in the above analysis restricts its application to turning on lathes. Any machining process that allows regeneration after a time interval τ is covered. τ is simply the time since the surface was last cut. Other machining operations will be considered in Chap. 2.

For turning, with the given assumptions, the diameter of the work does not affect the stability boundary.

Prior to Tlusty, it was known [11] that the resonant frequencies of machines were important but not why. Figure 1.11 shows the in-phase component of the chatter receptance for a structure with a single mode of vibration. It is clear that in this case, $G_{R,\max}(\omega)$ occurs just above the undamped natural frequency, i.e. just above resonance. Tlusty used his model of chatter with great effect and applied his method to a wide range of machine tools [13].

It was Tlusty's simple cutting force model that allowed him to relate chatter performance to the chatter receptance of a machine. Independently, others were making advances but were using more complex force models and including the effect of the phase between $x(t)$ and $x(t-\tau)$. Their results would indicate that the omission of phase had practical implications.

1.3.4 Graphical Approach

The development of more complex cutting force models will be considered later as it is considered helpful to introduce each major advance in chatter modelling one at a time. This section will introduce the effect of rotational speed that is significant when the phase between $x(t)$ and $x(t-\tau)$ is included. The method used is the

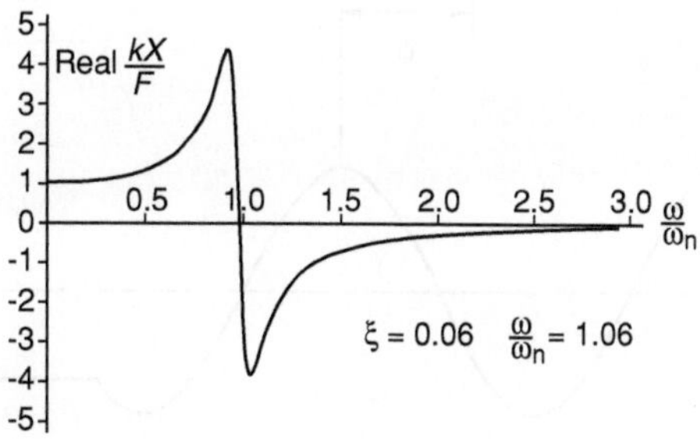

Fig. 1.11 In-phase response of a single-mode system (Program 1.6)

graphical approach of Guerney [14] as this is used extensively in later chapters when considering some methods of preventing chatter.

For this analysis, again consider the situation at the boundary of stability where the machine vibrates sinusoidally. It will continue to do so with no increase or decrease in the amplitude of vibration. The oscillation of the tool and the regenerating wave on the surface must result in an oscillating force, which acts on the structure to produce the cutting tool oscillation. This situation was considered previously (Fig. 1.10), and it was shown how the oscillating force arises. For a constant amplitude oscillation, let

$$x(t) = X_o \sin(\omega t - \phi) \tag{1.16}$$

The phase lag ϕ has the same significance as when used in the vibration theory developed in Appendix A.4. Thus, the response has a phase ϕ relative to the resultant of the exciting forces $F \sin \omega t$.

Thus, substituting in Eq. 1.10 gives the oscillating force as

$$F = -Rb[X_o \sin(\omega t - \phi) - X_o \sin(\omega(t - \tau) - \phi)] \tag{1.17}$$

We may now plot the two components of this force on the polar locus of the chatter receptance of the machine tool. However, before doing this, it is appropriate, by referring to Fig. 1.12, to recall the concept of rotating vectors (phasors) and the response locus (chatter receptance). The response locus has the force phasor located and fixed on the positive real axis. The displacement phasor is then at an angle ϕ to this.

Now, consider the two components of the force. The non-regenerative force F_n and the regenerative force F_r are located as shown in Fig. 1.13 where from Eq. 1.17

$$F_n = -RbX_o \sin(\omega t - \phi) \tag{1.18}$$

$$F_r = RbX_o \sin(\omega(t - \tau) - \phi) \tag{1.19}$$

If these force components are now plotted on the response locus, the diagram shown in Fig. 1.14 is obtained. Following the principle of one major step at a time,

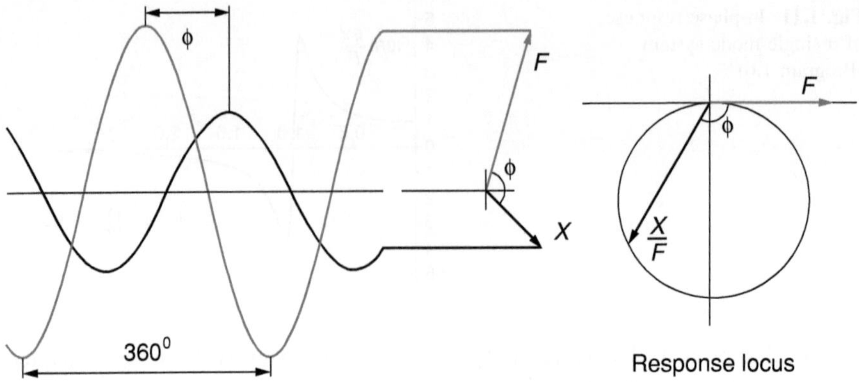

Fig. 1.12 Rotating vectors, sinusoidal excitation and the connection to the response locus (Program 1.7)

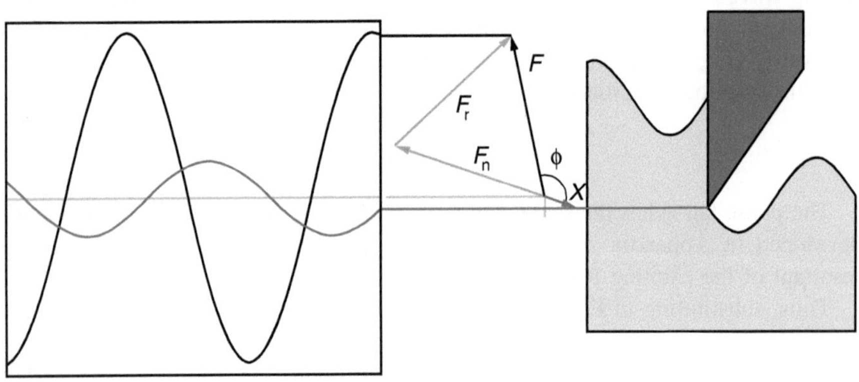

Fig. 1.13 Force components, regenerative F_r and non-regenerative F_n

the chatter receptance used is that of a structure with a single mode of vibration (see Appendix A.6) and is given by

$$\frac{kX_o}{F} = \frac{1}{1 - \frac{\omega^2}{\omega_n^2} + i2\xi\frac{\omega}{\omega_n}} = \frac{1 - \frac{\omega^2}{\omega_n^2} - i2\xi\frac{\omega}{\omega_n}}{\left(1 - \frac{\omega^2}{\omega_n^2}\right)^2 + 4\xi^2\frac{\omega^2}{\omega_n^2}}$$

The chatter frequency is ω, and the various vectors are in the directions shown. Note that the magnitude of both the force components is RbX_o so that from simple geometry, the force F has a magnitude $2RbX_o\cos(180 - \alpha)$.

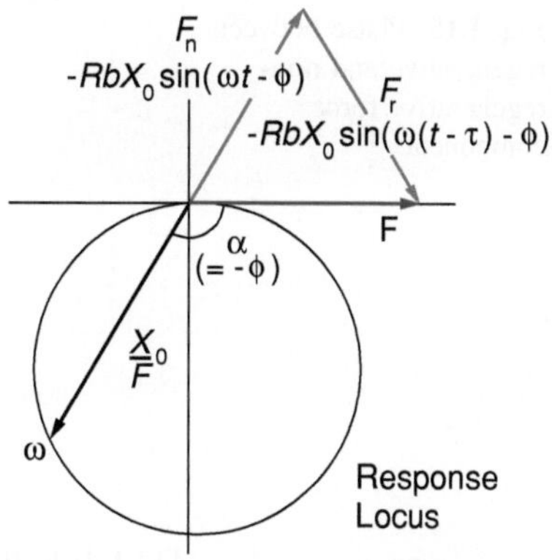

Fig. 1.14 Response locus with force vectors

For the situation shown in Fig. 1.14, let $S = X_o/F$ be the response at frequency ω. Thus,

$$S = \frac{X_o}{F} = \frac{X_o}{2RbX_o \cos(180 - \alpha)}$$

so that by rearranging, we obtain

$$b = b_{\text{lim}} = \frac{1}{2RS\cos(180 - \alpha)} \tag{1.20}$$

Note that $S\cos(180 - \alpha)$ is the negative in-phase component of the chatter receptance at the chatter frequency. It follows from Eq. 1.20 that the smallest value of b_{lim} is when the chatter frequency is at the maximum negative in-phase component of the chatter receptance $G_{R,\max}(\omega)$. This is as determined by Tlusty [12]. However, chatter does not always occur at the frequency that corresponds to the maximum negative in-phase component of the chatter receptance. This is because there is another condition that has to be met. The regenerative force component F_r lags behind the non-regenerative force F_n with a phase $\omega\tau$ as shown in Fig. 1.15.

From the sum of the angles in a triangle,

$$2\pi - \omega\tau + 2(\pi - \alpha) = \pi$$
$$\therefore \omega\tau = 3\pi - 2\alpha$$

Also, it should be noted that the regenerative force vector may have the same phase relationship if we add $2n\pi$ to the phase $\omega\tau$.

$$\therefore \omega\tau = 3\pi - 2\alpha + 2n\pi \tag{1.21}$$

Fig. 1.15 Phase between regenerative and non-regenerative force components

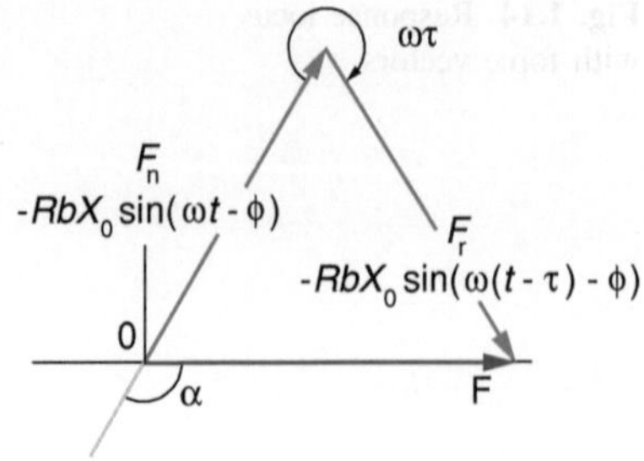

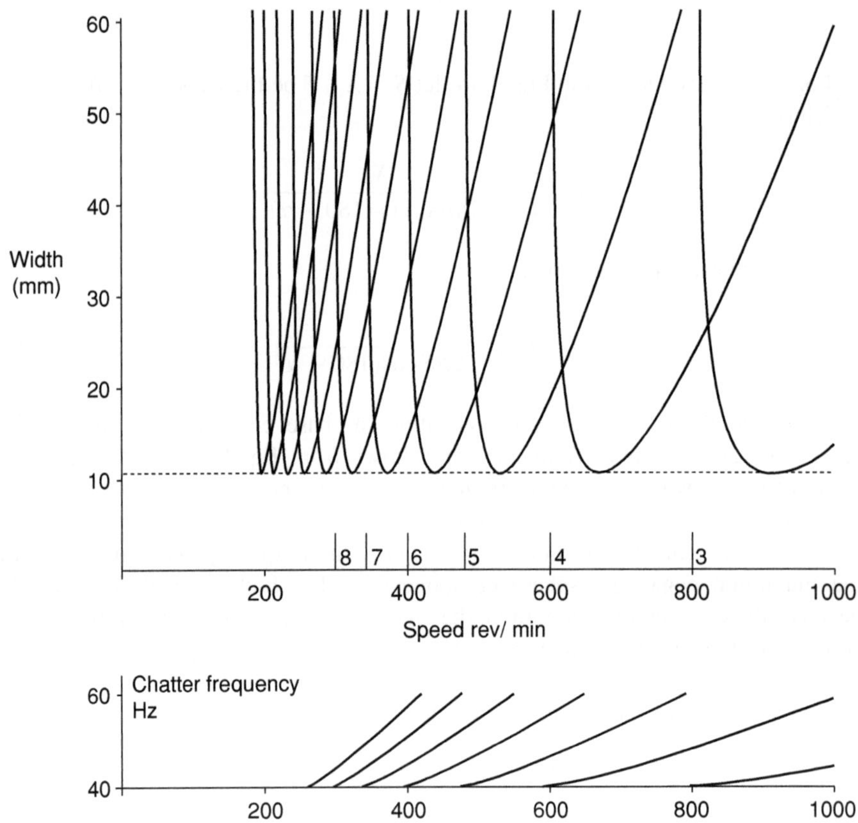

Fig. 1.16 Stability chart (Program 1.8)

where n is an integer. For different values of n, we get different values of τ and hence different values of rotational speed (=60/τ rev/min if τ is in seconds). In fact, for any ω, we can calculate b_{lim} from Eq. 1.20, and for each value of $n = 0$, 1, 2, 3, 4, etc., the associated rotational speed can be calculated. A plot of b_{lim} against rotational speed is conventionally called a stability chart, and an example

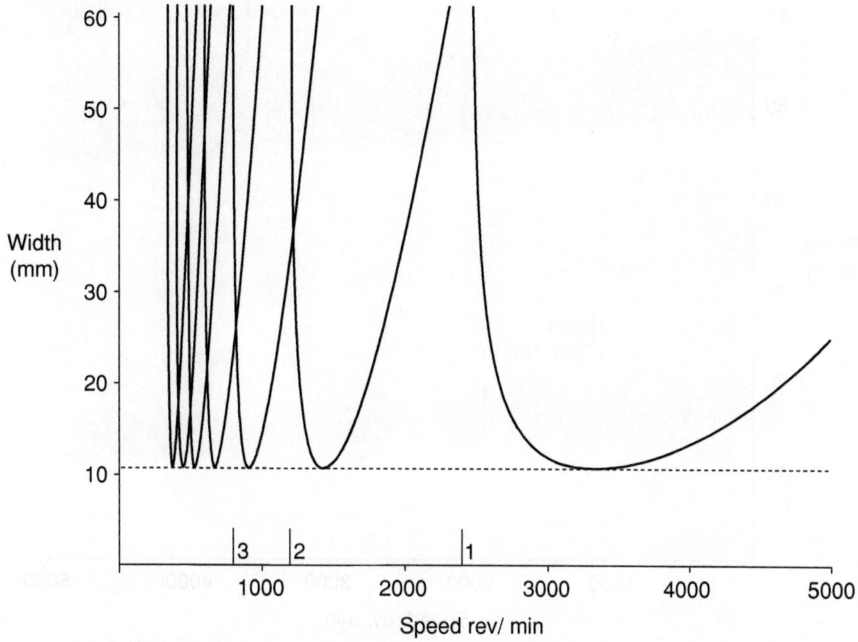

Fig. 1.17 Stability chart showing last lobe

is shown in Fig. 1.16. This example is for a lathe with a chatter receptance represented by a single mode with a stiffness $k = 4.0 \times 10^7$ N/m, an undamped natural frequency of 40 Hz and a damping ratio $\xi = 0.05$. Figure 1.16 also shows the frequency at which chatter will occur. It should be noted that for the single degree-of-freedom system considered, the chatter frequency is always above the natural frequency and that it is not proportional to the speed (as would be the case for forced vibration). For all speeds, the chatter frequency remains within a limited band and is not proportional to the speed.

In Fig. 1.16, the integer numbers above the speed axis indicate the conditions when an integer number of waves exist around the work circumference. When this is the case, $x(t) = x(t-\tau)$ and the chip thickness (and hence the cutting force) will remain constant. There is then no possibility of regenerative chatter, and so the value of b_{lim} tends to infinity. Also, it should be noted that there is a "last" lobe where there is less than one wave left on the work circumference. Figure 1.17 shows such a lobe on a stability chart that has the same conditions as for Fig. 1.16 but with the maximum value of speed increased to 5,000 rev/min.

To illustrate the significance of the stability lobes, a Program 1.9 was written that simulates turning. This program is a time simulation of the equation

$$m\frac{\mathrm{d}^2x(t)}{\mathrm{d}t^2} + c\frac{\mathrm{d}x(t)}{\mathrm{d}t} + kx(t) = Rb[h - x(t) + x(t - \tau)] \tag{1.22}$$

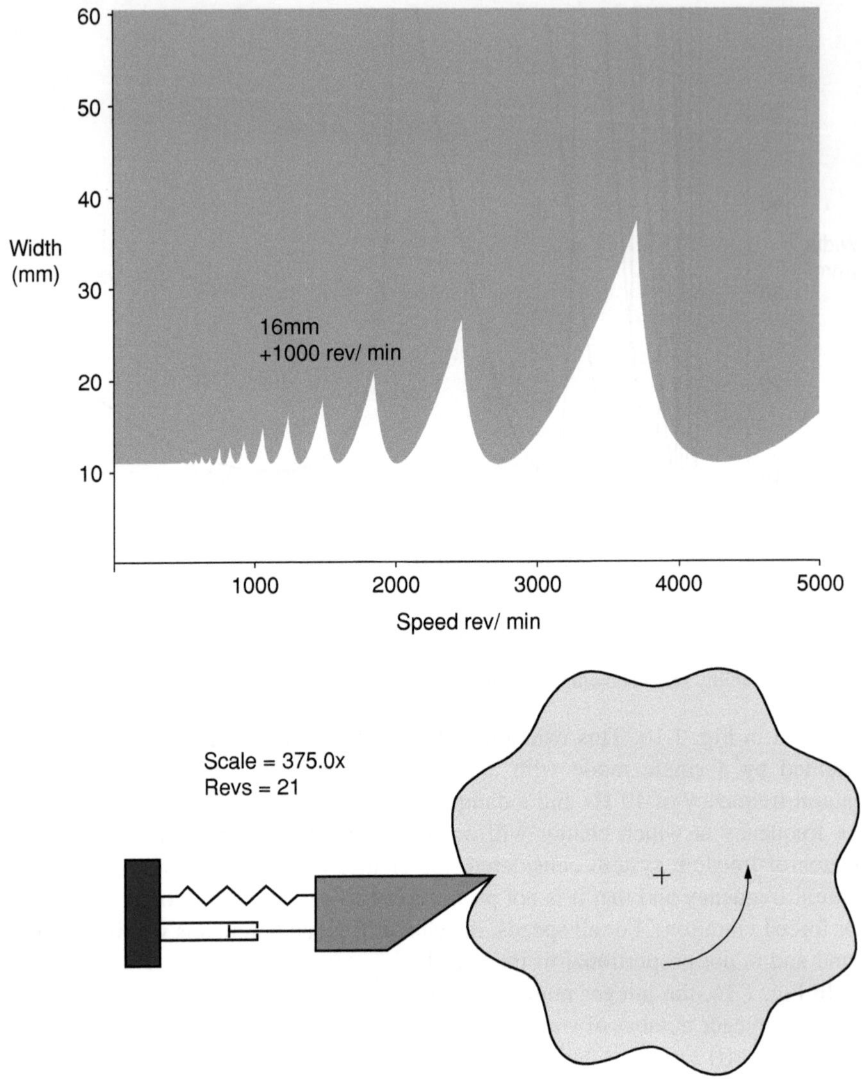

Fig. 1.18 Stability chart showing animation of chatter (Program 1.9)

The left-hand side represents the chatter receptance of the lathe by a single mode and the right-hand side the cutting force as described in Eq. 1.9. The simulation was written using a fourth-order Runge–Kutta approximation, and it is possible to check whether turning is stable or not for any particular width of cut and speed by simply clicking on the stability chart. A screenshot with the start-up values of $b = 16$ mm and speed 100 rev/min, after the work has completed 21 revolutions, is shown in Fig. 1.18.

The graphical analysis has confirmed that

- There is a width of cut below which chatter will not occur.
- Machines with improved chatter performance can be made by reducing $G_{R,\max}(\omega)$.
- Nothing in the above analysis restricts its application to turning on lathes. Any machining process that allows regeneration after a time interval τ is covered. τ is simply the time since the surface was last cut.
- For turning (with the cutting force model used), the diameter of the work does not affect the stability boundary.
- Additionally, at certain speeds, a greatly increased value for $b_{\lim}$ may be found by running between stability lobes or indeed beyond the last lobe.

1.3.5 Experimental Stability Charts

Now, we need to consider whether the stability charts predicted by the model are found in practice. Tobias [15] presented one of the earliest experimental charts for a vertical milling machine (see Fig. 1.19). We shall consider milling in the next chapter, but for now, it is necessary to accept that the depth of cut in vertical milling relates to the width of cut used in the above analyses. This experimental work confirmed the existence of the stability lobes but indicated a significant difference to those predicted by the theory that has been developed. As shown in Fig. 1.19, the stability lobes at lower speeds have increased values of $b_{\lim}$. Earlier work by Tobias and Fishwick [16–18] had predicted both stability lobes and the improvement at low speeds. They used a more complex model for the cutting force than that used by Tlusty [12]. Many of the early researchers on chatter had focused on the cutting force and attempted to use cutting force models alone to explain chatter, for example Doi [19]. In later years, experimental results were obtained for the forces arising under oscillating conditions while cutting [20–22]. The general consensus is that the improved chatter performance at low speeds is the result of what is known as “process damping” or “penetration rate damping”.

As the purpose of this part of this book is to describe methods of avoiding chatter, it is important to consider the source of process damping. The major cause is agreed to be the velocity of penetration of the tool into the work and hence the phrase “penetration rate damping”. The source of this damping is considered to be the result of an additional cutting force component that arises from the interference of the tool with the cut surface as the tool vibrates. There are two plausible mechanisms considered to be in play. Figure 1.20 shows the one that results from the fact that the cutting edge is not sharp but has a finite nose radius and also a wear land. It is apparent that the wear land would tend to prevent a surface being left of the kind shown and also resists motion into the work. This is because the short wavelength would not easily pass through the contact zone caused by the

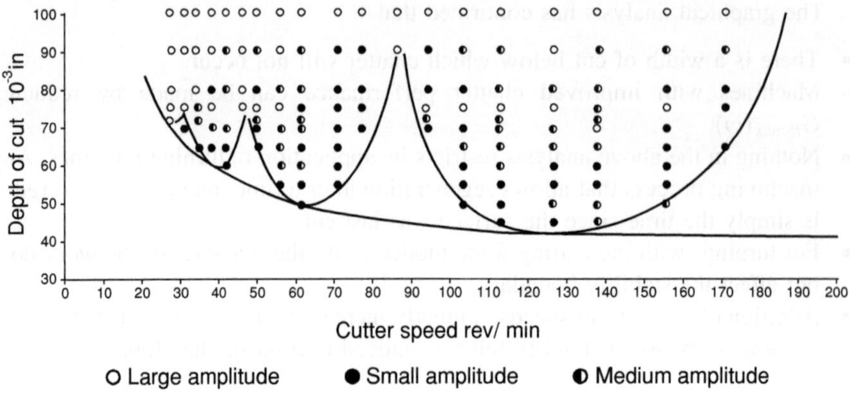

Fig. 1.19 Experimental stability for milling (after Tobias [15])

wear land. It is important to note that this effect arises for even small amplitudes of vibration. The second mechanism is that shown in Fig. 1.21 where the clearance face of the tool interferes with the cut surface. This effect is dependent on the amplitude of the vibration and the wavelength and might not be significant for very small amplitudes of vibration.

The model for the process damping force presented here is adapted from the work of Wallace and Andrew [21, 22]. The process damping force f_{pr} was expressed as follows:

$$f_{\mathrm{pr}} = k_{\mathrm{pr}} b \frac{\partial x}{\partial z} \tag{1.23}$$

where k_{pr} is a constant of proportionality and $\frac{\partial x}{\partial z}$ is the instantaneous slope of the surface being cut by the tool. For turning,

$$\frac{\partial x}{\partial z} = \frac{\partial x}{\partial t} \frac{1}{\left(\frac{\partial z}{\partial t}\right)} = \frac{\mathrm{d}x(t)}{\mathrm{d}t} \frac{1}{R_{\mathrm{w}} \Omega} \tag{1.24}$$

where Ω is the work rotational speed (rad/s) and R_{w} is the radius of the work.

The substitution of Eq. 1.24 into Eq. 1.23 gives

$$f_{\mathrm{pr}} = k_{\mathrm{pr}} b \frac{\mathrm{d}x(t)}{\mathrm{d}t} \frac{1}{R_{\mathrm{w}} \Omega} = k_{\mathrm{pr}} b \frac{\mathrm{d}x(t)}{\mathrm{d}t} \frac{2\pi\tau}{R_{\mathrm{w}}} \tag{1.25}$$

where τ is the time for one revolution of the work.

As f_{pr} acts to oppose the motion of the cutting tool, it can be subtracted from the cutting force caused by the chip cross section so that the equation of motion Eq. 1.22, for a machine modelled by a single mode, becomes

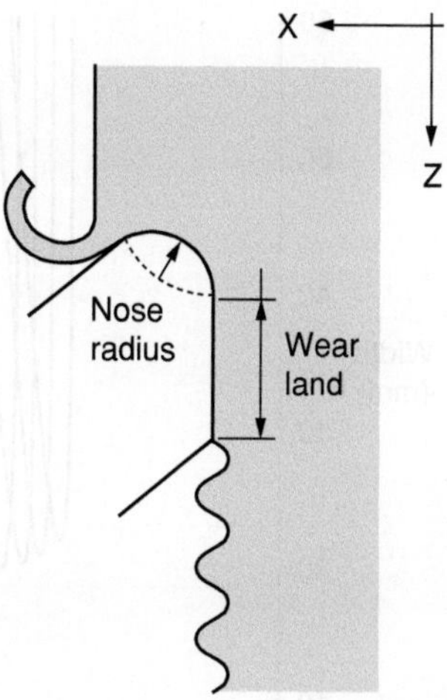

Fig. 1.20 Geometry of cutting edge greatly magnified

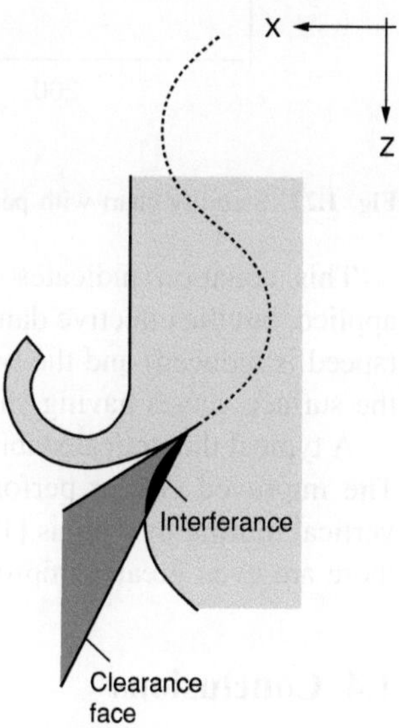

Fig. 1.21 Interference of tool flank and machined surface

$$m\frac{\mathrm{d}^2x(t)}{\mathrm{d}t^2} + c\frac{\mathrm{d}x(t)}{\mathrm{d}t} + kx(t) = b\left[R(-x(t) + x(t-\tau)) - k_{\mathrm{pr}}\frac{\mathrm{d}x(t)}{\mathrm{d}t}\frac{2\pi\tau}{R_{\mathrm{w}}}\right]$$

and rearranging

$$m\frac{\mathrm{d}^2x(t)}{\mathrm{d}t^2} + \left(c + k_{pr}\frac{2\pi\tau}{R_w}\right)\frac{\mathrm{d}x(t)}{\mathrm{d}t} + kx(t) = b[R(-x(t) + x(t-\tau))] \qquad (1.26)$$

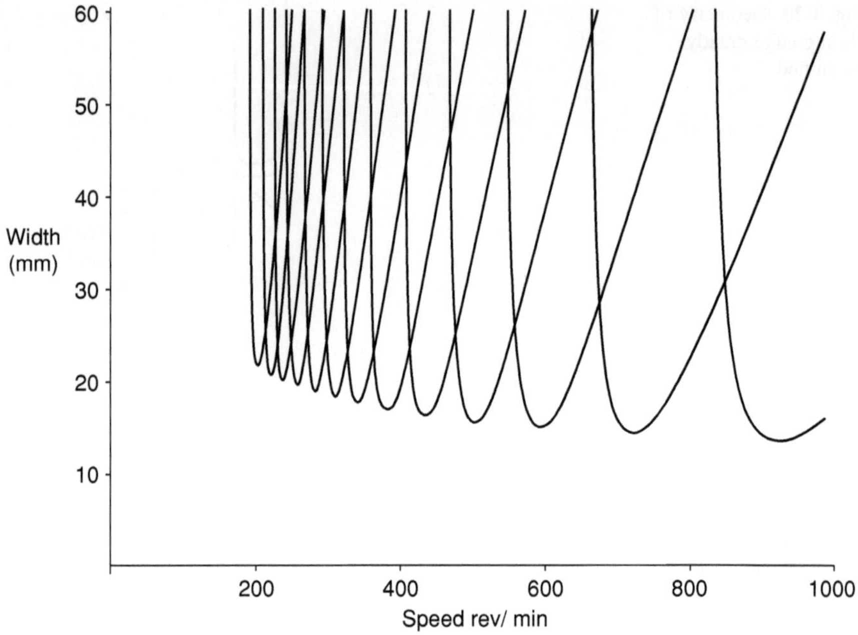

Fig. 1.22 Stability chart with penetration rate damping (Program 1.10)

This equation indicates that the model previously developed may still be applied, but the effective damping in the machine response increases as τ increases (speed is reduced) and the work radius is reduced. Both of these changes result in the surface waves having shorter wavelengths.

A typical theoretical stability chart derived from Eq. 1.26 is shown in Fig. 1.22. The improved chatter performance at low speed is similar to that measured for vertical milling by Tobias [15]. If the program is run and the work radius reduced, there are even greater improvements at low speed.

1.4 Conclusions

For the particular machining operation that has been modelled, i.e. turning, and for the assumptions made, the following points are of note.

- The width of cut determines whether chatter occurs. There is a width, for any speed, below which chatter will not occur.
- The mean chip thickness (feed rate) does not affect the onset of chatter.
- The machine characteristic that is of prime importance is the maximum negative in-phase component of the chatter receptance.
- The limiting width of cut varies with speed, and a stability chart shows stability lobes. It is possible to avoid chatter by choosing a speed between lobes or above the last lobe [23].

- At low speeds, the cutting force gives rise to penetration rate damping and the stability lobes have increased values of limiting width.

The question now arises as to whether the same effects occur in other machining operations. This will require the assumptions made in this chapter to be reconsidered in the next chapter.

References

1. Arnold RN (1946) The mechanism of tool vibration in the cutting of steel. Proc Inst Mech Eng 154(1):261–284
2. Taylor FW (1907) On the art of cutting metals. ASME, New York, p 30
3. Smith D (1911) Cutting tools. Trans Manchester Assoc Eng, p 133
4. Brooks JP (1905) Photographs of cutting tools in action. In: Proceedings of the institution of mechanical engineers, p 365
5. Doi S (1937) Chatter of lathe tool. J Soc Mech Eng Jpn 3(10):94
6. Doi S (1940) Chatter caused by lateral bending of work. J Soc Mech Eng Jpn 6(22):4–10
7. American Society of Mechanical Engineers (1939) Manual on cutting of metals
8. Schmidt W (1941) Zur dynamik der werkzeugmaschine. ZVDI 85(11):249
9. Holtzapffel JJ (1881) Hand or simple turning—principles and practice. In: Volume IV of the five volume work turning and mechanical manipulation, Dover
10. Entwistle RD (1997) Torsional compliance and chatter in grinding. PhD Dissertation, The University of Western Australia
11. Hahn RS (1952) Metal-cutting chatter and its elimination. Trans ASME J Eng Indus Ser B 74:1073–1080
12. Tlusty J, Spacek L (1954) Self-excited vibrations in machine tools. Czech Naklad CSAV, Prague
13. Tlusty J, Polacek M (1968) Experience with analyzing stability of machine tool against chatter. In: Proceedings of 9th MTDR Conference, vol 1, pp 521–570
14. Gurney JP, Tobias SA (1961) A graphical method for the determination of the dynamic study of machine tools. Int J Mach Tool Des Res 1:148–156
15. Tobias SA (1959) The vibrations of vertical milling machines under test and working conditions. Proc Inst Mech Eng 173(18):474–510
16. Tobias SA, Fishwick W (1956) The vibrations of radial drilling machines under test and working conditions. Proc Inst Mech Eng 170:232–247
17. Tobias SA, Fishwick W (1958) The chatter of lathe tools under orthogonal cutting conditions. Trans ASME 80:1079–1088
18. Tobias SA, Fishwick W (1958) Theory of regenerative machine tool chatter. Engineering 205:199–203 and 238–239
19. Doi S, Kato S (1956) Chatter vibration of lathe tools. Trans ASME 78:1127–1134
20. Smith JD, Tobias SA (1961) The dynamic cutting of metals. Int J Mach Tool Des Res 4:283–292
21. Wallace PW, Andrew C (1965) Machining forces: some effects of tool vibration. J Mech Eng Sci 7(2):152–162
22. Wallace PW, Andrew C (1966) Machining forces: some effects of removing a wavy surface. J Mech Eng Sci 8(2):129–140
23. Bediaga I, Muñoa J, Hernández J, López de Lacalle LN (2009) An automatic spindle speed selection strategy to obtain stability in high-speed milling. Int J Mach Tools Manuf 49(5):384–394

Chapter 2
Extension of Chatter Theory

2.1 Introduction

In the previous chapter, regenerative chatter theory was developed for turning and in particular for a facing operation. It is appropriate to summarise the assumptions made in the theory that was presented.

- A linear model of the process was assumed; that is, nonlinearities were ignored. It was argued that the onset of chatter will initially involve small amplitudes of vibration, and so, a linear model could reasonably be assumed when investigating the onset of chatter.
- Both the average and oscillating components of the cutting force were assumed to act in the same fixed direction.
- The cutting force was initially assumed to be proportional to the undeformed chip cross-sectional area and not dependent on the cutting speed. Subsequently, an additional force component was added that depended on the penetration rate of the tool into the work and this resulted in process damping.
- Only vibration in the direction that affected chip thickness was considered.
- A graphical method [1] of determining the stability boundary was presented, and the machine response used as an example had a single mode of vibration representing a one-degree-of-freedom system.

There are therefore several questions that need to be raised. Do the assumptions made give rise to any serious errors when solutions to chatter problems are attempted? What about real machine responses that involve multiple modes of vibration rather than a single mode? Does the chatter theory developed in the previous chapter apply to other machining operations? This chapter addresses these questions.

2.2 Multiple Modes of Vibration

Real machines have multiple modes of vibration. Though there may be a predominant mode, so that the machine response can be represented by the response of a single-degree-of-freedom system, this is not always the case. To illustrate the

B. Stone, *Chatter and Machine Tools*, DOI: 10.1007/978-3-319-05236-6_2,

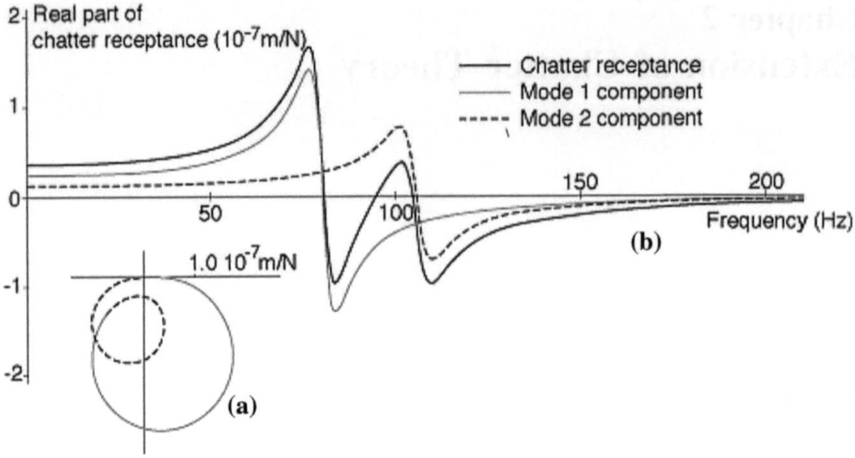

Fig. 2.1 Chatter receptance for a structure with two modes of vibration, **a** chatter receptance and **b** real part (Program 2.1)

effect on chatter of a machine with more than one significant mode, consider a lathe with two modes of vibration. The conventional wisdom is to assume that the response can be represented by the superposition of the responses of two one-degree-of-freedom systems. Thus, the chatter receptance is given by

$$\frac{X_o}{F} = \frac{1}{k_1\left(1 - \frac{\omega^2}{\omega_{n1}^2} + i2\xi_1\frac{\omega}{\omega_{n1}}\right)} + \frac{1}{k_2\left(1 - \frac{\omega^2}{\omega_{n2}^2} + i2\xi_2\frac{\omega}{\omega_{n2}}\right)}$$

Figure 2.1 shows the chatter receptance for some typical values of the mode parameters chosen to give equal values of the maximum negative in-phase component of the chatter receptance at two frequencies.

Mode 1: $k_1 = 4 \times 10^7$ N/m, $\omega_{n1} = 80$ Hz, $\xi_1 = 0.044$
Mode 2: $k_2 = 8 \times 10^7$ N/m, $\omega_{n2} = 105$ Hz, $\xi_2 = 0.04$

The graphical method [1] for predicting the stability chart is readily applied to this chatter receptance. For a facing operation, as considered in the previous chapter, a stability chart as shown in Fig. 2.2 is obtained. This chart does not include the process damping resulting from penetration rate effects. If these are included, the predicted stability chart is as shown in Fig. 2.3. The process damping has a proportionally greater effect on the mode with the higher natural frequency as this mode produces shorter wavelength surface waves for the same rotational speed.

Even at this stage, it is clear that chatter theory involves some complexity and it is to be extended even further in this chapter. To avoid being overwhelmed by the increasing complexity, we will adopt the approach initially developed by Tlusty [2] and simply examine the maximum negative in-phase component of the chatter

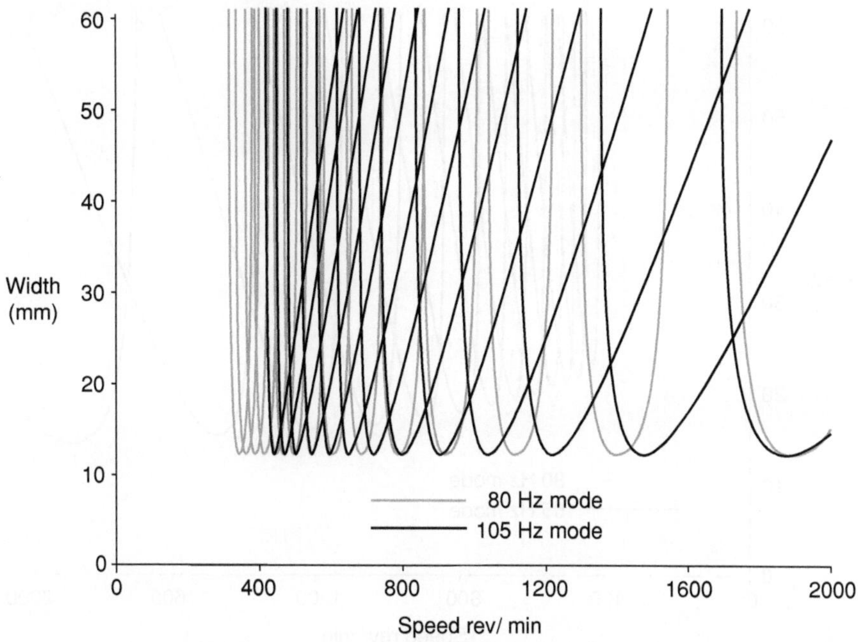

Fig. 2.2 Stability chart with no process damping (Program 2.2)

receptance, knowing that this yields the unconditionally stable (i.e. speed independent) width of cut. In practice, it will be expected that there will be stability lobes and the effects of process damping will give increased stability at lower speeds when there are shorter wavelength surface waves. These effects need to be borne in mind when only the unconditional stable width is calculated.

2.3 Vibration Direction

Thus far, only vibration in the direction that affects chip thickness has been considered. However, in practice, the vibration, resulting from an oscillating cutting force, may occur in a different direction. If the facing operation is reconsidered, then vibration could occur having components in each of the three directions shown in Fig. 2.4 (x, y, z).

Regenerative chatter occurs when a vibration causes a surface wave to be left that one revolution later causes a variation in the cutting force that in turn maintains the vibration or causes it to increase in amplitude. If the vibration is in the x direction as shown in Fig. 2.5a (and as considered in the previous chapter), then such a surface wave is left. However, suppose the vibration was in the z direction, the direction that would change the width of cut as shown in Fig. 2.5b. Any surface wave left on the faced surface will not regenerate one revolution later

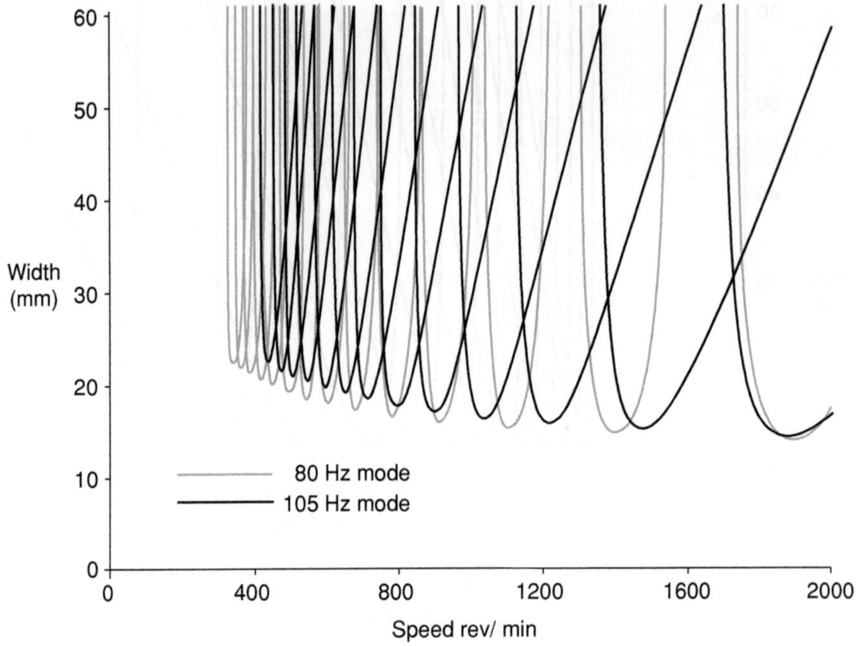

Fig. 2.3 Stability chart with process damping (Program 2.3)

Fig. 2.4 Principal directions of vibration when facing

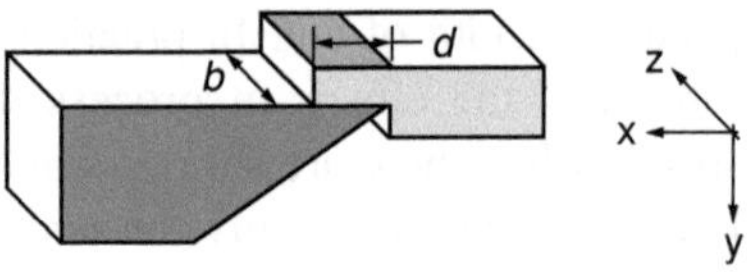

Fig. 2.5 Effective width of cut for vibration in different directions

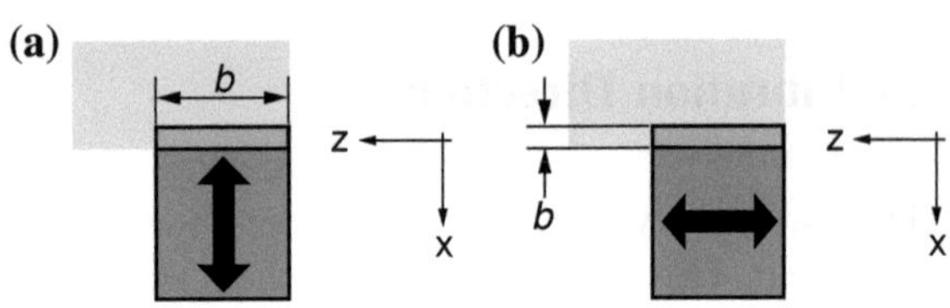

as the tool has moved in the x direction. However, even if such a wave did regenerate, because the width of cut b will normally be much greater than the depth of cut d, the variation in chip cross-sectional area would be an order of magnitude less than for vibration in the x direction with the same amplitude of vibration. For vibration in the x direction, we have shown that the cutting force F is given by

$F = Rbd = Rb[h - x(t) + x(t - \tau)]$ with an oscillating component $Rb[-x(t) + x(t - \tau)]$

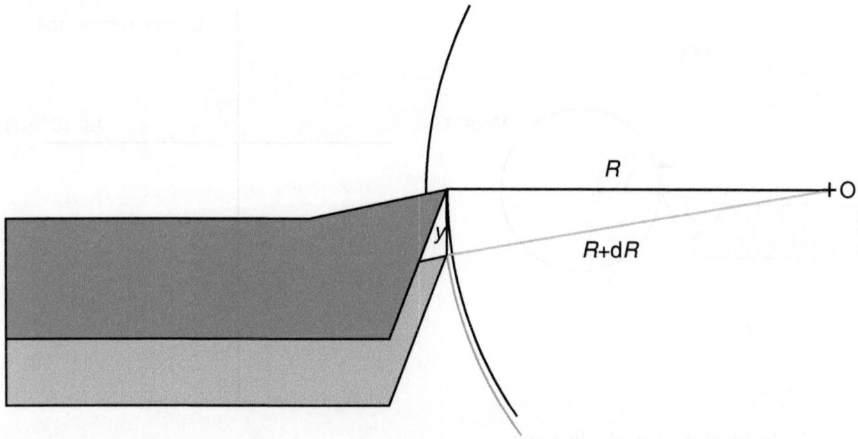

Fig. 2.6 Change in cutting radius for displacement in the y direction

For vibration in the z direction, the equivalent equation would be

$F = Rbd = Rd[b_{\text{mean}} - z(t) + z(t-\tau)]$ with an oscillating component $Rd[-z(t) + z(t-\tau)]$

For identical amplitudes of vibration, i.e. $x(t) = z(t)$, the ratio of the oscillating forces is

$$\frac{F_{x\text{vibration}}}{F_{z\text{vibration}}} = \frac{Rb[-x(t) + x(t-\tau)]}{Rd[-z(t) + z(t-\tau)]} = \frac{b}{d}$$

So that vibration in the z direction causes a far smaller variation in the force, and hence, the prospect of regeneration occurring is greatly reduced.

Now, consider vibration in the y direction (the direction that would cause a change in the cutting speed). This may cause Arnold type B chatter under some circumstances but for the model presented thus far, not regenerative chatter. The amplitude of the wave left on the machined surface is extremely small compared to the amplitude of vibration. For example, if the cutting tool moves down a distance y, then the cutting radius (R) increases to ($R + dR$) and the depth of cut (see Fig. 2.6) is decreased by an amount dR given by

$$(R + dR)^2 = R^2 + y^2 \text{ so that } R^2 + 2RdR + dR^2 = R^2 + y^2$$

And since dR^2 is negligible $dR = y^2/2R = y \cdot y/2R$ and as $y/2R$ is very small, dR is much smaller than y. Hence, the regenerating wave resulting from a vibration in the y direction has a very small amplitude compared to the vibration amplitude, and therefore, the prospect of regeneration occurring is very unlikely.

In practice, any mode of vibration of a machine tool has an associated mode shape and this defines the direction in which the vibration will occur. It is thus

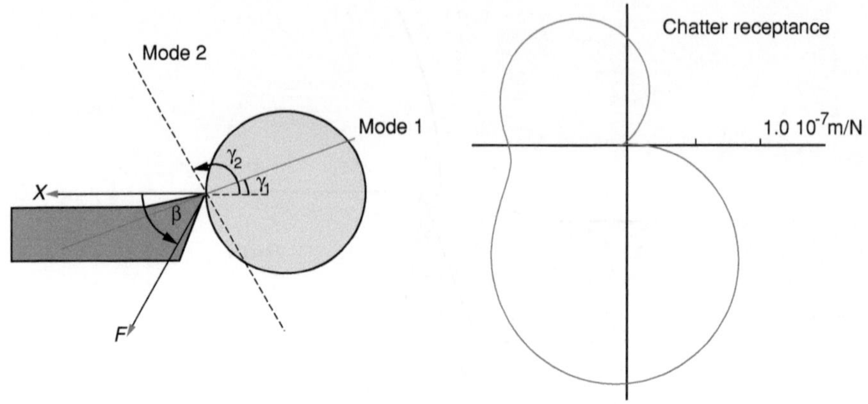

Fig. 2.7 Inclination of mode direction

possible for the mode direction to be, for example, in the x–y plane of Fig. 2.4 and at some angle to the x direction. The effect of the mode direction needs to be considered as it has significant practical application for improving chatter performance.

2.4 Mode Direction

Consider the facing operation with the structural response represented by a single mode that has a mode direction inclined at an angle γ to the horizontal x-axis as shown in Fig. 2.7. Note that X represents the component of the relative displacement between the work and cutting edge, and though F is shown only applied to the cutting edge, there is an equal and opposite force on the work.

The direction of the cutting force F relative to the x direction is β. Let the mode have a direct response (i.e. displacement and force in the mode direction) defined by stiffness k, modal mass m and viscous damping c. The direct response, where both force and displacement are in the mode direction, as shown in Appendix A.6, is given by

$$\frac{X_{\text{direct}}}{F_{\text{direct}}} = \frac{1}{k - m\omega^2 + i\omega c} \tag{2.1}$$

The chatter receptance is the response in the x direction divided by the force (F) in the cutting force direction. The component of F in the mode direction is $F_{\text{direct}} = F\cos(\gamma - \beta)$. The response in the x direction is $X_{\text{direct}} = X/\cos\gamma$. Substituting in Eq. 2.1 gives

$$\frac{X_{\text{direct}}}{F_{\text{direct}}} = \frac{X}{\cos\gamma F\cos(\gamma - \beta)} = \frac{1}{k - m\omega^2 + i\omega c}$$

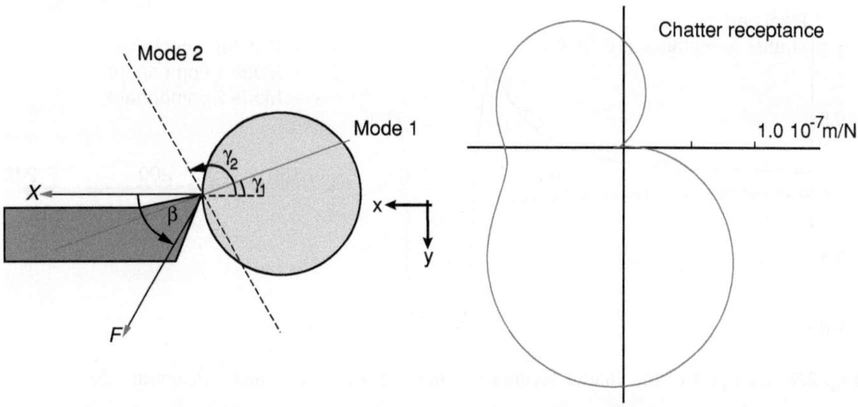

Fig. 2.8 Chatter receptance for a machine with two inclined modes (Program 2.4)

so that the chatter receptance is given by

$$\frac{X}{F} = \frac{\cos \gamma \cos(\gamma - \beta)}{k - m\omega^2 + i\omega c} \tag{2.2}$$

If either $\cos \gamma$ or $\cos(\gamma - \beta)$ is zero, the chatter receptance becomes zero and regenerative chatter will not occur. This is the case when the mode direction is perpendicular to either the x direction (the chip thickness direction) or the cutting force direction.

In practice, the machine response involves several modes, and each has its associated mode direction. In general, when the x direction or cutting force direction is perpendicular to one mode, it will not be so for all the other modes. Thus, the resultant chatter receptance will not be zero. As an example, consider a machine response that has two modes, each of which can be represented by the response of a single-degree-of-freedom system with stiffness k, undamped natural frequency ω_n and damping ratio ξ. Figure 2.8 shows the chatter receptance for a machine with two modes with the following characteristics:

Mode 1: $k_1 = 4 \times 10^7$ N/m, $\omega_{n1} = 100$ Hz, $\xi_1 = 0.05$, mode direction $\gamma = 20°$
Mode 2: $k_2 = 6 \times 10^7$ N/m, $\omega_{n2} = 120$ Hz, $\xi_2 = 0.02$, mode direction $\gamma = 120°$

The cutting force direction (β) was assumed to be at 60° to the horizontal x direction. Thus, neither of the modes is perpendicular to the x or F directions and both contribute to the chatter receptance.

The stability chart has stability lobes, but when considering improvements to chatter performance, the unconditional width of cut serves as a suitable indicator. In this example, using a cutting force coefficient of $R = 4.0 \times 10^8$ N/m^2, the unconditional width of cut is 12 mm for the chatter receptance shown.

If on resolving X and F into any mode direction they lie in opposite directions, then the effective mode stiffness (k) is negative. Figure 2.8 illustrates this

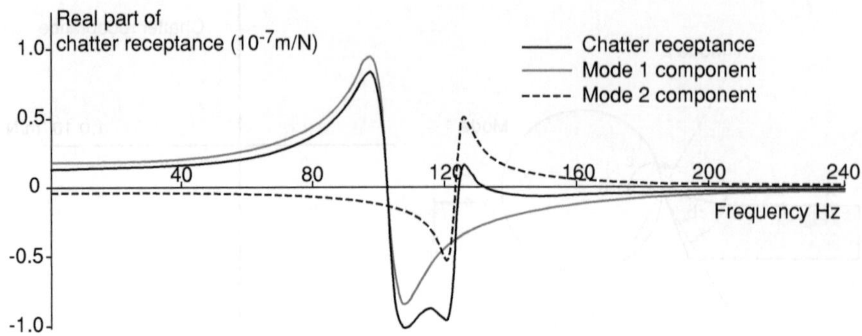

Fig. 2.9 Real part of the chatter receptance showing a negative mode (Program 2.5)

significant feature—it is possible for the chatter receptance to include such a negative mode. This is more clearly seen if the real part of the chatter receptance is plotted against frequency (see Fig. 2.9) and the modal contributions examined. If Mode 2 in Fig. 2.9 was the only mode present, the mean cutting force would cause the tool to move into the work.

Tlusty [2] used negative modes to cancel positive ones and investigated changing the circumferential position of the tool relative to the work for a turning operation. For the mode characteristics and directions used for Fig. 2.8, the tool position may be changed and the unconditional width found. Figure 2.10 shows the variation of the unconditional width of cut with tool position. For the example shown, the conventional tool position has an unconditional width of cut of 12 mm. However, this width would be increased to 45 mm if the tool were positioned at an angle 34° above the horizontal. In the worst position (almost vertically under or over the work), the width is 9 mm. Tlusty [2] measured similar variations experimentally.

Changing the tool position to advantage was noted in the previous chapter. During the discussion of Arnold's [3] paper

> Mr. Alfred Harrison, A.I.Mech.E., commenting on Mr. Pamacott's remarks, said he had found that a great improvement could be achieved by down-cutting, either by putting the tool on the opposite side of the slide or by reversing the direction of the lathe and inverting the tool.

It can be shown that inverting the tool and reversing the rotation may also be advantageous confirming what was known experimentally in 1946. Figure 2.11 shows the variation of unconditional width of cut with position for both an inverted tool and the standard configuration. The mode characteristics listed below have been chosen to show this:

Mode 1: $k_1 = 4 \times 10^7$ N/m, $\omega_{n1} = 100$ Hz, $\xi_1 = 0.05$, mode direction $\gamma = 20°$
Mode 2: $k_2 = 6 \times 10^7$ N/m, $\omega_{n2} = 140$ Hz, $\xi_2 = 0.02$, mode direction $\gamma = 100°$

The cutting force direction (β) was again assumed to be at 60° to the x direction.

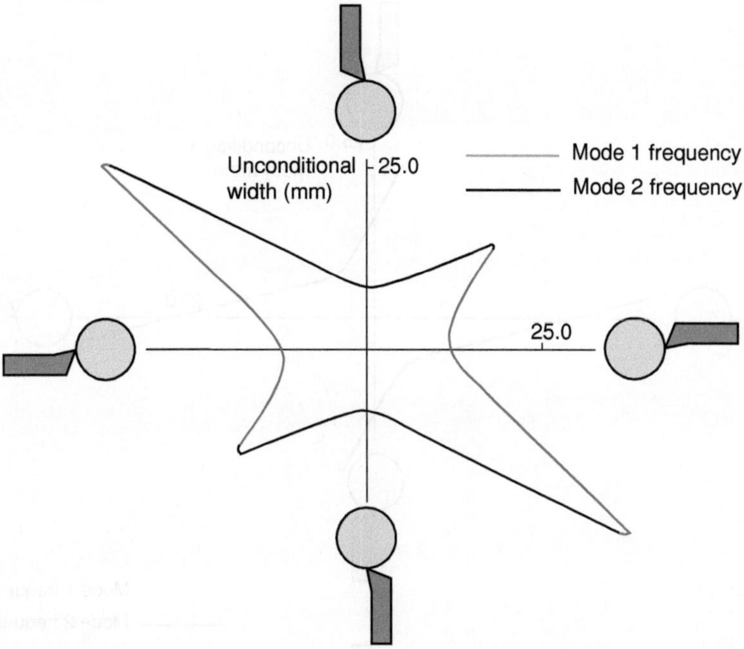

Fig. 2.10 Variation of unconditional width with circumferential position of the tool (Program 2.6)

For the inverted tool in the conventional location (with reverse rotation), the unconditional width is 30 mm, whereas for the not inverted, the width is 13 mm. The total variation of unconditional width with tool position varies between 9 and 45 mm as before.

2.5 Turning, Boring, Drilling and Milling

Thus far, only the specific turning operation of facing has been considered. It is necessary to consider other turning operations and also different machining operations such as boring, drilling and milling. Grinding will be examined in Chap. 5.

For facing on a lathe, the onset of chatter was governed by the width of cut. This width was the length of the cutting edge in contact with the work, and it was perpendicular to the direction that most affected the chip thickness, i.e. the feed direction, see Fig. 2.12a. The question to be answered is does this apply to all machining operations?

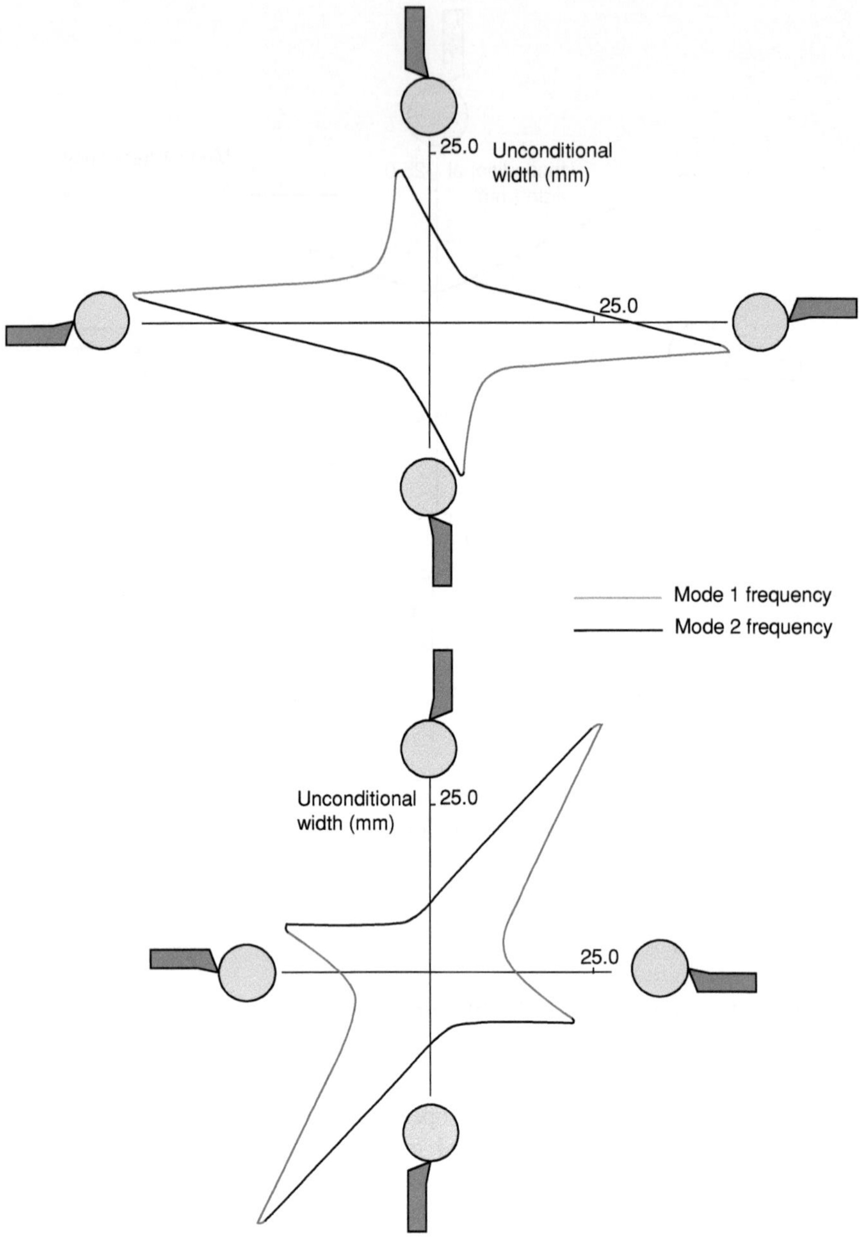

Fig. 2.11 Variation of unconditional width with circumferential position for inverted and conventional tool mounting (Program 2.6)

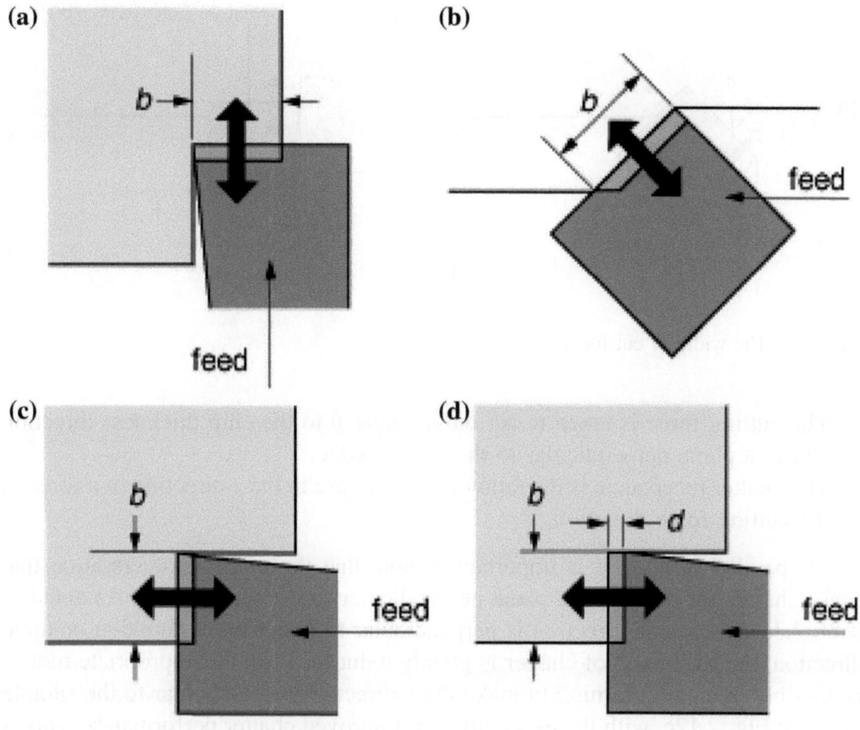

Fig. 2.12 The width of cut for various turning operations

2.5.1 *Turning*

If the case of parallel turning is considered, see Fig. 2.12b, the chip cross section is as shown. For the chatter model that has been developed, the variation in oscillating force depends in the first place on the varying area of this chip. The width of the chip is still the contact length of the cutting edge with the work. However, the direction (x) that most affects the chip thickness is not the feed direction. It is, however, perpendicular to the cutting edge and the cut surface. Thus, facing and parting off, which have the chip thickness direction the same as the feed direction, do not represent all machining operations. Also, for facing, the cutting force was considered to act in a plane that was perpendicular to the cutting edge. The direction was assumed to be at an angle β to the chip thickness direction. For parallel turning, the direction of the cutting force is assumed to be subject to the same rules. Thus, to begin to generalise,

1. The width of cut is the length of the cutting edge in contact with the work.
2. The chip thickness direction (x) is perpendicular to the cutting edge and the cut surface.

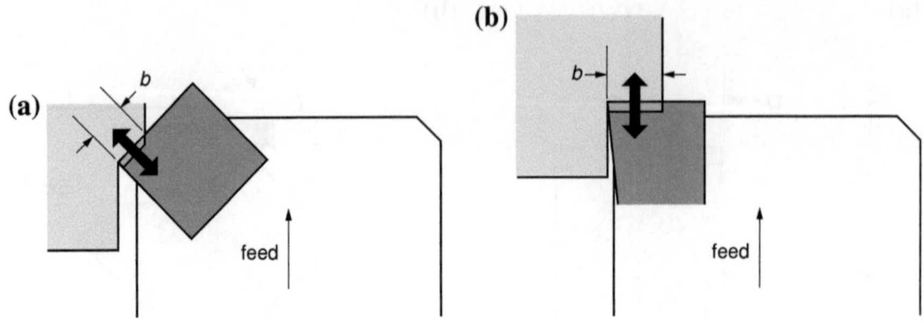

Fig. 2.13 The width of cut for boring

3. The cutting force is taken to act on an angle β to the chip thickness direction and in a plane perpendicular to the cutting edge.
4. The chatter receptance is the ratio of the response in the x direction to a force in the cutting force direction.

For parallel turning, it is important to note that the mode(s) of vibration that cause chatter are usually in a plane perpendicular to the spindle axis. As noted in Sect. 2.3, if the mode direction is perpendicular to either the x direction or force direction, the likelihood of chatter is greatly reduced. With the appropriate tool, it is possible in parallel turning to make the x direction perpendicular to the spindle axis, see Fig. 2.12c, with the expectation of improved chatter performance. This is found experimentally to be the case. However, if the tool has a sharp corner as shown, a spiral "thread" will be left on the cut surface. This may be overcome by using a tool shaped as in Fig. 2.12d. However, if the dimension d is too large, then vibration (and chatter) may occur in a direction perpendicular to the spindle axis.

2.5.2 Boring

Boring has great similarities to parallel turning. Figure 2.13a shows the width of cut and chip cross section for a typical operation. The width of the chip is the contact length of the cutting edge with the work, and the direction (x) that most affects the chip thickness is perpendicular to the cutting edge. Thus, boring may be considered as internal turning.

In practice, it is found that chatter is very common with long boring bars. This is because they have very little damping and easily vibrate in a transverse direction, i.e. perpendicular to their axes. Thus, it can be helpful to ensure that the chip width is perpendicular to the preferred vibration direction as shown in Fig. 2.13b. Methods of increasing the damping of boring bars will be considered further in Chap. 4.

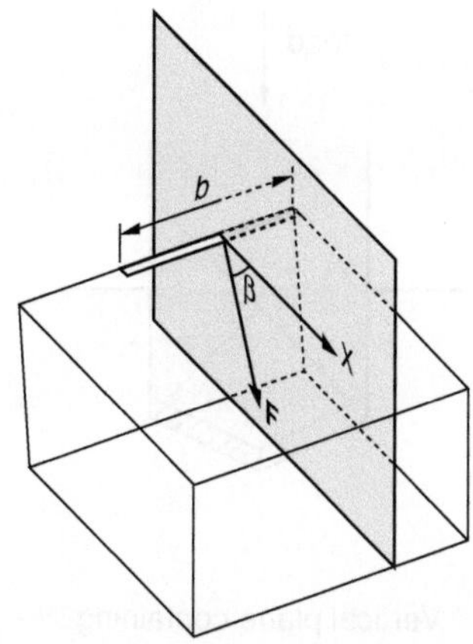

Fig. 2.14 Force and displacement directions for turning or boring

At this stage, it is possible to show the x direction and the force direction for any turning or boring operation, see Fig. 2.14. The x and F directions do not depend on the feed direction.

The cutting edge has so far been considered only to feed in a straight line, whether radial or axially. It is now necessary to consider what happens when the cutting edge is rotating as in drilling.

2.5.3 Drilling

For turning and boring, the cutting edge moved, but its orientation did not change, and the cutting edge did not rotate. In drilling, the complexity of chatter theory is now increased as there are two cutting edges and they are rotating and inclined to each other. Figure 2.15 shows, for each cutting edge, the chip cross sections. The x and F directions for each chip are also shown using the information from Fig. 2.14. The combination of the two forces produces a resultant force that acts along the axis of the drill. The question to be answered is "What is the direction of vibration that will most likely cause chatter by producing a variation in the chip thickness?" Transverse vibration of the drill (perpendicular to the feed) may be a problem on entry, but once drilling is in process, it is not normally a problem. An axial vibration will cause a variation of the same amount in both chips. It follows that chatter may occur in the axial direction of the drill. In this case, the effective total width, perpendicular to the direction affecting the chip thickness, is the diameter of the drill, see Fig. 2.16a.

The governing chatter receptance for drilling is thus the axial response resulting from an axial force. It is appropriate to restate that this receptance is that resulting from equal and opposite forces on drill and work and the response is the relative axial displacement between drill and work. Some of the earliest work on regenerative chatter was on radial arm drilling machines [4]. The theory developed previously is

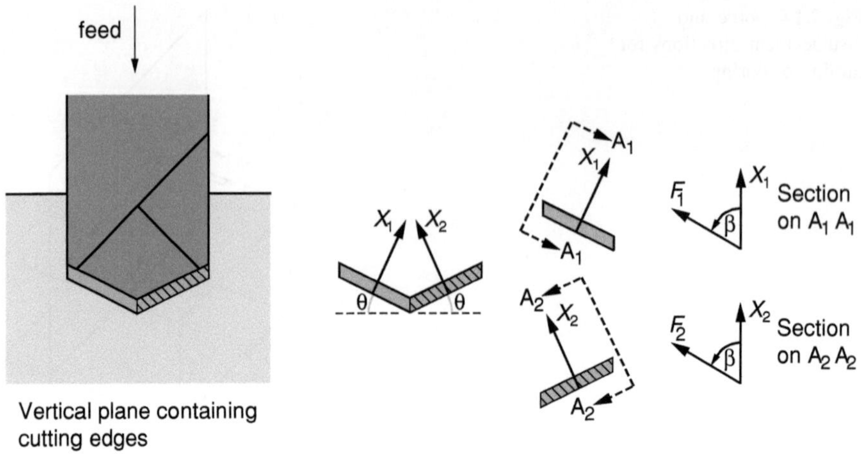

Fig. 2.15 Chip cross sections and forces for drilling

applicable with the value of b representing the diameter of the drill, but with one major change. For turning and boring, the time before a surface wave is cut again (τ) was the time for one revolution of the work. For drilling, it is the time between the two cutting edges. This is half the time for one revolution of the drill ($\tau/2$). As a result, Eq. 1.21 expressing the phase between the regenerative and non-regenerative components of the force in turning/boring is modified for drilling to

$$\omega\tau/2 = 3\pi - 2\alpha + 2n\pi \tag{2.3}$$

Hence, if the chatter receptance was the same, the stability lobes will occur at half the speed compared to turning/boring. However, the unconditional width of cut (i.e. drill diameter) is not affected—only the location of the stability lobes along the speed axis.

2.5.3.1 Enlarging Holes/Spot Facing

It is often the case when drilling that a hole is being enlarged (see Fig. 2.16b). In this case, the total width of cut b is the difference between the diameters of the drill and the existing hole.

Spot facing is very similar to drilling but with the angle θ in Fig. 2.15 zero. However, spot facing cutters often have four teeth resulting in a total width of cut b that is four times the width of cut for one tooth. Also the time between teeth is one quarter of the time for one revolution. Thus, Eq. 2.3 becomes

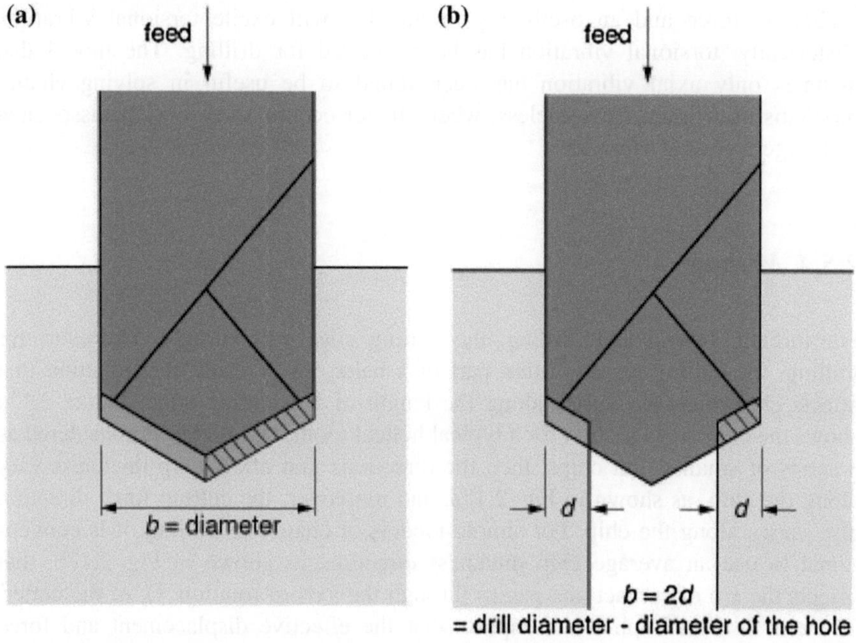

Fig. 2.16 The effective width of cut for drilling

$$\omega\tau/4 = 3\pi - 2\alpha + 2n\pi \tag{2.4}$$

As a result, the stability lobes occur at one quarter of the speed compared to turning/boring. In summary, the speed at which a stability lobe resides is inversely proportional to the number of cutting edges on the rotating tool.

2.5.3.2 Torsional Vibration

It was noted above that the resultant force in drilling acted along the axis of the drill. However, the forces acting on the two cutting edges also produce a torque. The question therefore arises as to whether it is possible to have torsional chatter in drilling? For chatter to involve torsional vibration, some form of regeneration must occur. For purely torsional vibration, without axial vibration, there would be no force variation unless the cutting force depends on speed. While any torsional vibration would cause a variation in speed, it would cause no variation in force and hence no chatter.

The practical situation is far more complicated. The cutting force is more complex than the model that simply relates force to the chip cross-sectional area. Also it is possible for torsional vibration and the axial vibration to be coupled by a structural mode. Even when not so coupled, an axial vibration causes both an oscillating force and an oscillating torque that will excite torsional vibration. Historically, torsional vibration has been ignored for drilling. The model that assumes only axial vibration has been found to be useful in solving chatter problems in drilling. Nevertheless, when chatter occurs, there may be associated and large torsional vibrations.

2.5.4 Milling

For turning, boring and drilling, the cutting edge was straight. However, for milling, the cutting edge is often part of a helix. As a result, the direction that affects chip thickness varies along the length of the cutting edge. Figure 2.17a shows the chip cross section for a typical helical tooth. If this chip is considered as a series of small-width chips, then the directions that affect chip thickness vary along the chip, as shown in Fig. 2.17a, and moreover, the cutting force direction also varies along the chip. For simple models of chatter in milling, it is conventional to use an average chip thickness direction, as shown in Fig. 2.17b, that bisects the arc of contact and passes through the axis of rotation, O, of the cutter. There is also an implicit assumption that the effective displacement and force direction are in a plane perpendicular to the axis of rotation of the cutter. However, the helix angle causes both the displacement and force to have components along the axis of rotation. Nevertheless, when looking for solutions to chatter in milling, these conventional assumptions have produced practical solutions.

In passing, it is important to note why a helix angle is employed on milling cutters. The width of contact with the work is often large, and a straight tooth would engage the full width at the same instant. This would give rise to large impacts on engagement and, when leaving the cut, exciting significant forced vibration in the drive train. The helix angle allows each tooth to gradually come into engagement and gradually leave the cut.

Returning to chatter, it is necessary to determine not only the effective displacement and force directions but also the effective width of cut. For practical reasons, the width of the work in contact with the cutter is defined as b, see Fig. 2.18. However, in the equations developed previously, b has been the length of the cutting edge in contact with the work. From Fig. 2.18, it may be seen that as the cutter rotates each cutting edge comes in contact with the arc of cut in such a way that the length in contact increases from zero to a maximum and then reduces to zero. For the example shown, the maximum length of contact remains constant for a short time. Also at any instant, more than one cutting edge may be in contact with the work. To use the theory developed previously, it is necessary to estimate an average length in contact. An approximation may be obtained by determining the average number of teeth in contact via the angle, θ, subtended by the arc of cut. From Fig. 2.18 $\cos\theta = \frac{R-d}{R}$, where R is the radius of the cutter and d the depth of cut.

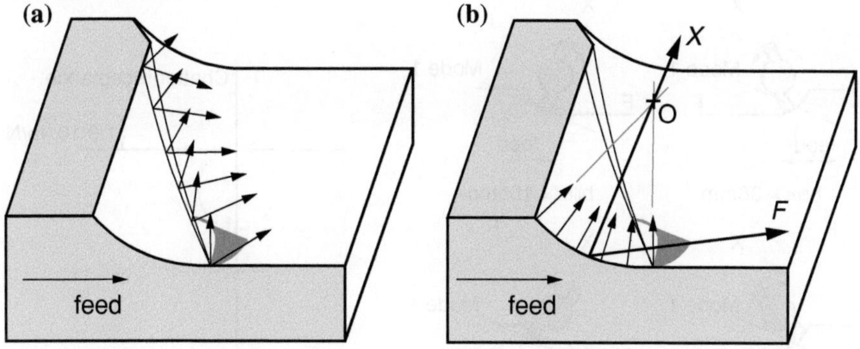

Fig. 2.17 Variation of displacement and force directions along a helical tooth

Fig. 2.18 Width of cut for milling

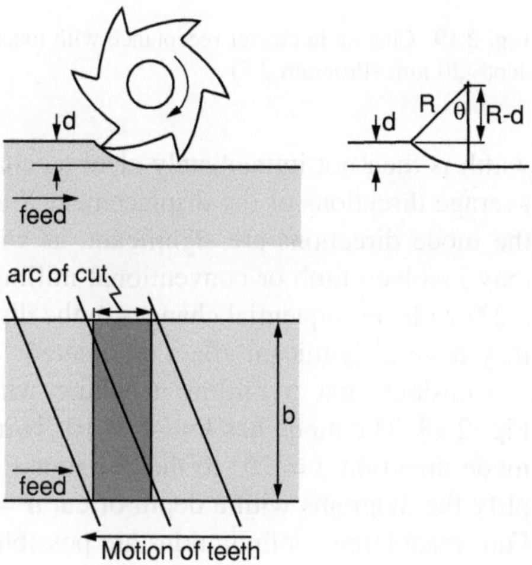

If there are N teeth on the cutter, the average number of teeth, z_c, in contact with the arc of cut will be $z_c = \frac{N\theta}{2\pi}$, where θ is in radians.

Thus, the average effective width in contact is $z_c b$ and substituting in Eq. 1.15 gives $z_c b = \frac{-1}{2RG_R(\omega)}$ so that at the stability boundary $b_{\text{lim}} = \frac{-1}{2Rz_c\, G_R(\omega)}$.

And similarly substituting in Eq. 1.20 gives

$$b_{\text{lim}} = \frac{1}{2Rz_c S\cos(180 - \alpha)} \tag{2.5}$$

Thus, the limiting work width, b_{lim}, above which chatter may occur is modified by the reciprocal of z_c. Provided all of the assumptions made are reasonable, this means that b_{lim} is reduced by having a larger number of teeth on the cutter. z_c is also increased when the depth of cut is increased. However, the effect on the stable

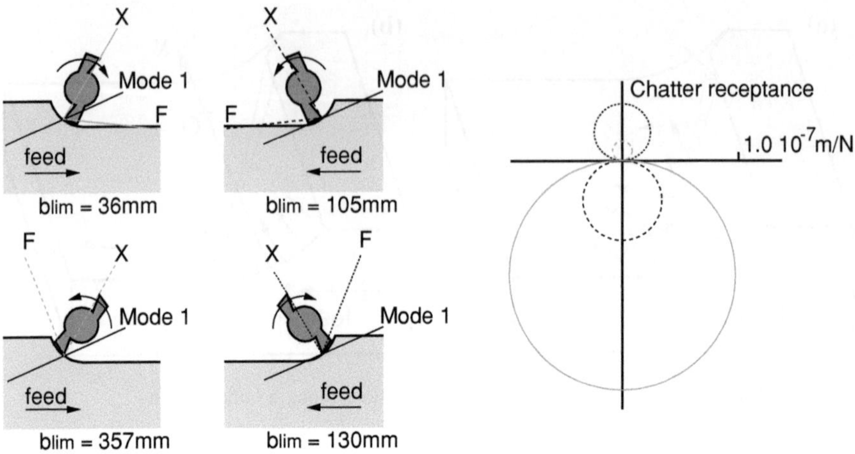

Fig. 2.19 Change in chatter receptance with machining configuration for a single mode, cutting depth 20 mm (Program 2.7)

width is then not immediately clear as changing the depth of cut will change the average directions of the displacement X and the force F. This in turn means that the mode directions are significant, as will be shown below. Milling operations may involve climb or conventional milling (as depicted in Figs. 2.19, 2.20, 2.21, 2.22) with consequential changes in the displacement and force directions, and this may have a significant effect on chatter.

Consider first a milling machine with one significant mode as shown in Fig. 2.19. The mode has $k = 4 \times 10^7$ N/m, $\omega_n = 100$ Hz, $\xi = 0.05$ and with the mode direction $\gamma = 20°$ to the horizontal. A two-tooth cutter was chosen to simplify the diagrams with a depth of cut $d = 20$ mm and a cutter radius of 30 mm. Four machining configurations are possible, and the chatter receptance for each is shown. The displacement direction, X, is normal to the arc of cut and the force at 60° to the displacement. For each configuration, the unconditional width of the work is shown for a cutting force coefficient $R = 4.0 \times 10^8$ N/m^2. Recall that there would be stability lobes on the associated stability charts. It is important to note that this simulation predicts a large improvement to be obtained by using the best configuration $b_{\text{lim}} = 357$ mm when climb milling feed left to right, compared to the worst, $b_{\text{lim}} = 36$ mm when conventional milling feed left to right.

The predicted variation in performance changes with the depth of cut. Figure 2.20 shows the variation when the depth of cut is increased to 50 mm, with all of the other variables remaining unchanged. Note that not only do the displacement and force directions change, but the average number of teeth in contact increases. For this case, the best configuration gives $b_{\text{lim}} = 74$ mm while climb milling feed right to left, and the worst gives $b_{\text{lim}} = 20$ mm while conventional milling feed right to left.

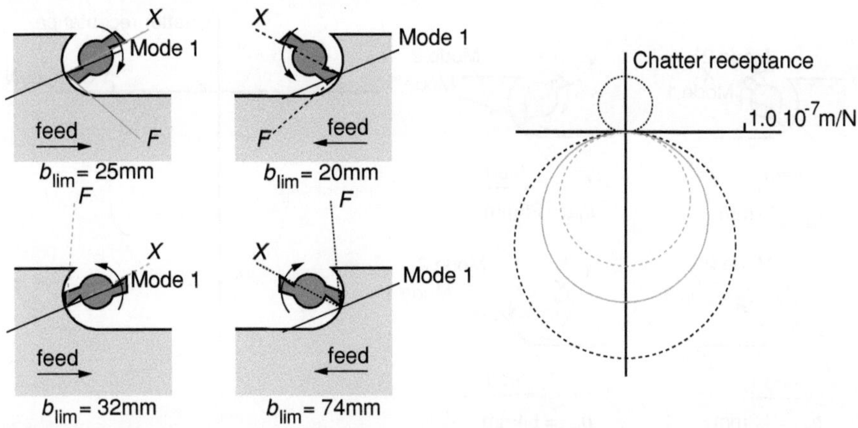

Fig. 2.20 Change in chatter receptance with machining configuration for a single mode, cutting depth 50 mm (Program 2.7)

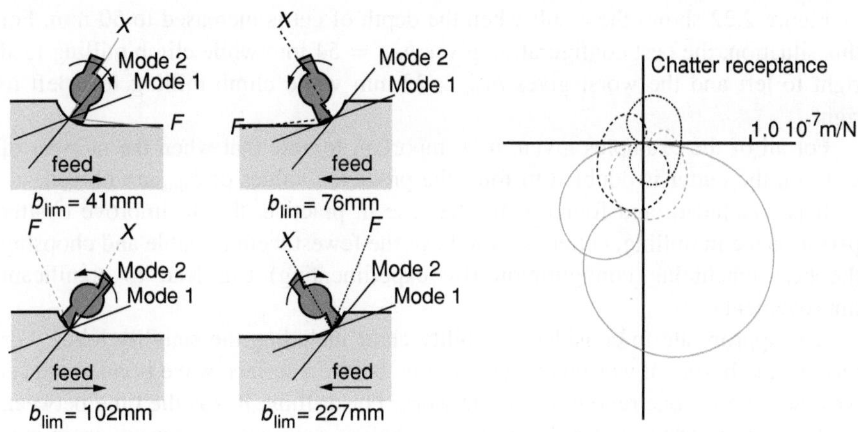

Fig. 2.21 Change in chatter receptance with machining configuration for two modes, cutting depth 20 mm (Program 2.8)

When there are several vibration modes with different directions, the situation is more involved. Figure 2.21 shows the comparative performance for a chatter receptance consisting of two modes with the following characteristics:

Mode 1: $k_1 = 4 \times 10^7$ N/m, $\omega_{n1} = 100$ Hz, $\xi_1 = 0.05$, mode direction $\gamma = 20^{\text{o}}$
Mode 2: $k_2 = 6 \times 10^7$ N/m, $\omega_{n2} = 120$ Hz, $\xi_2 = 0.05$, mode direction $\gamma = 50^{\text{o}}$

The other parameters are the same as those used to generate Fig. 2.19. For this situation, the best configuration gives $b_{\text{lim}} = 227$ mm while climb milling feed right to left and the worst gives $b_{\text{lim}} = 41$ mm while conventional milling feed left to right.

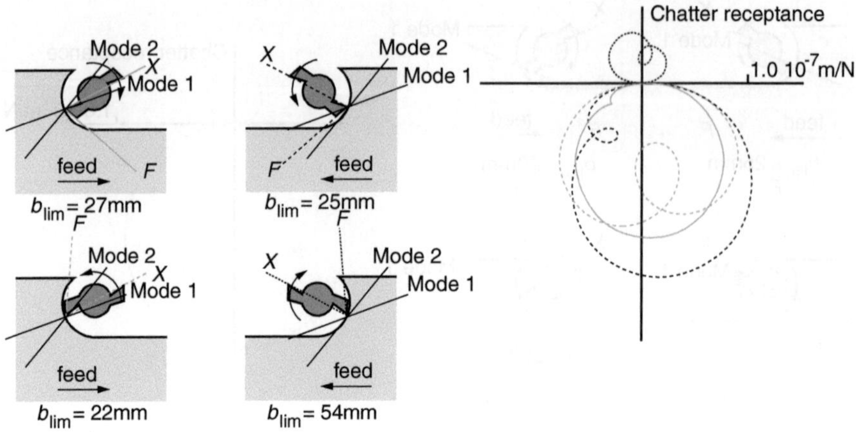

Fig. 2.22 Change in chatter receptance with machining configuration for two modes, cutting depth 50 mm (Program 2.8)

Figure 2.22 shows the result when the depth of cut is increased to 50 mm. For this situation, the best configuration gives $b_{\text{lim}} = 54$ mm while climb milling feed right to left and the worst gives $b_{\text{lim}} = 22$ mm while climb milling feed left to right.

For all of the examples given, it is important to note that when the number of teeth on the cutter is doubled to four, the predicted values of b_{lim} are halved.

It is concluded, and found to be the case in practice, that to improve chatter performance in milling, cutters should have the fewest teeth possible and choosing the best machining configuration (by experimenting) can lead to significant improvements.

It is appropriate to consider a stability chart including the stability lobes. For turning and boring, it was noted that the time before a surface wave is cut again (τ) was the time for one revolution of the work. For drilling, it was the time between the two cutting edges. For milling, the time before a surface wave is cut again is τ/N where N is the number of teeth and so, in general, the phase condition equation becomes

$$\omega\tau/N = 3\pi - 2\alpha + 2n\pi \tag{2.6}$$

Thus, the number of teeth on a cutter affects the unconditional stable width and also the location of the stability lobes along the speed axis of the stability chart. An example of a stability chart (without penetration rate damping) is shown in Fig. 2.23. The same parameter values were used as for Fig. 2.19 but for a cutter with four teeth. It is important to note that as the number of teeth is increased the last lobe occurs at a lower speed. This is significant for high-speed milling where machining may occur close to or above the last lobe.

The milling operation used for the examples given applies to slab milling or milling with an end mill. The operation known as face milling has some differences.

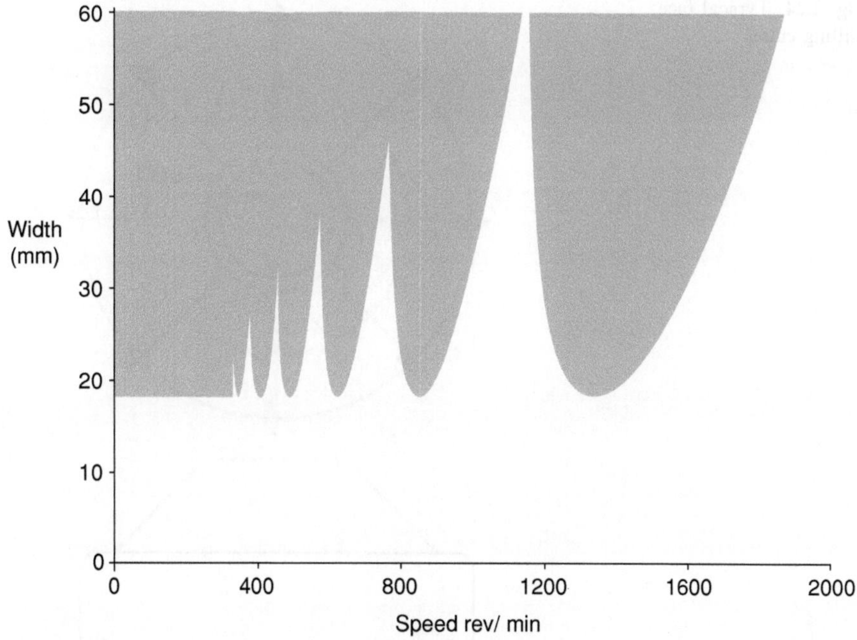

Fig. 2.23 Stability chart for milling with one significant mode (Program 2.9)

2.5.5 Face Milling

Figure 2.24 shows a typical face milling cutter. It is conventional for this application to determine the depth of cut (*d*) at the stability boundary. This depth is directly related to the width of the contact with each cutting edge, which is normally an insert. The motion of the inserts is directly analogous to that of the teeth during slab milling and end milling. Thus, there is an arc of cut and the average displacement direction and force direction are determined in the same way. The depth of cut, *d*, shown in Fig. 2.24, is here equivalent to the work width used above for slab milling.

If the width of the work while face milling is smaller than the cutter diameter as shown in Fig. 2.25, then there is the possibility of changing the directions of the average force and displacement relative to the mode directions of the machine. Figure 2.25 shows three possible machining configurations and the associated displacement and force directions. These three configurations show two extreme locations (a) and (c), where the cutter just overlaps the work. The configuration (b) is when the cutter is central to the work. Depending on the mode/s causing chatter and the mode directions, the three configurations may have significantly different stable depths of cut.

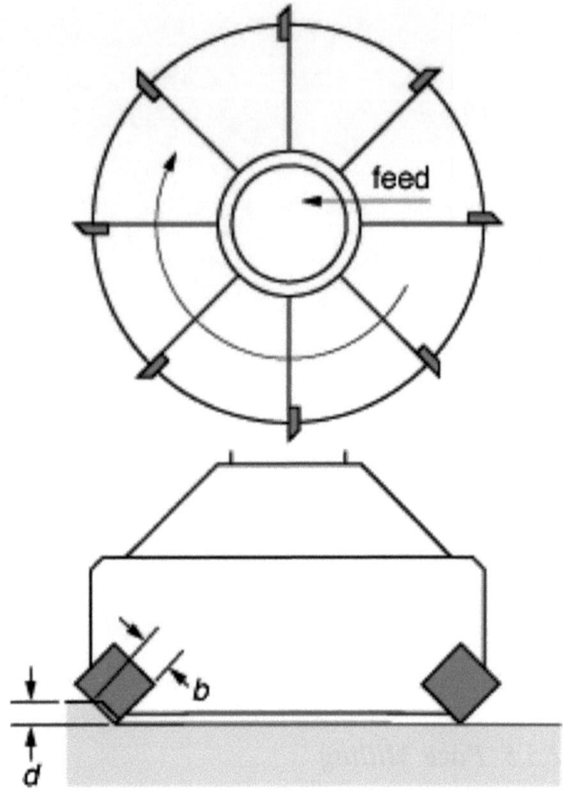

Fig. 2.24 Typical face milling cutter

2.6 Mode-Coupling Chatter

Now that multiple modes with associated mode directions have been considered, it is possible to describe another form of chatter. This is called mode-coupling chatter and occurs without any regeneration. For the machining operations considered thus far, the wave left by a cutting edge was subsequently cut by the same cutting edge or another tooth on the same tool. When turning a thread or turning with a large feed, any surface wave left by the cutting edge is not machined one revolution later, and so, there is no regeneration. Figure 2.26 shows this form of chatter when using a boring bar with a feed equal to the contact length of the tool. The feed/rev is the contact length b, and the depth of cut d.

Typically, boring bars have cross sections with different principal second moments of area that result in two vibration modes with mode directions at right angles. These modes may each be modelled as single-degree-of-freedom systems to illustrate this type of chatter. Figure 2.27 shows the orientation of the two modes relative to cutting edge.

For the simplified analysis that follows, it is assumed that the effective cutting speed of the work does not affect the force. Hence, the force is the cutting force

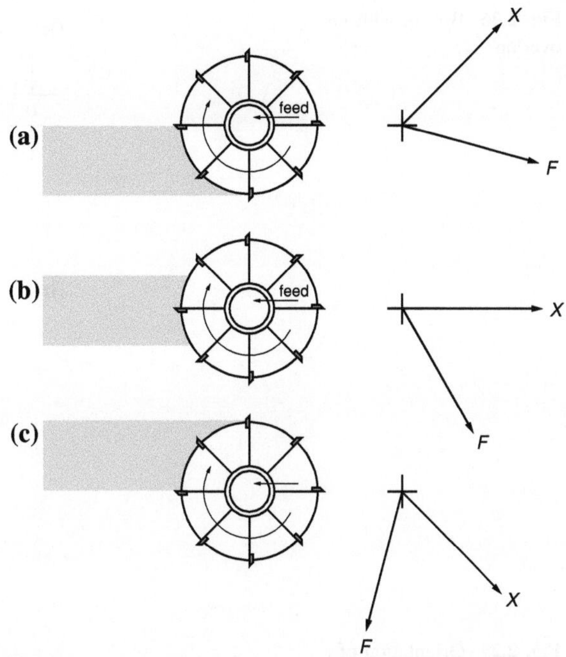

Fig. 2.25 Effect of position of face mill with respect to the work

coefficient (R) times the undeformed chip cross section (bd), i.e. Rbd. Let the cutting force act at an angle β from the horizontal as shown in Fig. 2.27.

Now, let Mode 1 be at an angle α from the horizontal. Then, for "thread cutting" with no overlap, i.e. no regeneration, for Mode 1, the equation of motion is

$$m_1 \frac{d^2 x_1(t)}{dt^2} + c_1 \frac{dx_1(t)}{dt} + k_1 x_1(t) = -Rb(x_1(t)\cos\alpha + x_2(t)\sin\alpha)\cos(\beta - \alpha), \tag{2.7}$$

and for Mode 2,

$$m_2 \frac{d^2 x_2(t)}{dt^2} + c_2 \frac{dx_2(t)}{dt} + k_2 x_2(t) = Rb(x_1(t)\cos\alpha + x_2(t)\sin\alpha)\sin(\beta - \alpha) \tag{2.8}$$

Assume solutions of the form $x_1(t) = A_1 e^{(\sigma + i\omega)t}$ and $x_2(t) = A_2 e^{(\sigma t + i[\omega t - \phi])}$ so that the motion is sinusoidal and growing exponentially. Substituting in Eqs. 2.7 and 2.8,

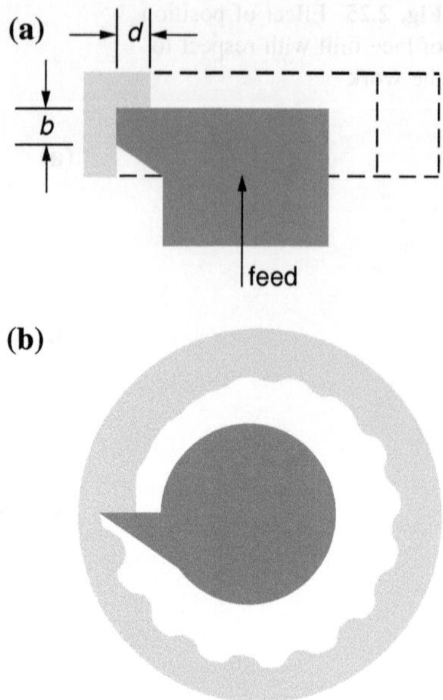

Fig. 2.26 Boring with no overlap

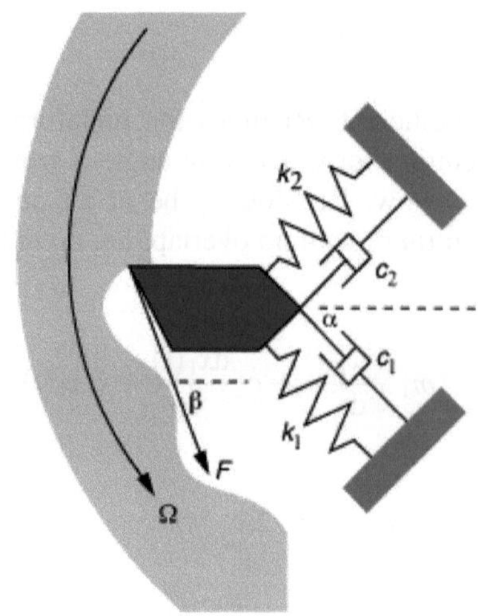

Fig. 2.27 Orientation of modes relative to the cutting position

$$m_1(\sigma + i\omega)^2 A_1 e^{(\sigma+i\omega)t} + c_1(\sigma + i\omega)A_1 e^{(\sigma+i\omega)t} + k_1 A_1 e^{(\sigma+i\omega)t} = -Rb\left(A_1 e^{(\sigma+i\omega)t} \cos\alpha + A_2 e^{(\sigma t+i[\omega t-\phi])} \sin\alpha\right) \cos(\beta - \alpha)$$

$$m_2(\sigma + i\omega)^2 A_2 e^{(\sigma t+i[\omega t-\phi])} + c_2(\sigma + i\omega)A_2 e^{(\sigma t+i[\omega t-\phi])} + k_2 A_2 e^{(\sigma t+i[\omega t-\phi])} = Rb\left(A_1 e^{(\sigma+i\omega)t} \cos\alpha + A_2 e^{(\sigma t+i[\omega t-\phi])} \sin\alpha\right) \sin(\beta - \alpha)$$

Expanding and dividing throughout by A_2, putting $A_1/A_2 = p$ and multiplying the second equation throughout by $e^{i\phi}$ gives

$$m_1(\sigma + i\omega)^2 p + c_1(\sigma + i\omega)p + k_1 p = -Rb\left(p\cos\alpha + e^{-i\phi}\sin\alpha\right)\cos(\beta - \alpha)$$
$$m_2(\sigma + i\omega)^2 + c_2(\sigma + i\omega) + k_2 = Rb\left(pe^{i\phi}\cos\alpha + \sin\alpha\right)\sin(\beta - \alpha)$$

Collecting real and imaginary parts of each equation,

$$m_1(\sigma^2 - \omega^2)p + c_1\sigma p + k_1 p = -Rb(p\cos\alpha + \cos\phi\sin\alpha)\cos(\beta - \alpha)$$
$$m_1 2\sigma\omega p + c_1\omega p = Rb\sin\phi\sin\alpha\cos(\beta - \alpha)$$
$$m_2(\sigma^2 - \omega^2) + c_2\sigma + k_2 = Rb(p\cos\phi\cos\alpha + \sin\alpha)\sin(\beta - \alpha)$$
$$m_2 2\sigma\omega + c_2\omega = Rbp\sin\phi\cos\alpha\sin(\beta - \alpha)$$

These represent four equations with four unknowns σ, ω, p and ϕ. To find the stability boundary, put $\sigma = 0$ with b as an unknown. Then,

$$(k_1 - m_1\omega^2)p = -Rb(p\cos\alpha + \cos\phi\sin\alpha)\cos(\beta - \alpha) \tag{2.9}$$

$$c_1\omega p = Rb\sin\phi\sin\alpha\cos(\beta - \alpha) \tag{2.10}$$

$$\left(k_2 - m_2\omega^2\right) = Rb(p\cos\phi\cos\alpha + \sin\alpha)\sin(\beta - \alpha) \tag{2.11}$$

$$c_2\omega = Rbp\sin\phi\cos\alpha\sin(\beta - \alpha) \tag{2.12}$$

Dividing Eq. 2.10 by Eq. 2.12 gives

$$\frac{\sin\alpha\cos(\beta - \alpha)}{p\cos\alpha\sin(\beta - \alpha)} = \frac{c_1\omega p}{c_2\omega}$$

and rearranging gives

$$p = \sqrt{\frac{c_2\sin\alpha\cos(\beta - \alpha)}{c_1\cos\alpha\sin(\beta - \alpha)}} \tag{2.13}$$

Divide Eq. 2.11 by Eq. 2.9

$$\frac{(p\cos\phi\cos\alpha + \sin\alpha)\sin(\beta - \alpha)}{-(p\cos\alpha + \cos\phi\sin\alpha)\cos(\beta - \alpha)} = \frac{(k_2 - m_2\omega^2)}{(k_1 - m_1\omega^2)p}$$

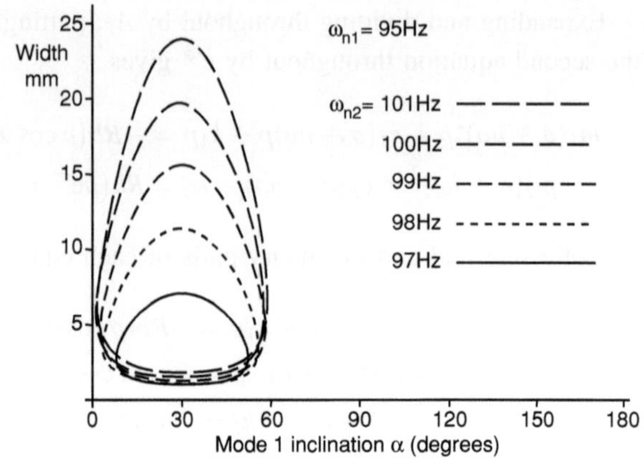

Fig. 2.28 Stability boundary for mode-coupling chatter (Program 2.10)

and rearranging to find cos ϕ

$$\cos\phi = -\frac{(k_1 - m_1\omega^2)p\sin\alpha\sin(\beta-\alpha) + (k_2 - m_2\omega^2)p\cos\alpha\cos(\beta-\alpha)}{((k_1 - m_1\omega^2)p^2\cos\alpha\sin(\beta-\alpha) + (k_2 - m_2\omega^2)\sin\alpha\cos(\beta-\alpha))} \tag{2.14}$$

Thus, we have p from Eq. 2.13 and for a given ω, we have ϕ from Eq. 2.14. From Eqs. 2.10 and 2.11, we can obtain two independent equations for b

$$b = \frac{c_1\omega p}{R\sin\phi\sin\alpha\cos(\beta-\alpha)}$$
$$b = \frac{(k_2 - m_2\omega^2)}{R(p\cos\phi\cos\alpha + \sin\alpha)\sin(\beta-\alpha)}$$

With a computer program, it is possible to search for value(s) of ω that give the same values of b from both equations. Figure 2.28 shows results for b against α for a boring bar with the following characteristics, $K_1 = 1.0 \times 10^7$ N/m, $\omega_{n1} = 95$ Hz, $\xi = 0.01$, cutting force direction $\beta = 60°$.

The different curves are for different values of the second natural frequency ω_{n2}. It is assumed that the effective mass for each mode is the same as a boring bar is being considered. Thus, $\omega_{n1} = \sqrt{\frac{k_1}{m}}$ and $\omega_{n2} = \sqrt{\frac{k_2}{m}}$ so that eliminating m gives $k_2 = \frac{\omega_{n2}^2}{\omega_{n1}^2}k_1$.

The viscous damping ratio is considered to be the same for each mode so that $c_1 = 2\xi\sqrt{mk_1}$ and $c_2 = 2\xi\sqrt{mk_2}$.

It is found that mode-coupling chatter may only occur when the direction of the lower-frequency mode lies between the displacement and force directions. As shown in Fig. 2.28, there are, for any particular value of α in the required range,

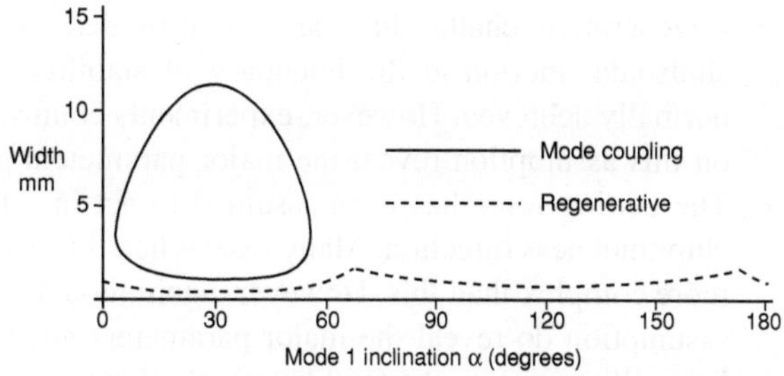

Fig. 2.29 Stability boundary for mode-coupling and regenerative chatter (Program 2.11)

normally two solutions for b. Instability occurs for values of b between these two values. Note that the solutions are independent of the rotational speed of the work.

The question arises as to how the limiting widths for mode-coupling chatter compare with those for regenerative chatter with the same structural model. Figure 2.29 shows a comparison for $k_1 = 1.0 \times 10^7$ N/m, $\omega_{n1} = 95$ Hz, $\omega_{n2} = 100$ Hz, $\xi = 0.01$ and the cutting force direction $\beta = 60^\circ$. The limiting width of cut for regenerative chatter is always significantly less than for mode-coupling chatter. For the model presented, mode-coupling chatter does not depend on speed. The boundary for regenerative chatter shown in Fig. 2.29 is the unconditional width.

Any particular machining operation can be much more complex than those chosen above to illustrate a single major effect. For example, in some machining operations, e.g. grinding (see Chap. 5), part of the surface wave regenerates but not the full width. Mode directions will not always be at right angles, as for the boring bar example above. Process damping will also have an effect when the surface waves have a short wavelength.

To illustrate the fact that mode-coupling chatter has a lower and upper stability boundary, a program (Program 2.12) was written that simulates boring. It is a time simulation of Eqs. 2.7 and 2.8. The simulation was written using a fourth-order Runge–Kutta approximation, and it is possible to vary the parameters and confirm that under certain conditions an increase in the width of cut will stop mode-coupling chatter. Using this animation, it will be found that when mode-coupling chatter occurs, the cutting edge moves anticlockwise around an ellipse. The consequence of this is that energy is input to the system and causes chatter.

2.7 Conclusions

The theory of regenerative chatter has been extended to a variety of machining operations, and several significant effects have been noted. In each case, assumptions were made so that the effect to be illustrated was highlighted. It is important to restate the assumptions that have been made and assess them.

- Regenerative chatter has been investigated assuming a constant-amplitude sinusoidal motion at the boundary of stability. In practice, no such state is normally achieved. However, experiments confirm that models of chatter based on this assumption reveal the major parameters governing chatter.
- The cutting force has been assumed to act in a fixed direction relative to the chip thickness direction. Many researchers have shown that the cutting force is more complex than this. However, again models of chatter based on the simple assumption do reveal the major parameters governing chatter.
- For milling cutters, the total length of all cutting edges in contact with the work varies with time as teeth enter and exit the arc of cut. Average values have therefore been assumed for both the displacement and the force directions. This assumption needs to be continuously borne in mind when considering solutions for chatter in milling.

This chapter has illustrated several methods that may usefully be employed to prevent chatter in addition to those presented in Chap. 1. Thus, the following solutions may be usefully employed.

- The total length of the cutting edge(s) in contact with the work should be reduced. For milling, this can be achieved by reducing the number of teeth on the cutter.
- The machining configuration should be chosen so that either the displacement or force direction is close to being perpendicular to the direction of the most significant mode.
- Machining between stability lobes may be possible. This is especially the case for high-speed milling when operating at a speed close to or above the last stability lobe.
- Process damping can be beneficial if the wavelength of chatter marks is small. This may be achieved by reducing the surface cutting speed.
- For mode-coupling chatter with a boring bar, chatter may be eliminated by ensuring that the direction of the lower-frequency mode does not lie between the displacement and force directions.

There is a considerable amount of literature on the topics covered in this chapter, a sample of which is given in the bibliography below. Many of these publications contain far more complex models of chatter, but it is found that the major parameters illustrated in this chapter remain the major ones affecting chatter.

Thus far, chatter theory has been developed to the stage where improvements can be made quite simply and experimentally. There are other ways of stopping chatter. In the next chapter, a range of chatter-resistant multi-tooth cutters will be presented and their performance assessed.

References

1. Gurney JP, Tobias SA (1961) A graphical method for the determination of the dynamic study of machine tools. Int J Mach Tool Des Res 1:148–156
2. Tlusty J, Polacek M (1968) Experience with analyzing stability of machine tool against chatter. In: Proceedings of 9th MTDR conference, vol 1, pp 521–570
3. Arnold RN (1946) The mechanism of tool vibration in the cutting of steel. Proc Inst Mech Eng 154(1):261–284
4. Tobias SA, Fishwick W (1956) The vibrations of radial drilling machines under test and working conditions. Proc Inst Mech Eng 170:232–247

Bibliography

Opitz H, Bernardi F (1969) Investigation and calculation of the chatter behaviour of lathes and milling machines. Annals of CIRP 18(1):335–343

Tlusty J (1986) Dynamics of high-speed milling. ASME J Eng Ind 108:509–687

Smith S, Tlusty J (1993) Efficient simulation programs for chatter in milling. Ann CIRP 42(1):463–466

Altintas Y, Budak E (1995) Analytical prediction of stability lobes in milling. Ann CIRP 44(1):357–362

Budak E, Altintas Y (1998) Analytical prediction of chatter stability in milling—part I: general formulation. J Dyn Sys Meas Control 120(1):22–30

Budak E, Altintas Y (1998) Analytical prediction of chatter stability in milling—part II: application of the general formulation to common milling systems. J Dyn Sys Meas Control 120(1):31–36

Altintas Y, Shamoto E, Lee P, Budak E (1999) Analytical prediction of stability lobes in ball end milling. Trans ASME J Manufact Sci Eng 121:586–592

Insperger T, Stepan G (2000) Stability of the milling process. Periodica Polytech Mech Eng 44(1):47–57

Li Z, Li XP (2000) Modelling and simulation of chatter in milling using predictive force model. J Mach Tools Manuf 40(14):2047–2071

Altintas Y (2001) Analytical prediction of three dimensional chatter stability in milling. Jpn Soc Mech Eng Int 44(3):717–723

Gasparetto A (2001) Eigenvalue analysis of mode-coupling chatter for machine-tool stabilization. J Vib Control 7(2):181–197

Altintas Y, Weck M (2004) Chatter stability in metal cutting and grinding. Ann CIRP 53(2):619–642

Merdol D, Altintas Y (2004) Multi frequency solution of chatter stability for low immersion milling. Trans ASME J Manufact Sci Eng 126(3):459–466

Li H, Shin YC (2006) Comprehensive dynamic end milling simulation model. ASME J Manuf Sci Eng 128(1):86–95

Altintas Y, Stepan G, Merdol D, Dombovari Z (2008) Chatter stability of milling in frequency and discrete time domain. CIRP J Manufact Sci Technol 1:35–44

Khachan S, Ismail F (2009) Machining chatter simulation in multi-axis milling using graphical method. Int J Mach Tools Manuf 49:163–170

Chapter 3
Chatter-Resistant Multi-Tooth Cutters

3.1 Introduction

It has been shown that for multi-tooth cutters, regenerative chatter arises when a small oscillation of a cutting edge results in a wave being left on the cut surface. The following tooth has to remove this wave and, depending on the width of the work, leaves a wave of smaller or greater amplitude. When the latter is the case, each succeeding tooth leaves a wave of greater amplitude, resulting in a vibration of large amplitude. The situation shown in Fig. 3.1 is a simplified representation with the circular motion of each tooth represented by a linear motion. Figure 3.1 shows the conditions existing at the boundary of stability when each tooth leaves a wave of the same amplitude. Any increase in the width of the work would result in the vibration amplitude increasing. It must be stressed again that though this neutrally stable situation would not normally be obtained in practice, it nevertheless predicts the stability boundary quite accurately.

In this chapter, various means of improving chatter performance by reducing the regenerative force will be considered. These methods involve modification of standard multi-tooth cutters. Clearly, if the regenerative force was reduced to zero, then regenerative chatter would not be possible. In order to indicate the order of magnitude of the improvement in chatter performance, a greatly simplified model of the cutting process will be used with a reference set of conditions. The chatter receptance will be represented by a single mode of vibration with the following characteristics.

$$k = 4.0 \times 10^7\,\mathrm{N/m};\quad \omega_n = 120\,\mathrm{Hz};\quad \xi = 0.05;$$

For any particular machine, the chatter receptance will be more complex than the single mode used here, but the improvements indicated by using a single mode will also indicate the order of magnitude likely to be achieved on a real machine. As shown in Chap. 2, the chatter receptance depends on the directions of the cutting force and the displacement direction (that affects the chip thickness). In milling, these directions change with the depth of cut, etc. However, in this

B. Stone, *Chatter and Machine Tools*, DOI: 10.1007/978-3-319-05236-6_3,

Fig. 3.1 Simplified stability boundary conditions for conventional milling

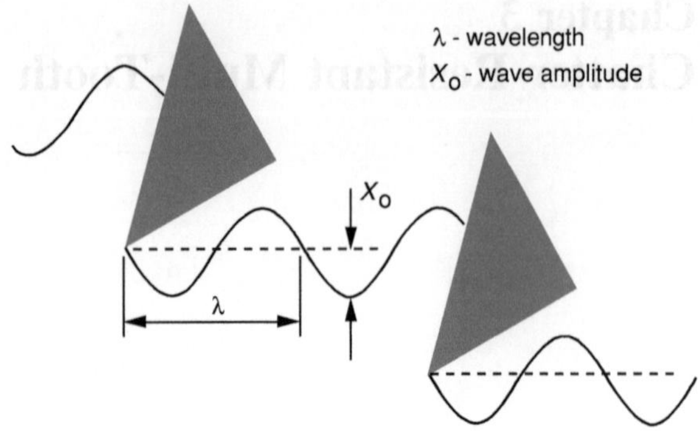

chapter, such changes will be ignored as the purpose is to investigate each type of cutter by fixing as many parameters as possible while varying just the one under investigation. Also, the effects of process damping will not be included though it must always be remembered that, for short wavelength chatter marks, process damping may be significant.

The cutting force coefficient will be assumed to be a constant $R = 4 \times 10^8$ N/m^2. This simplified model will indicate the order of magnitude likely to be achieved with the actual (more complex) forces present. The reference cutter has four teeth, a diameter of 60.0 mm and the depth of cut, d, is 20 mm as shown in Fig. 3.2.

Using the model of chatter developed in Chaps. 1 and 2, the stability chart for this reference cutter is as shown in Fig. 3.3.

Chatter-resistant cutters fall into four main types each of which reduce the resultant regenerative force and hence improve chatter performance. The first varies the pitch spacing of the teeth around the circumference and hence is called an alternating pitch cutter. The operating principle is to attempt to have the regenerative force on one tooth cancelled by the regenerative force on the next.

3.2 Alternating Pitch Cutters

The earliest suggestion of varying the pitch to prevent chatter appears to have been made by Hahn [1]. This was before models of regenerative chatter were more fully developed and later Slavicek [2] and Opitz et al. [3] presented models involving chatter theory. The significant parameter for these cutters is the amount by which the pitch of alternate teeth varies from the average pitch p. The pitch of one will be increased by an amount s and the next decreased by s. For any particular machining operation, s will be a proportion of the wavelength, λ, of the chatter marks left on the surface, see Fig. 3.4. To determine a quantitative measure of the improvement, it is helpful to add the regenerative components of chip variation on two adjacent teeth to obtain

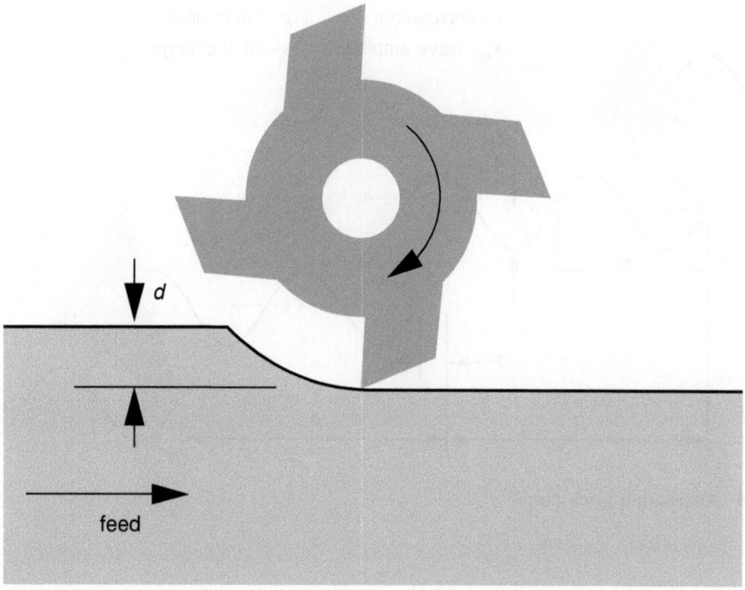

Fig. 3.2 Reference machining conditions for comparison of cutters

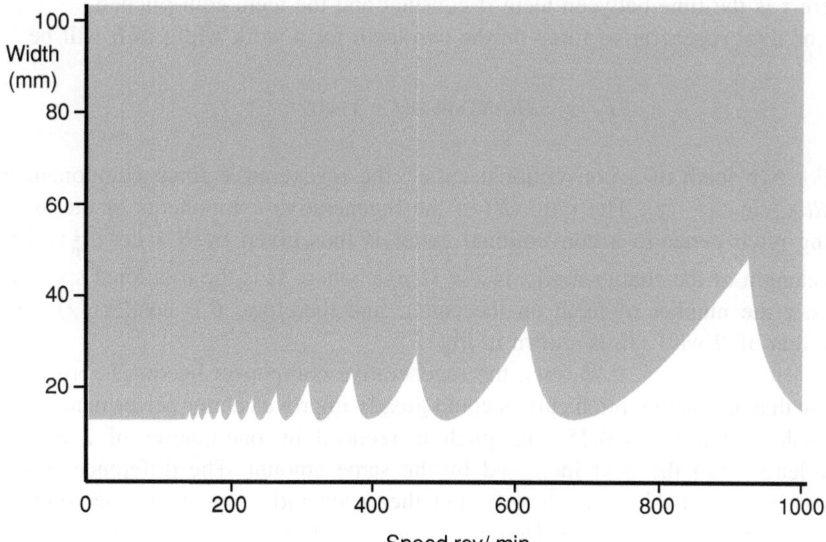

Fig. 3.3 Stability chart for the reference machining operation

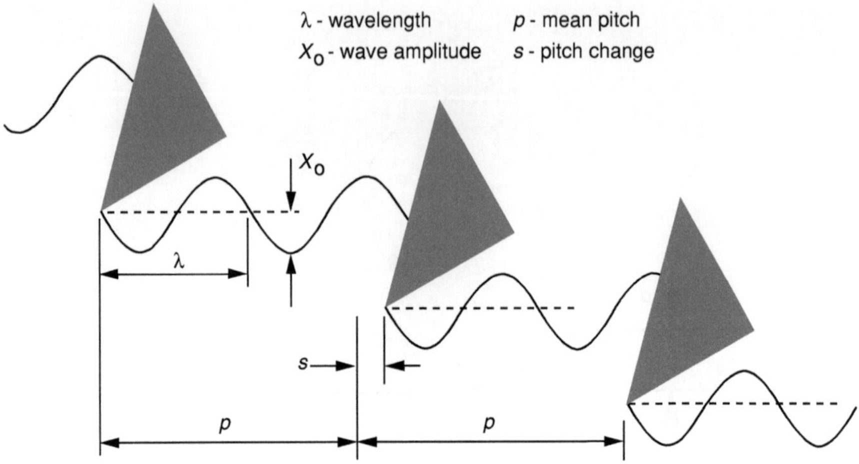

Fig. 3.4 Alternating pitch cutter

$$
\begin{aligned}
&- X_o \sin\omega\left(t - \frac{(p+s)}{p}\tau\right) - X_o \sin\omega\left(t - \frac{(p-s)}{p}\tau\right)\\
&= -2X_o \sin\omega(t-\tau)\cos\omega\frac{s}{p}\tau
\end{aligned}
$$

where τ is the time between teeth if constant and the teeth equi-pitched.

The total regenerative force on the two teeth for a work width of b will be

$$
F_r = -2RbX_o \sin\omega(t-\tau)\cos\omega\frac{s}{p}\tau
$$

For two teeth of a conventional cutter, the regenerative force component is $-2RbX_o \sin\omega(t-\tau)$. The ratio (*B*) of the regenerative components of the alternating pitch cutter to a conventional cutter is thus given by $B = \cos\omega\frac{s}{p}\tau$. The wavelength of the chatter marks is $\lambda = \Omega zp/\omega$ where Ω is the rotational speed in rad/s, z the number of teeth on the cutter, and therefore, $B = \cos(2\pi s/\lambda)$. The variation of B with s/λ is shown in Fig. 3.5.

When s/λ is 0.25, 0.75, etc., the regenerative component becomes zero indicating that alternating pitch cutters could greatly improve chatter performance. For example, when $s/\lambda = 0.25$, one pitch is reduced by one-quarter of a chatter wavelength and the next increased by the same amount. The difference in the pitches is then half a wavelength, and the regenerative force on one tooth is opposite to that on the other. However, for this to happen, it is necessary for the chatter frequency to not change when the alternating pitch cutter is used instead of its regular pitch counterpart. It is possible that chatter could occur at a different frequency. To investigate whether this is the case a stability chart needs to be

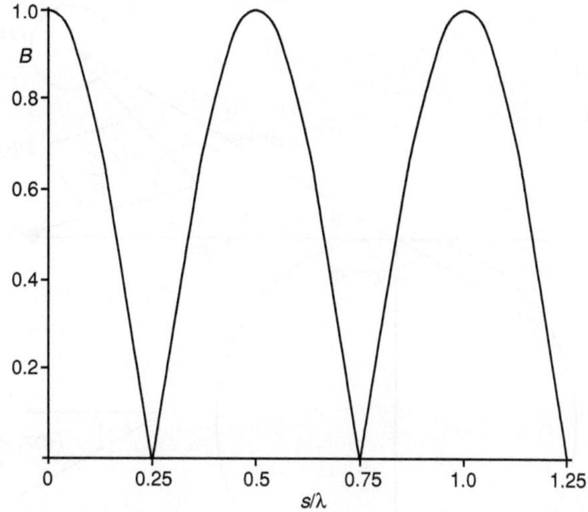

Fig. 3.5 Regenerative force ratio as a function of s/λ

derived to determine whether such cutters are effective over a wide speed range. The graphical method (described in Chap. 1) is very helpful in achieving this. Figure 3.6 shows the single-mode chatter receptance and the oscillating forces when the average number of teeth in contact is z_c. The non-regenerative force will be $-Rbz_cX_o \sin \omega t$. In the analysis above, the sum of the regenerative force on two teeth has been found. When the regenerative forces on the two teeth are adjusted to accommodate the fact that the average number of teeth in contact is z_c, their magnitudes both become $Rbz_cX_o/2$ but their phases are different. To improve the clarity of Fig. 3.6, these phases are not drawn to scale.

The computer program used to find the variation of the limiting stable width of cut with rotational speed of the cutter used the following steps.

First: the amount of alternating pitch was chosen and represented by defining s/p.

Second: values of $\omega\tau$ were stepped through to construct the chart. For each $\omega\tau$, the value of β was found using the geometry of Fig. 3.6 from

$$\tan\beta = \frac{\frac{Rbz_cX_o}{2}\sin\left(\frac{(p-s)}{p}\omega\tau - \pi\right) + \frac{Rbz_cX_o}{2}\sin\left(\frac{(p+s)}{p}\omega\tau - \pi\right)}{Rbz_cX_o + \frac{Rbz_cX_o}{2}\cos\left(\frac{(p-s)}{p}\omega\tau - \pi\right) + \frac{Rbz_cX_o}{2}\cos\left(\frac{(p+s)}{p}\omega\tau - \pi\right)}$$

$$\therefore \tan\beta = \frac{\sin\left(\left[1-\frac{s}{p}\right]\omega\tau - \pi\right) + \sin\left(\left[1+\frac{s}{p}\right]\omega\tau - \pi\right)}{2 + \cos\left(\left[1-\frac{s}{p}\right]\omega\tau - \pi\right) + \cos\left(\left[1+\frac{s}{p}\right]\omega\tau - \pi\right)}$$

Third: for the value of β obtained, the frequency ω, at which chatter must be occurring for the vector A to have the location shown in Fig. 3.6, is found from the phase angle equation for a single-mode system.

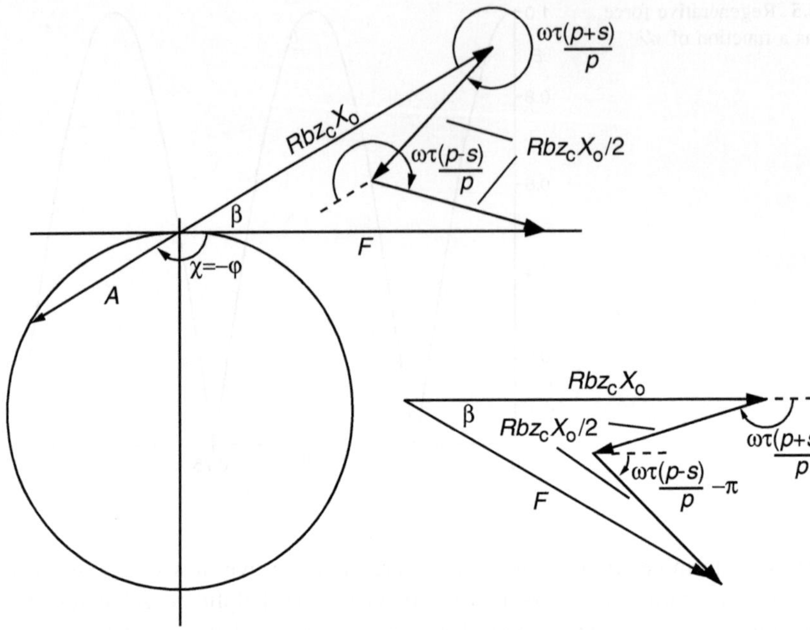

Fig. 3.6 Graphical solution of stability boundary at a chatter frequency ω

$$\tan\phi = \frac{-2\xi\omega/\omega_n}{\left(1-\frac{\omega^2}{\omega_n^2}\right)} \text{ (See Appendix A.4)} \quad \text{where } \phi = \beta - \pi$$

Rearranging this equation,

$$\tan\phi\frac{\omega^2}{\omega_n^2} - 2\xi\frac{\omega}{\omega_n} - \tan\phi = 0$$

and therefore,

$$\frac{\omega}{\omega_n} = \frac{2\xi \pm \sqrt{4\xi^2 + 4\tan^2\phi}}{2\tan\phi} = \frac{\xi}{\tan\phi} \pm \sqrt{1 + \frac{\xi^2}{\tan^2\phi}}$$

Only the positive root will give a positive value for ω. Now, the response for a single-mode system is, from Appendix A.4,

$$\frac{kX}{F} = \frac{1}{\left(\left(1-\frac{\omega^2}{\omega_n^2}\right) + 4\xi^2\frac{\omega^2}{\omega_n^2}\right)^{1/2}}$$

In the present case, $X = X_o$, and from the geometry of Fig. 3.6, the resultant force magnitude is

$$F = \frac{Rbz_cX_o}{2}\left[2 + \cos\left(\frac{(p-s)}{p}\omega\tau - \pi\right) + \cos\left(\frac{(p+s)}{p}\omega\tau - \pi\right)\right] \Big/ \cos\beta$$

Then,

$$\frac{X_o}{F} = \frac{1}{k\left(\left(1-\frac{\omega^2}{\omega_n^2}\right) + 4\xi^2\frac{\omega^2}{\omega_n^2}\right)^{1/2}}$$

$$= \frac{X_o\cos\beta}{\frac{Rbz_cX_o}{2}\left[2 + \cos\left(\frac{(p-s)}{p}\omega\tau - \pi\right) + \cos\left(\frac{(p+s)}{p}\omega\tau - \pi\right)\right]}$$

Solving for the width b_{lim} at the stability boundary gives

$$b_{\mathrm{lim}} = \frac{k\left(\left(1-\frac{\omega^2}{\omega_n^2}\right)^2 + 4\xi^2\frac{\omega^2}{\omega_n^2}\right)^{1/2} 2\cos\beta}{Rz_c\left[2 + \cos\left(\frac{(p-s)}{p}\omega\tau - \pi\right) + \cos\left(\frac{(p+s)}{p}\omega\tau - \pi\right)\right]}$$

As examples of the improvements in chatter performance that may be made, consider the reference set of conditions with $s/p = 5\ \%$. The corresponding stability chart is shown in Fig. 3.7 with the results for a standard cutter presented shaded for comparison.

At low speed, the lobes are very close together and so only their lower envelope has been shown (dotted). For the three lobes shown at the higher speeds, there are sometimes two solutions for b_{lim} at a given speed. In practice, chatter would occur above the lower width and hence the dotted vertical line shows the effective stability lobe. Note that there are ranges of speed where the performance is greatly improved. These correspond to the conditions where the regenerative force will be small because s/λ is 0.25, 0.75, etc., where λ is the wavelength of the chatter marks that would have been left by a standard equal-pitch cutter while vibrating at the frequency of the maximum negative real part of the chatter receptance. For different values of s/p, the ranges of speed where the performance is greatly improved are different. For example, Fig. 3.8 shows the stability chart when $s/p = 2.5\ \%$.

Using Program 3.1, the effect of varying the value of s/p may be investigated. In this program, the chatter receptance is not changed when the cutter diameter and depth of cut are changed (as it would in a comprehensive model) as it is the intention to isolate the alternating pitch effect. However, the effect of having more teeth in contact with increasing depth of cut is modelled in the program. It is clear that greatly improved stability can be achieved over a range of speeds. However, the stability charts shown in Figs. 3.7 and 3.8 are for a machine with a chatter receptance that can be represented by a single mode. In practice, machines have

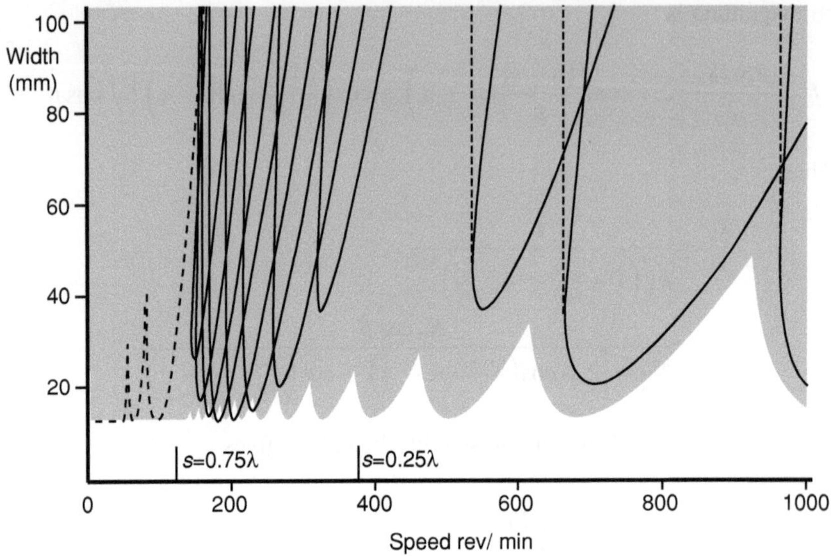

Fig. 3.7 Stability chart for an alternating pitch cutter with *s/p* = 5 % (Program 3.1)

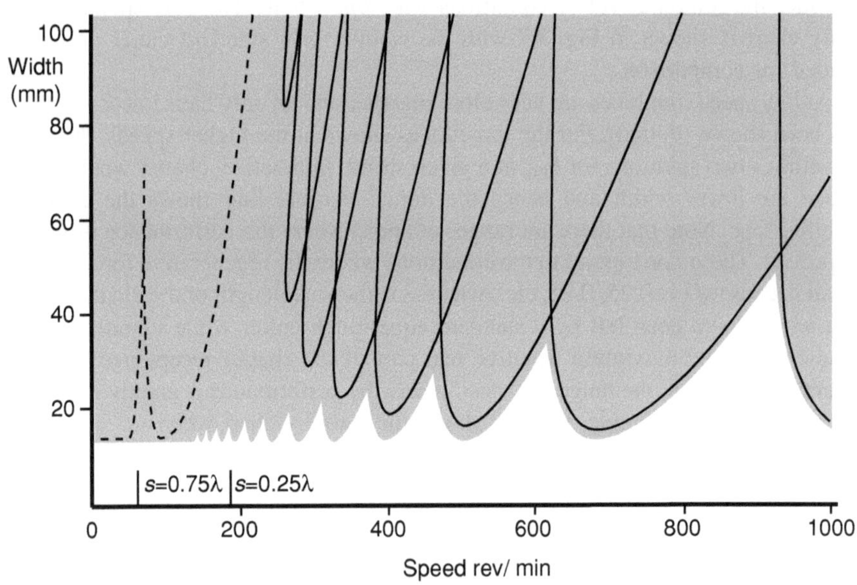

Fig. 3.8 Stability chart for an alternating pitch cutter with *s/p* = 2.5 % (Program 3.1)

many modes with different natural frequencies. Thus, though chatter close to one of the mode's natural frequencies may be avoided, it is not certain that another mode will have been improved at the desired rotational speed.

There is also the practical difficulty of producing the alternating pitch cutter. If a standard cutter is modified by removing material from the rake face of the cutter, then the radius of the cutting edge, from the centre of the cutter, is reduced because of the clearance angle. For all teeth to be cutting, additional material has to be removed from the clearance face of the teeth that are not ground back.

It is evident that it would be desirable to have a cutter with improved performance over the whole speed range and be effective for a machine with multiple modes. Such a cutter is the Strasman cutter.

3.3 Strasman (Ripping or Roughing) Cutters

Strasman cutters are named after their inventor but in different parts of the world are known as ripping or roughing cutters. An example of such a cutter is shown in Fig. 3.9. These cutters have what appears to be a thread machined around their periphery. The thread has a sinusoidal form. In fact, the profile is more complex than a simple thread cut around the periphery. This is because each cutting edge has to have a clearance angle. Considering the additional cost of manufacturing such a cutter, it is appropriate to ask why such a cutter would have been developed. As there is no record of aiming to improving chatter performance, the most likely objective was to break up the chip into many smaller widths rather than have one very wide chip that is difficult to handle. It was claimed and known [4] that these cutters required less power to remove metal than conventional cutters. It was only when making experimental comparisons of the chatter performance of multi-tooth cutters that it was observed that Strasman cutters gave significant chatter improvements. These improvements can be understood by considering the instantaneous chip thickness on one of the teeth. First, it is necessary to define some major parameters of these cutters. Figure 3.10 shows the amplitude and wavelength (L) of the sinusoidal form of the thread.

For the reference machining operation (a single mode of vibration with $k = 4.0 \times 10^7$ N/m; $\omega_n = 120$ Hz; $\xi = 0.05$; $R = 4 \times 10^8$ N/m^2, cutter with four teeth, diameter 60.0 mm and depth of cut 20 mm), the instantaneous chip cross section, as the cutting edge exits the cut, is shown in Fig. 3.11 for a Strasman cutter with a wavelength of 6 mm, an amplitude of 2 mm and a feed of 2 mm/rev. Due to the profile left by previous cutting teeth, the width in cut on each tooth is $0.36L$. Also, the surface of the chip being cut was machined in part by the tooth in cut one revolution previously (i.e. -4τ previously), partly by the previous tooth (i.e. $-\tau$ previously) and partly by the tooth which is in fact the tooth behind the tooth in cut (i.e. -3τ previously). The widths of these three surfaces are $0.11L$ for teeth -1 and -3 and $0.14L$ for the tooth in cut one revolution earlier (tooth -4).

Fig. 3.9 Strasman (or roughing/ripping) cutter showing the sinusoidal form of the cutting edge

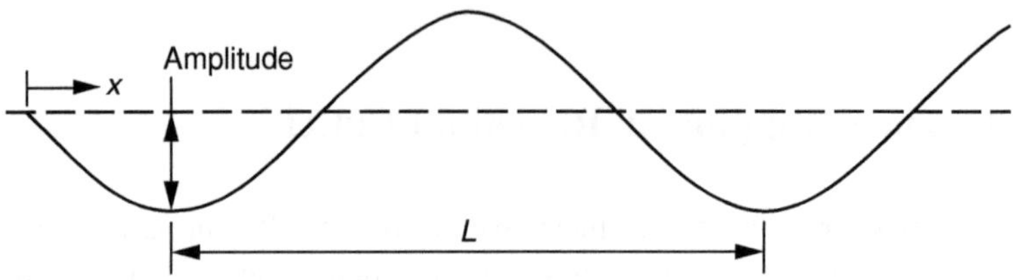

Fig. 3.10 Major parameters of a Strasman cutter

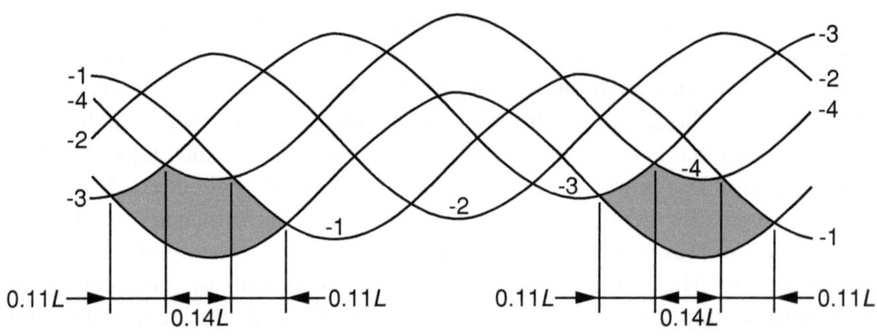

Fig. 3.11 Instantaneous chip cross section for a Strasman cutter with feed of 2 mm/rev

It is apparent that there are two main reasons why these cutters have improved chatter performance. First, the total width in cut is greatly reduced compared to the apparent width of cut. Second, the varying time delays, for the three parts of the surface profile, may have similar effects to those of the alternating pitch cutters.

Consider the regenerative force on one tooth for the conditions described. This is formed by three components with different time lags depending on which tooth last machined the surface. Thus, the regenerative force is

$$F_r = RbX_o[0.11 \sin\omega(t-\tau) + 0.14 \sin\omega(t-4\tau) + 0.11 \sin\omega(t-3\tau)]$$

where τ is the time between teeth.

Expanding,

$$F_r = RbX_o\begin{bmatrix} \sin\omega\tau(0.11 \cos\omega\tau + 0.14 \cos 4\omega\tau + 0.11 \cos 3\omega\tau) \\ -\cos\omega\tau(0.11 \sin\omega\tau + 0.14 \sin 4\omega\tau + 0.11 \sin 3\omega\tau) \end{bmatrix}$$

so that the magnitude of this regenerative force is

$$F_r = RbX_o\sqrt{\begin{matrix}(0.11\cos\omega\tau + 0.14\cos 4\omega\tau + 0.11\cos 3\omega\tau)^2 \\ +(0.11\sin\omega\tau + 0.14\sin 4\omega\tau + 0.11\sin 3\omega\tau)^2\end{matrix}}$$

For a conventional cutter with four teeth, the regenerative force on one tooth is $RbX_o \sin\omega(t-\tau)$ and this has an amplitude RbX_o. Thus, the ratio (B) of the regenerative force amplitude of the Strasman cutter to a conventional cutter is given by

$$B = \sqrt{\begin{matrix}(0.11\cos\omega\tau + 0.14\cos 4\omega\tau + 0.11\cos 3\omega\tau)^2 \\ +(0.11\sin\omega\tau + 0.14\sin 4\omega\tau + 0.11\sin 3\omega\tau)^2\end{matrix}}$$

The variation of B with $\omega\tau$ is shown in Fig. 3.12.

It is important to note that Fig. 3.12 is a specific case showing the reduction in the regenerative force for a particular Strasman cutter at the exit of cut with a particular depth of cut. As the cutting edge moves around the arc of cut, the proportions of the surface that were cut by the previous teeth vary. For the example above, this variation is shown in Fig. 3.13 where the cutting edge exits the arc of cut when θ just exceeds 70°.

The cutting forces at the exit condition are drawn on the response locus (see Fig. 3.14). The same graphical method of solution was adopted as was used for the alternating pitch cutters. The average number of teeth in contact was taken to be z_c. A value of $\omega\tau$ was chosen and the value of β found using the geometry of Fig. 3.14,

$$\tan\beta = \frac{0.11 \sin(\omega\tau-\pi) + 0.14 \sin(4\omega\tau-\pi) + 0.11 \sin(3\omega\tau-\pi)}{1 + 0.11 \cos(\omega\tau-\pi) + 0.14 \cos(4\omega\tau-\pi) + 0.11 \cos(3\omega\tau-\pi)}$$

As for the alternating pitch cutter,

$$\tan\phi = \frac{-2\xi\omega/\omega_n}{\left(1-\frac{\omega^2}{\omega_n^2}\right)}, \quad \text{where } \phi = \beta - \pi \text{ and}$$

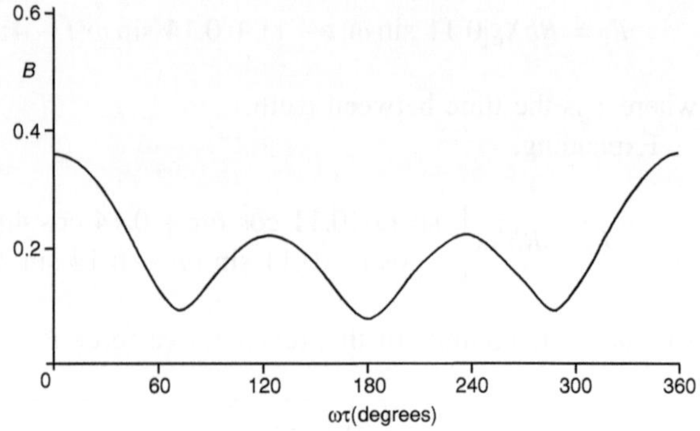

Fig. 3.12 Regenerative force ratio as a function of $\omega\tau$, for a Strasman cutter

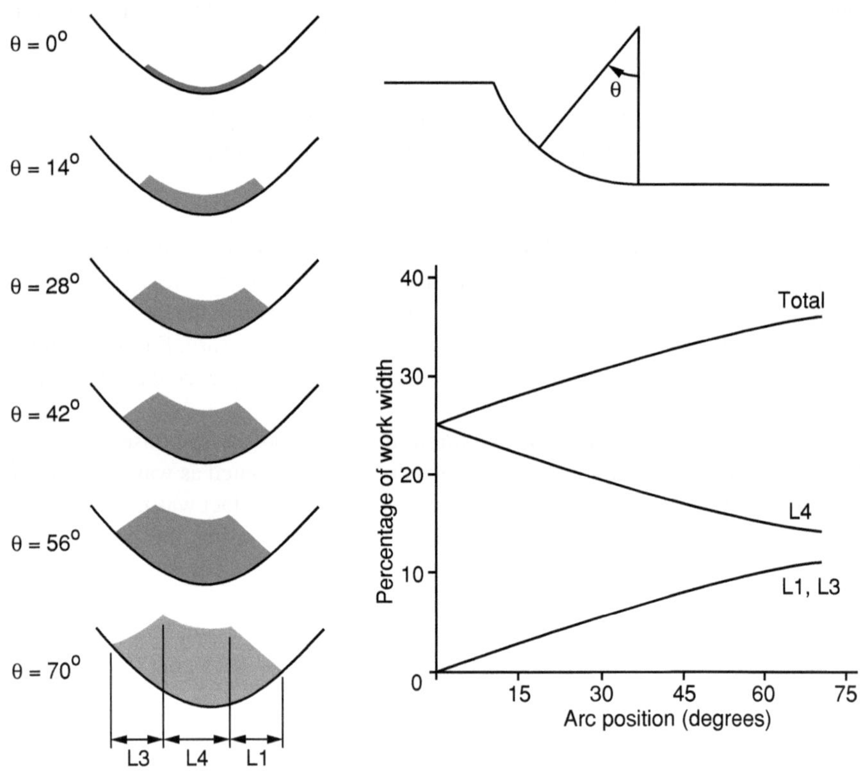

Fig. 3.13 Variation of chip cross section with rotation

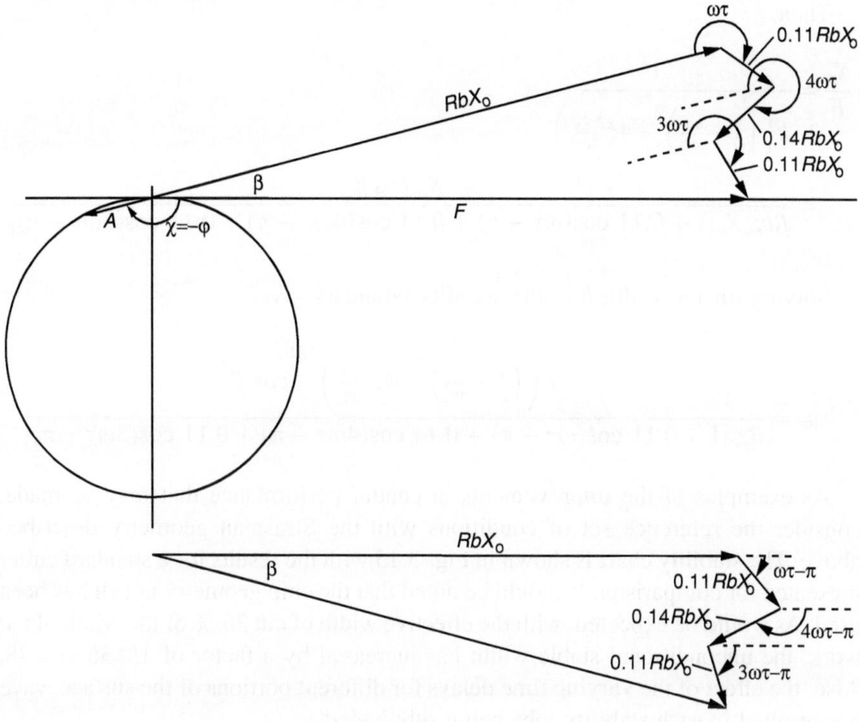

Fig. 3.14 Graphical solution of stability boundary at a chatter frequency ω

$$\frac{\omega}{\omega_n} = \frac{2\xi \pm \sqrt{4\xi^2 + 4\tan^2\phi}}{2\tan\phi} = \frac{\xi}{\tan\phi} \pm \sqrt{1 + \frac{\xi^2}{\tan^2\phi}}$$

Only the positive root will give a positive value for ω, and response is

$$\frac{kX}{F} = \frac{1}{\left(\left(1 - \frac{\omega^2}{\omega_n^2}\right)^2 + 4\xi^2\frac{\omega^2}{\omega_n^2}\right)^{1/2}}$$

Noting that $X = X_o$ and that from the geometry of Fig. 3.14, the force

$$F = \frac{Rbz_cX_o}{\cos\beta}[1 + 0.11\cos(\omega\tau - \pi) + 0.14\cos(4\omega\tau - \pi) + 0.11\cos(3\omega\tau - \pi)]$$

Then,

$$\frac{X_o}{F} = \frac{1}{k\left(\left(1-\frac{\omega^2}{\omega_n^2}\right)^2 + 4\xi^2\frac{\omega^2}{\omega_n^2}\right)^{1/2}}$$

$$= \frac{X_o \cos\beta}{Rbz_c X_o[1 + 0.11\cos(\omega\tau - \pi) + 0.14\cos(4\omega\tau - \pi) + 0.11\cos(3\omega\tau - \pi)]}$$

Solving for the width b at the stability boundary gives

$$b_{\lim} = \frac{k\left(\left(1-\frac{\omega^2}{\omega_n^2}\right)^2 + 4\xi^2\frac{\omega^2}{\omega_n^2}\right)^{1/2}\cos\beta}{Rz_c[1 + 0.11\cos(\omega\tau - \pi) + 0.14\cos(4\omega\tau - \pi) + 0.11\cos(3\omega\tau - \pi)]}$$

As examples of the improvements in chatter performance that may be made, consider the reference set of conditions with the Strasman geometry described above. The stability chart is shown in Fig. 3.15 with the results for a standard cutter presented for comparison. It should be noted that the chip geometry at exit has been used. As would be expected, with the effective width of cut 36 % of the width of the work, the unconditional stable width has increased by a factor of 1/0.36 = 2.78. Also, the effect of the varying time delays for different portions of the surface wave has resulted in each stability lobe being subdivided.

If Program 3.2 is run and the feed and number of teeth varied, it will be found that the chip cross section can become far more complex than the ones presented above. This is because the surface wave that is being machined may have portions left by more than three teeth. The program does not calculate for this situation, and no solution will be displayed.

The actual machining conditions are far more complex than those assumed above. The chip cross section varies around the arc of cut, as does the chatter receptance. The cutting forces also include penetration rate effects. However, the simplified model that has been presented indicates the major improvements that may be expected when using Strasman cutters. As predicted, in practice, these cutters do give greatly improved chatter performance and also have reduced power consumption [4].

A possible drawback to the use of Strasman cutters is that the tools have to be ground on the rake face when being re-sharpened. The major disadvantage from the use of these cutters is the quality of the surface finish that is far from flat because of the sinusoidal form on the cutting edge. Consequently, the profile of the cutting edge was changed by some manufacturers in an attempt to produce an improved surface finish. Such cutters are called trapezoidal cutters because of their tooth profile.

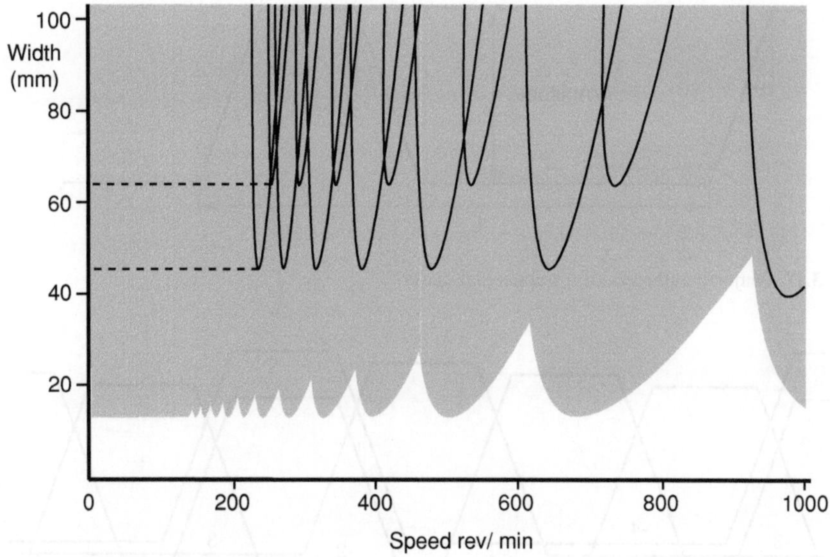

Fig. 3.15 Stability chart for a Strasman cutter (Program 3.2)

Fig. 3.16 Trapezoidal cutter showing the form of the cutting edge

3.4 Trapezoidal Cutters

An example of a trapezoidal cutter is shown in Fig. 3.16. This is similar to a Strasman cutter but with a significantly different form given to the profile of the cutting edge. As with the Strasman cutters, it is necessary to define some major parameters of these cutters. Figure 3.17 defines the amplitude and length (L) of the form of the thread. A significant parameter is ε, which in part determines the percentage of the profile that is not cutting. The angle θ defines the angle of the

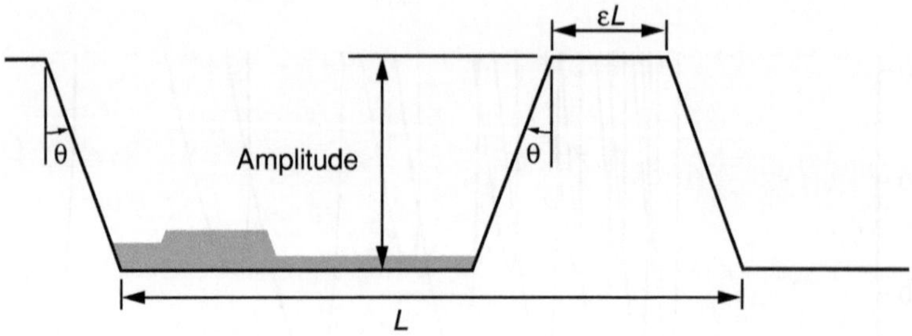

Fig. 3.17 Major parameters of a trapezoidal cutter

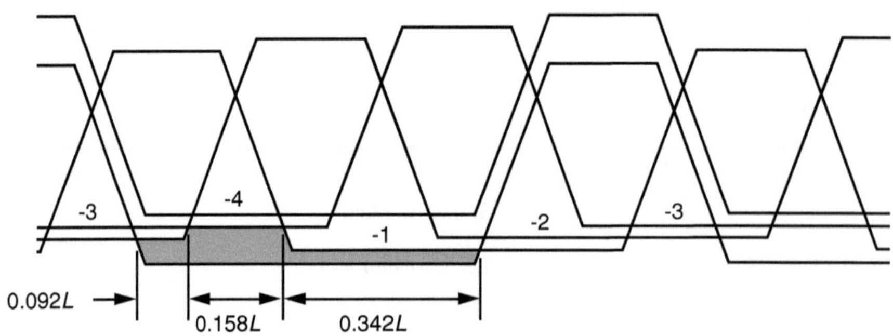

Fig. 3.18 Instantaneous chip cross section for a trapezoidal cutter with ε = 0.25 and a feed of 2 mm/rev

trapezoid's flanks. Most available trapezoidal cutters have a small value for ε. Figure 3.17 shows a case when $\varepsilon = 0.25$ and $\theta = 22.5°$.

Adopting the same approach as used for the Strasman cutter, an example cutter is chosen having an amplitude of 4 mm, $L = 12$ mm, $\varepsilon = 0.25$ and $\theta = 22.5°$ to investigate the standard conditions. The instantaneous chip cross section, as the cutting edge exits the cut, is shown in Fig. 3.18. Due to the profile remaining from previous teeth, the width in cut on each tooth is $0.592L$. As for the Strasman cutter, the surface of the chip being cut was machined in part by the tooth in cut one revolution previously (i.e. -4τ previously), partly by the previous tooth (i.e. $-\tau$ previously) and partly by the tooth behind the tooth in cut (i.e. -3τ previously). The widths of these three surfaces are $0.342L$ for tooth -1, $0.158L$ for tooth -3 and $0.092L$ for the tooth in cut one revolution earlier (tooth -4).

The regenerative force is

$$F_{\mathrm{r}} = RbX_{\mathrm{o}}[0.342 \sin \omega(t - \tau) + 0.158 \sin \omega(t - 4\tau) + 0.092 \sin \omega(t - 3\tau)]$$

This is very similar to a Strasman cutter but with different coefficients for each sine term. Therefore, following the same approach, the ratio (B) of the regenerative force amplitude of the trapezoidal cutter to that of a conventional cutter is given by

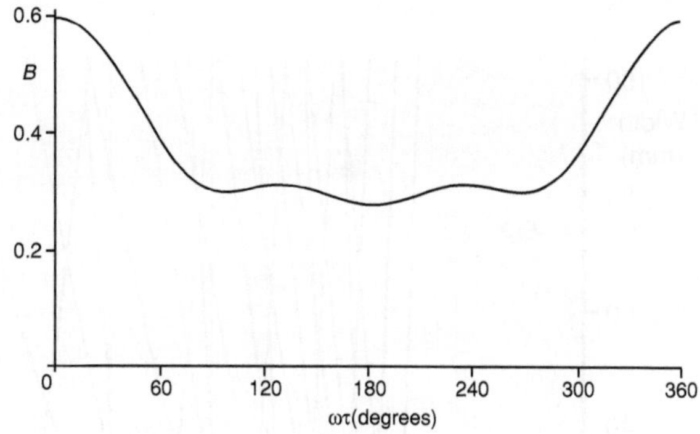

Fig. 3.19 Regenerative force ratio as a function of $\omega\tau$, for a Strasman cutter with $\varepsilon = 0.25$

$$B = \sqrt{\begin{aligned}&(0.342\cos\omega\tau + 0.158\cos 4\omega\tau + 0.092\cos 3\omega\tau)^2\\&+(0.342\sin\omega\tau + 0.158\sin 4\omega\tau + 0.092\sin 3\omega\tau)^2\end{aligned}}$$

The variation of B with $\omega\tau$ is shown in Fig. 3.19. It is clear that the reduction in the regenerative force is far less than for a Strasman cutter.

Using the same assumptions and method as for the Strasman chart shown in Fig. 3.15, a stability chart for the trapezoidal cutter may be calculated and is shown in Fig. 3.20. As expected, the unconditional width of cut has been increased by a factor of 1/0.592 = 1.69 because of the reduced length of the cutting edge in contact with the work.

Reconsidering the chip cross section of Fig. 3.18, it is apparent the width of cut may be reduced considerably by increasing ε. There is a practical limit to this reduction as the total width of the work must be machined by the cutter requiring each trapezoidal form to overlap the others. Figure 3.21 shows a trapezoidal cutter with the same parameter values as used previously but with $\varepsilon = 0.65$. The three values for the regenerating surface are now $0.0684L$ for teeth -1 and -3 and $0.1816L$ for the tooth in cut one revolution earlier (tooth -4). The total width of cut is now $0.318L$ compared to L for a conventional cutter. As the same amount of material is removed with $\varepsilon = 0.25$ and $\varepsilon = 0.65$, the maximum chip thickness is necessarily greater for $\varepsilon = 0.65$.

The ratio (B) of the regenerative force amplitude of this cutter to that of a conventional cutter is given by

$$B = \sqrt{\begin{aligned}&(0.0684\cos\omega\tau + 0.1816\cos 4\omega\tau + 0.0684\cos 3\omega\tau)^2\\&+(0.0684\sin\omega\tau + 0.1816\sin 4\omega\tau + 0.0684\sin 3\omega\tau)^2\end{aligned}}$$

The variation of B with $\omega\tau$ is shown in Fig. 3.22. It is clear that a further reduction in regenerative force has been achieved. It is therefore to be expected that the stability chart will show similar improvements and this is confirmed by Fig. 3.23. The unconditional stable width has been increased by a factor of $1/0.318 = 3.14$.

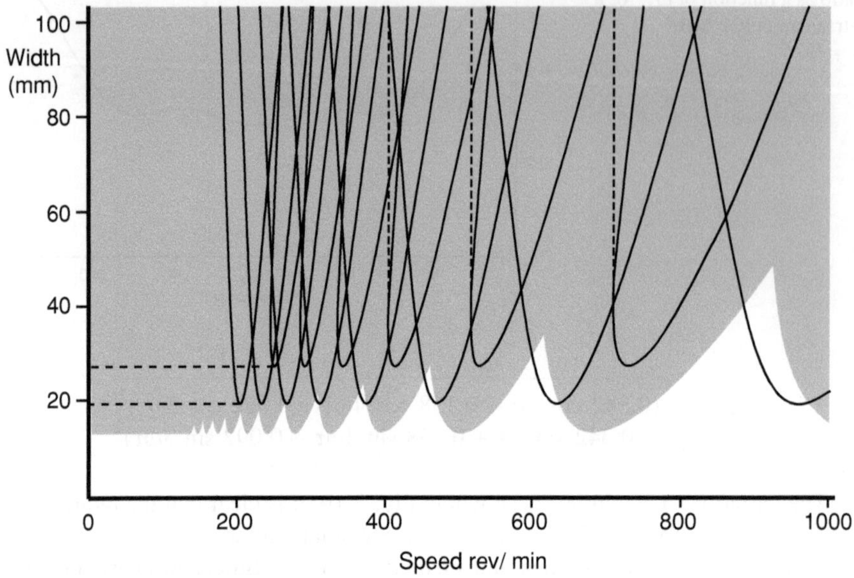

Fig. 3.20 Stability chart for a trapezoidal cutter with $\varepsilon = 0.25$ (Program 3.3)

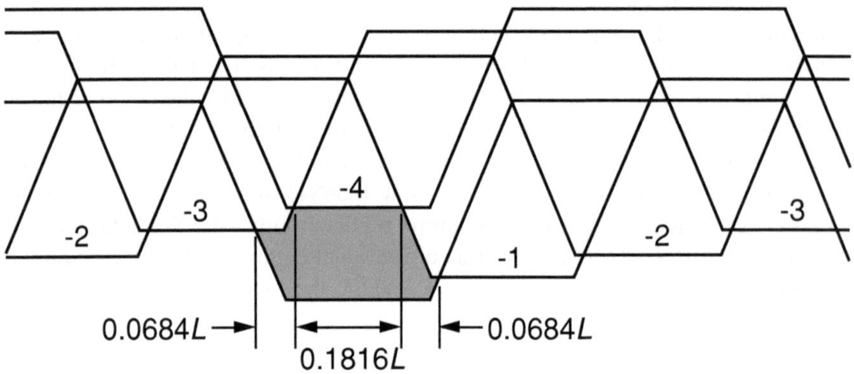

Fig. 3.21 Instantaneous chip cross section for a trapezoidal cutter with $\varepsilon = 0.25$

There is little in the literature about the chatter performance of trapezoidal cutters. They appear to have been first discussed by Pye and Stone [5], and it seems that most cutters of this type have not been optimised for chatter performance as the values of ε are normally small.

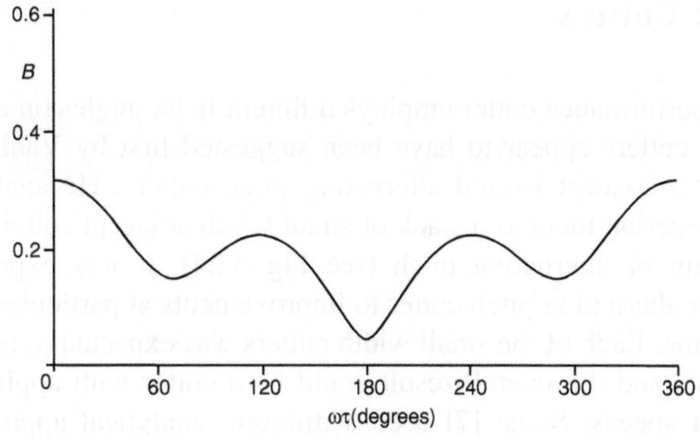

Fig. 3.22 Regenerative force ratio as a function of $\omega\tau$, for a Strasman cutter with $\varepsilon = 0.65$

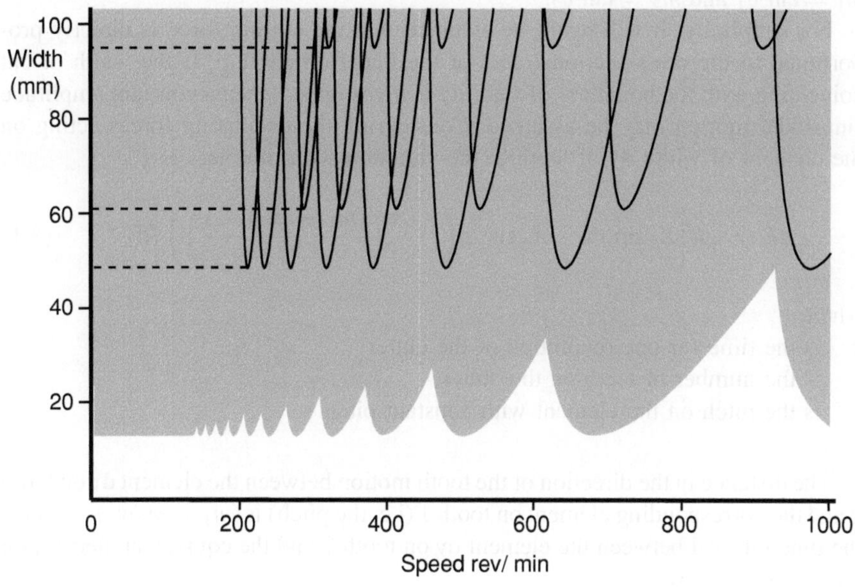

Fig. 3.23 Stability chart for a trapezoidal cutter with $\varepsilon = 0.65$ (Program 3.4)

The cutters used to explore the benefits of Strasman and trapezoidal cutters have had four teeth. In practice, such cutters may be found with more teeth and more than one "thread" is used around the periphery. If there are two threads, they are started on opposite sides of the cutter. They may be analysed in the same way as applied to the cutters that have been described with four teeth and one start on the thread.

3.5 Bi-helix Cutters

Another high-performance cutter employs different helix angles on adjacent teeth. These bi-helix cutters appear to have been suggested first by Vanherk [6] as an extension of the concept behind alternating pitch cutters. He analysed bi-helix cutters by considering them as a stack of small width adjacent cutters, each with a different amount of alternating pitch (see Fig. 3.24). It was expected that the limitation of an alternating pitch cutter to improvements at particular speeds could thus be overcome. Each of the small width cutters was expected to be effective for different speeds, and the overall result could be a cutter with application over a wider range of speeds. Stone [7] used a different analytical approach that was presented in a restricted publication [8]. The complete analysis is presented here as it shows that bi-helix cutters can deliver a greater improvement than that suggested by Vanherk's model that used a stack of alternating pitch cutters.

Consider a small width of the cutter dy, a distance y from the constant pitch line as shown in Fig. 3.25. Also, let the helix angles be θ_1 and θ_2 and also define $m_1 = \tan\,\theta_1$ and $m_2 = \tan\,\theta_2$.

For simplicity, it will again be assumed that the cutting force is directly proportional to the cross-sectional area of the undeformed chip. If the width of cut coinciding with the boundary of stability is investigated, then a constant amplitude sinusoidal motion may be assumed. Considering the oscillating forces acting on the element of width dy of tooth 2, the variation with time is

$$\mathrm{d}F = -R\left\lfloor X_\mathrm{o} \sin\,\omega t - X_\mathrm{o} \sin\,\omega\left(t - \frac{\tau}{z}\left(\frac{(m_1 - m_2)y + s}{s}\right)\right)\right\rfloor \mathrm{d}y \tag{3.1}$$

where

τ is the time for one revolution of the cutter

z is the number of teeth on the cutter

s is the pitch on the element with constant pitch.

The distance in the direction of the tooth motion between the element dy on tooth 2 and the corresponding element on tooth 1 (i.e. the pitch) is $(m_1 - m_2)y + s$. Thus, the time interval between the element dy on tooth 2 and the equivalent element on tooth 1 is given by

$$\frac{\tau}{z}\left(\frac{(m_1 - m_2)y + s}{s}\right)$$

and the total oscillating force on tooth 2 resulting from a chip width b normal to the direction of motion of the tooth is given by

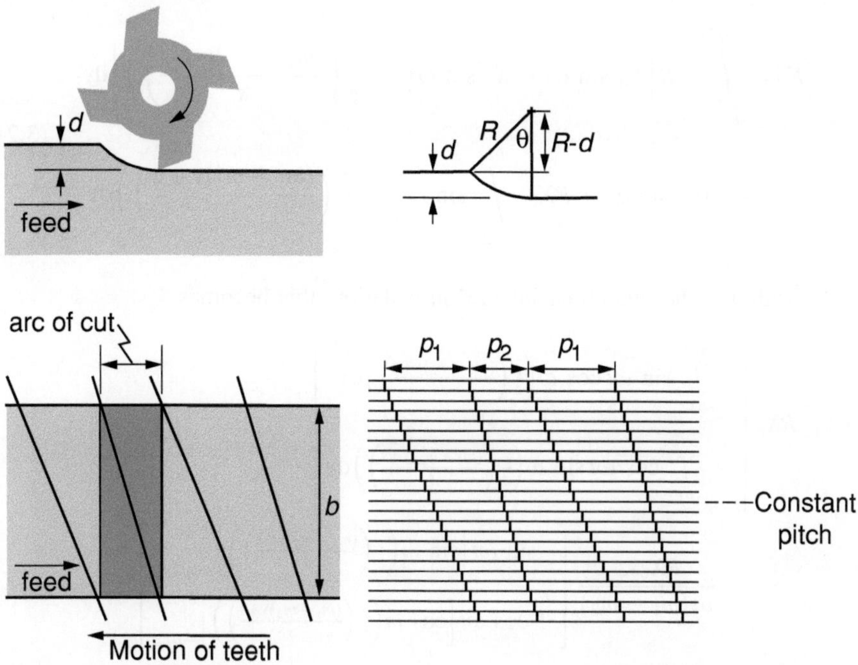

Fig. 3.24 Developed plan view of a bi-helix cutter and its representation by a series of alternating pitch cutters

Fig. 3.25 Two teeth of a bi-helix cutter (developed)

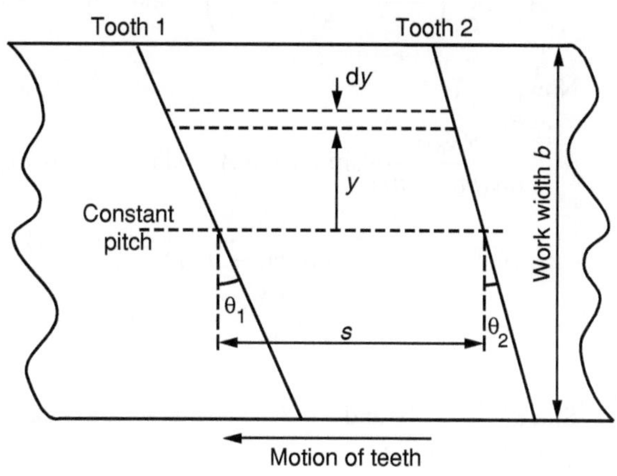

$$F = \int_{-b/2}^{b/2} -R\left[X_o \sin \omega t - X_o \sin \omega\left(t - \frac{\tau}{z}\left(\frac{(m_1 - m_2)y + s}{s}\right)\right)\right] dy$$

$$\therefore F = -RbX_o \sin \omega t + RX_o \int_{-b/2}^{b/2} \sin \omega\left(t - \frac{\tau}{z}\left(\frac{(m_1 - m_2)y + s}{s}\right)\right) dy \tag{3.2}$$

Considering the remaining integral in isolation, this becomes

$$RX_o \left[\begin{array}{l} \int_{-b/2}^{b/2} \sin \omega t \cos \omega\left(\frac{\tau}{z}\left(\frac{(m_1 - m_2)y+s}{s}\right)\right) dy \\ - \int_{-b/2}^{b/2} \cos \omega t \sin \omega\left(\frac{\tau}{z}\left(\frac{(m_1 - m_2)y+s}{s}\right)\right) dy \end{array}\right]$$

$$= \frac{RX_o sz}{\omega\tau(m_1 - m_2)} \left[\begin{array}{l} \sin \omega t\left[\sin \omega\left(\frac{\tau}{z}\left(\frac{(m_1 - m_2)y+s}{s}\right)\right)\right]_{-b/2}^{b/2} \\ + \cos \omega t\left[\cos \omega\left(\frac{\tau}{z}\left(\frac{(m_1 - m_2)y+s}{s}\right)\right)\right]_{-b/2}^{b/2} \end{array}\right]$$

$$= \frac{RX_o sz}{\omega\tau(m_1 - m_2)}[\sin \omega t(\sin A - \sin B) + \cos \omega t(\cos A - \cos B)]$$

where

$$A = \frac{\omega\tau}{z}\left(\frac{(m_1 - m_2)b/2 + s}{s}\right) \quad \text{and} \quad B = \frac{\omega\tau}{z}\left(\frac{-(m_1 - m_2)b/2 + s}{s}\right)$$

Now,

$$\frac{RX_o sz}{\omega\tau(m_1 - m_2)}[\sin \omega t(\sin A - \sin B) + \cos \omega t(\cos A - \cos B)]$$

$$= \frac{RX_o sz}{\omega\tau(m_1 - m_2)}[\cos(\omega t - A) - \cos(\omega t - B)]$$

$$= \frac{RX_o sz}{\omega\tau(m_1 - m_2)} 2 \sin\left[\frac{2\omega t - (A + B)}{2}\right] \sin\left[\frac{(A - B)}{2}\right]$$

Since $\frac{(A+B)}{2} = \frac{\omega\tau}{z}$ and $\frac{(A-B)}{2} = \frac{\omega\tau}{z}\frac{(m_1 - m_2)b}{2s}$, the total oscillating force acting on one tooth is, from Eq. 3.2,

$$F = -Rb\left[X_o \sin \omega t - \frac{2X_o sz}{\omega\tau(m_1 - m_2)b} \sin\left(\omega t - \frac{\omega\tau}{z}\right) \sin\left(\frac{\omega\tau}{z}\frac{(m_1 - m_2)b}{2s}\right)\right] \tag{3.3}$$

For a conventional cutter, $m_2 = m_1$ and hence, the regenerative term becomes

$$\frac{2sz}{\omega\tau(m_1 - m_2)b}\sin\left(\frac{\omega\tau}{z}\frac{(m_1 - m_2)b}{2s}\right) = 1$$

so that Eq. 3.3 reduces to the standard form for a milling operation

$$F = -RbX_o\left[\sin\,\omega t - \sin\left(\omega t - \frac{\omega\tau}{z_c}\right)\right] \tag{3.4}$$

Note that for the previous cutters in this chapter, τ was the time between teeth. Here, it is the time for one revolution of the cutter because the time between teeth varies along the cutting edge. It is clear that varying the helix angle of successive teeth modifies only the second term of Eq. 3.3, i.e. the regenerative force. At the boundary of stability, the ratio (B) of the regenerative force with varying helix angle to the regenerative force for a conventional cutter with equal helix angles is

$$B = \frac{2sz}{\omega\tau(m_1 - m_2)b}\sin\left(\frac{\omega\tau}{z}\frac{(m_1 - m_2)b}{2s}\right) \tag{3.5}$$

As would be expected as $m_1 \to m_2$, the ratio tends to unity.

The ratio of the regenerative forces, B, may be expressed in terms of the number of waves (λ) crossed by the second tooth. Consider tooth 1 leaving chatter marks as shown in Fig. 3.26.

Tooth 2 will cross some number λ of these over its length PQ. From the parallel-line intercept theorem, the number crossed on PQ is the same as the number crossed on QR. The length of QR $= b(m_1 - m_2)$ so that the length of one wave in the direction QR $= \frac{2\pi sz}{\omega\tau}$.

Therefore, the number of waves crossed $\lambda = \frac{b(m_1 - m_2)\omega\tau}{2\pi sz}$ and substituting in Eq. 3.5 gives a regenerative force ratio of $\frac{\sin(\pi\lambda)}{\pi\lambda}$.

The variation of this ratio (B) with the number of waves crossed is shown in Fig. 3.27 where it is seen that when an exact number of waves is crossed, the amplitude of the regenerative force becomes zero.

It is noted that by varying the helix angle of successive teeth and provided at least one wave is crossed, a reduction in the regenerative force by a factor of approximately five is assured. To illustrate this effect, the mechanism of the development of the regenerative force will be considered for both standard cutters and bi-helix cutters.

With constant helix angle cutting, the wave left by the preceding tooth will be parallel to the tooth in cut as may be seen in Fig. 3.28. The variation of the cross-sectional area of the chip as the tooth removes a complete wave may be determined by considering the tooth in each of the positions numbered 1–9. In the position drawn (i.e. position 1), the cross section of the chip is shown shaded and is the mean chip cross section. The chip cross sections at the positions 1–9 are shown in Fig. 3.28 where the area over or under the mean is shown shaded with 100 indicating the maximum variation from the mean. This is the important parameter

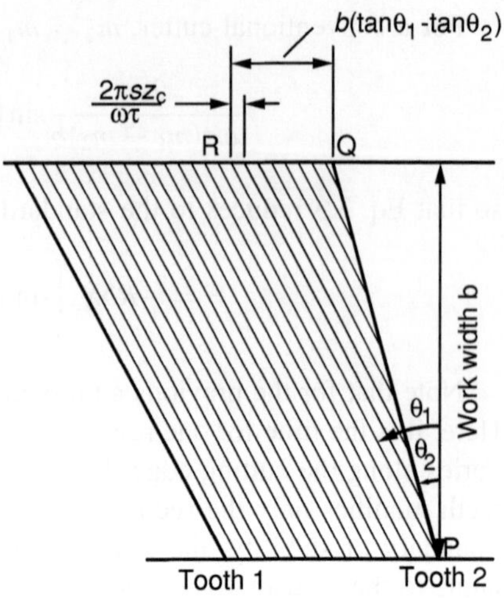

Fig. 3.26 Developed plan view of cutting operation, showing number of waves crossed by one tooth

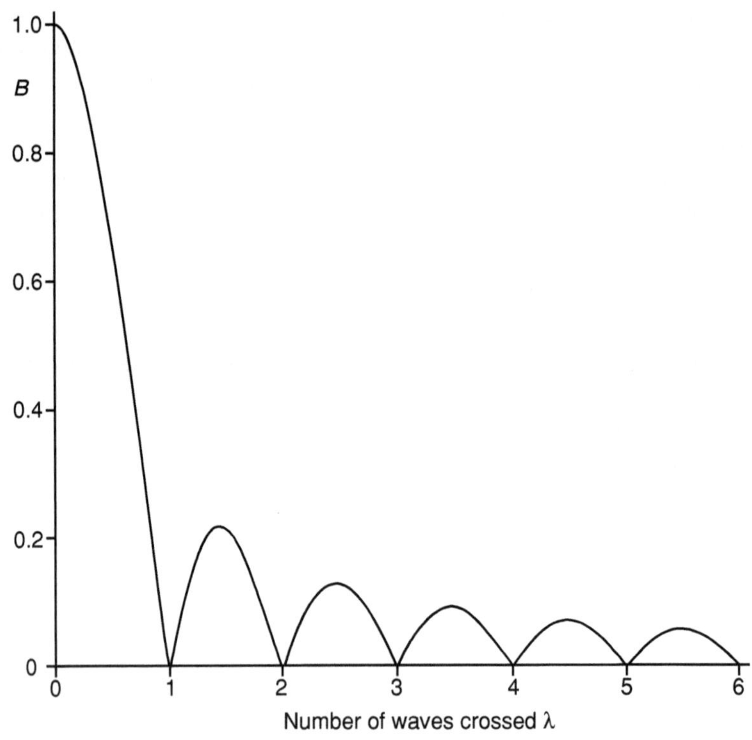

Fig. 3.27 Variation of amplitude of the generative force with number of waves crossed

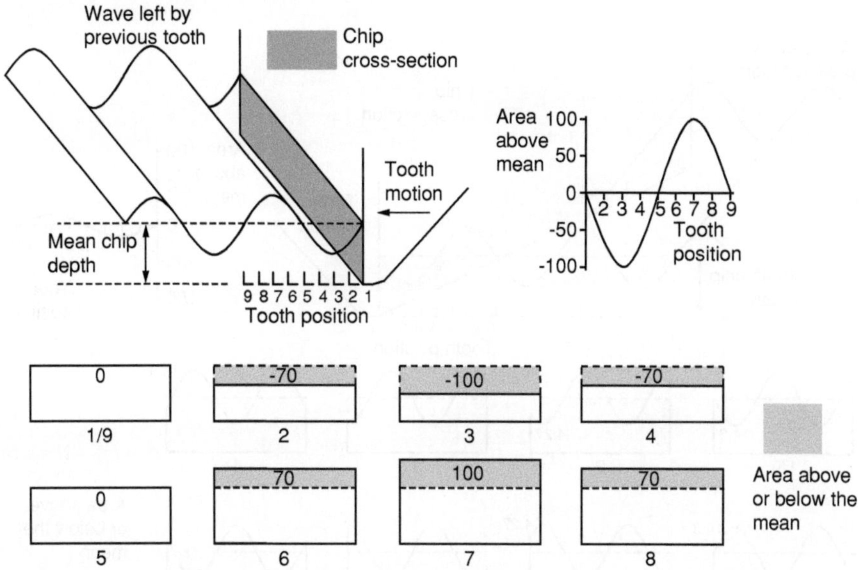

Fig. 3.28 Variation of chip cross section for constant helix cutting

as it is the oscillating forces that determine whether chatter occurs. If the force is proportional to the area of the cross section, an oscillating force of amplitude proportional to 100 acts on the tooth.

For the different-angle case, the removal of a wave of the same amplitude will be considered and, for comparison, the units of area will be the same. The difference in approach angle between sequential teeth results in the waves left by the preceding tooth being at an angle to the tooth in cut (Fig. 3.29). The tooth in cut will cross the waves left by the preceding tooth obliquely and may span more than one wave. As an example, consider a difference in angle that results in the tooth in cut crossing one and two-third waves. This condition is shown in Fig. 3.29 with the chip cross section shaded for the tooth position 1. The variation of the chip cross section as the tooth moves through positions 1–9 may be determined and is shown with the area over or under the mean shown shaded. It may be seen that at each position considered, the areas above and below the mean line are nearly equal. The residual area above or below the mean is greatly reduced compared with the constant helix case, and so the amplitude of the regenerative force has been greatly reduced.

Unlike the cutters considered previously, the bi-helix cutter poses several extra difficulties when attempting to see the effect on a stability chart. The main difficulty may be observed by considering the two cases shown in Fig. 3.30. If the work width is narrow and the arc of contact is long, then a situation similar to broaching exists as shown in Fig. 3.30a. However, if the reverse is true, then for a small arc of cut and a wide work piece, the cutting edges will appear to move

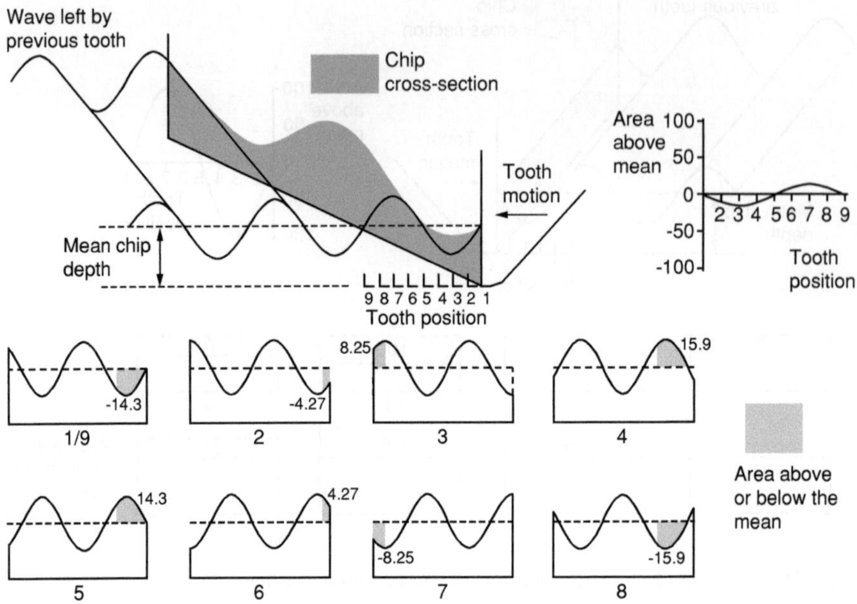

Fig. 3.29 Variation of chip cross section for bi-helix cutting

across the width of the work. This situation will be termed milling and is shown in Fig. 3.30b. A stability chart will be derived for both cases. Following the practice of examining effects in isolation, the effective chatter receptance for each case will be considered to be the same even though in practice this is unlikely.

3.5.1 Theoretical Stability Chart for Broaching

The same graphical approach will be used as for the other cutters already analysed. Also, as far as possible, the same standard conditions will be used. Figure 3.31 shows the chatter receptance locus and the force components when the average number of teeth in contact is z_c.

From Eq. 3.3, including the number of teeth in contact, the non-regenerative force is $-z_c RbX_o \sin \omega t$ and the regenerative force is

$$z_c Rb \frac{2X_o sz}{\omega\tau(m_1 - m_2)b} \sin\left(\omega t - \frac{\omega\tau}{z}\right) \sin\left(\frac{\omega\tau}{z}\frac{(m_1 - m_2)b}{2s}\right)$$

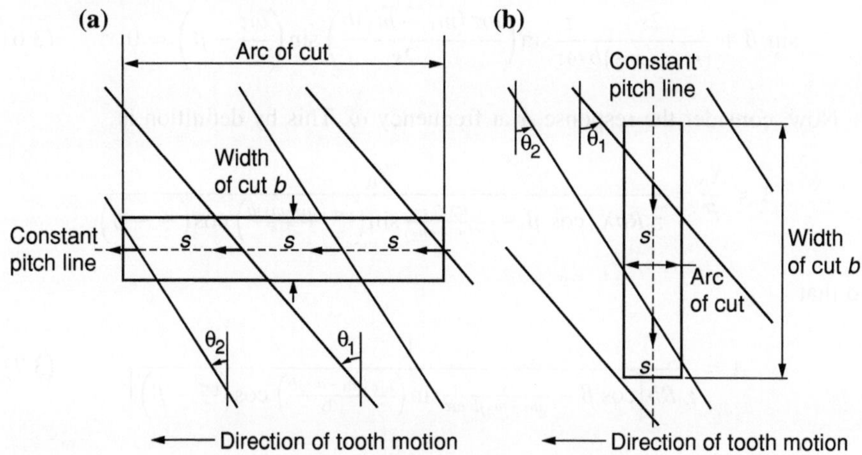

Fig. 3.30 Arc of cut and work width for **a** broaching and **b** milling

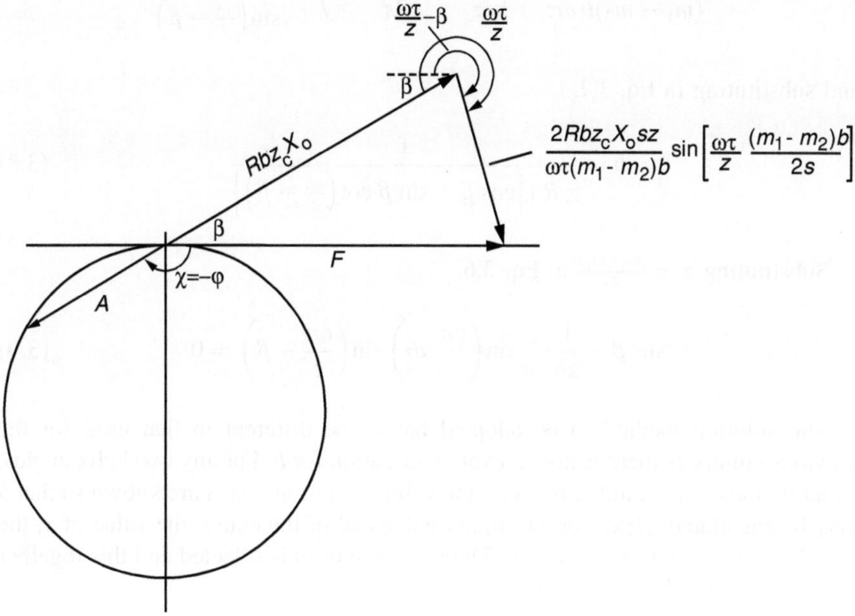

Fig. 3.31 Graphical solution of stability boundary at a chatter frequency ω

From the geometry of Fig. 3.31,

$$z_c R b X_o \sin\beta \frac{z_c R 2 X_o s z}{\omega\tau(m_1 - m_2)} \sin\left(\frac{\omega\tau}{z}\frac{(m_1 - m_2)b}{2s}\right)\sin\left(\frac{\omega\tau}{z} - \beta\right) = 0$$

and dividing by $z_c R b X_o$,

$$\sin\beta + \frac{2s}{(m_1 - m_2)b}\frac{z}{\omega\tau}\sin\left(\frac{\omega\tau}{z}\frac{(m_1 - m_2)b}{2s}\right)\sin\left(\frac{\omega\tau}{z} - \beta\right) = 0 \tag{3.6}$$

Now, consider the response A at frequency ω. This by definition is

$$A = \frac{X_o}{F} = \frac{X_o}{z_c R b X_o \cos\beta - \frac{z_c R 2 X_o s z}{\omega\tau(m_1 - m_2)}\sin\left(\frac{\omega\tau}{z}\frac{(m_1 - m_2)b}{2s}\right)\cos\left(\frac{\omega\tau}{z} - \beta\right)}$$

so that

$$A = \frac{1}{z_c R b\left[\cos\beta - \frac{2s}{(m_1 - m_2)b}\frac{z}{\omega\tau}\sin\left(\frac{\omega\tau}{z}\frac{(m_1 - m_2)b}{2s}\right)\cos\left(\frac{\omega\tau}{z} - \beta\right)\right]} \tag{3.7}$$

From Eq. 3.6,

$$\frac{2s}{(m_1 - m_2)b}\frac{z}{\omega\tau}\sin\left(\frac{\omega\tau}{z}\frac{(m_1 - m_2)b}{2s}\right) = \frac{-\sin\beta}{\sin\left(\frac{\omega\tau}{z} - \beta\right)}$$

and substituting in Eq. 3.7,

$$b = \frac{1}{z_c R A\left[\cos\beta + \sin\beta\cot\left(\frac{\omega\tau}{z} - \beta\right)\right]} \tag{3.8}$$

Substituting $\alpha = \frac{(m_1 - m_2)}{2s}$ in Eq. 3.6,

$$\sin\beta + \frac{1}{\alpha b}\frac{z}{\omega\tau}\sin\left(\frac{\omega\tau}{z}\alpha b\right)\sin\left(\frac{\omega\tau}{z} - \beta\right) = 0 \tag{3.9}$$

The solution method to be adopted has to be different to that used for the previous cutters as there is not an explicit equation for b. For any two helix angles, cutter diameter and number of teeth, the values of m_1, m_2 and s are known so that α may be calculated. Next, for any rotational speed of the cutter, the value of τ, the time for one revolution, is known. Then, a value of ω is selected and this together with the structural characteristics gives

$$\tan\phi = \frac{-2\xi\omega/\omega_n}{\left(1 - \frac{\omega^2}{\omega_n^2}\right)} \quad \text{and hence } \beta = \phi + \pi \text{ and}$$

$$A = \frac{X}{F} = \frac{1}{k\left(\left(1 - \frac{\omega^2}{\omega_n^2}\right)^2 + 4\xi^2\frac{\omega^2}{\omega_n^2}\right)^{1/2}}$$

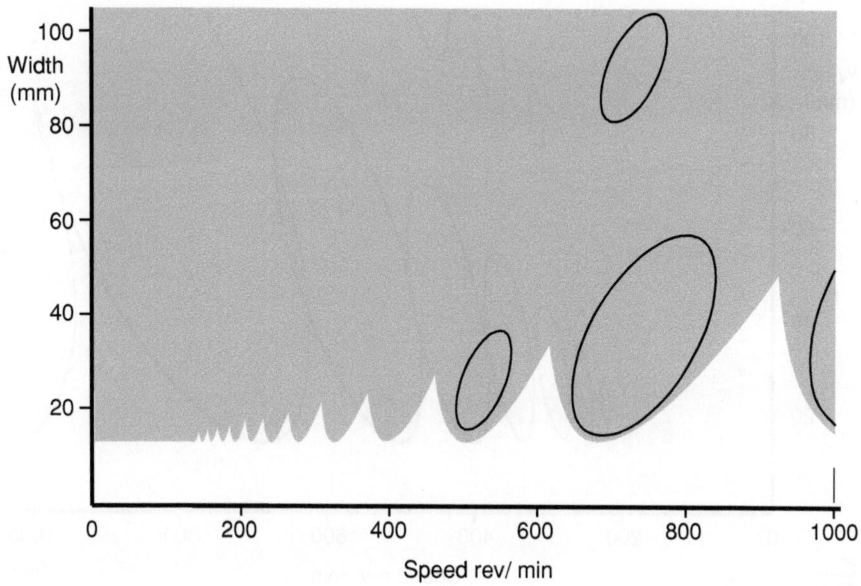

Fig. 3.32 Stability chart for broaching, helix angles 25° and 35° (Program 3.5)

The relevant values are substituted in Eq. 3.6, and the value of b is determined. This trial value and the values of the other parameters are then inserted into the left-hand side of Eq. 3.7. This process is repeated for different values of ω until Eq. 3.7 is satisfied. The value for speed is then incremented and the search algorithm repeated.

A typical stability chart for the reference conditions defined previously and using helix angles of 25° and 35° is shown in Fig. 3.32. It is evident that at low speed, chatter is eliminated. Also where chatter does occur, it is possible to increase the width of the work in cut and restore stability. This is because the regenerative force is reduced as the number of waves crossed (see Fig. 3.27) is increased. The wavelengths are shorter at low speeds, and so many waves are crossed. At higher speeds, fewer waves are crossed, reducing stability, but this may be mitigated by an increase in width of cut that leads to more waves being crossed.

It is interesting to observe the bi-helix effect in broaching for a smaller difference in helix angles. Figure 3.33 shows the stability chart when the helix angles are 27.5° and 32.5°.

The explanation for the shape of this stability chart is that due to the small difference in helix angle, insufficient waves are crossed to achieve stability unless lower speeds and/or increased widths of cut are employed.

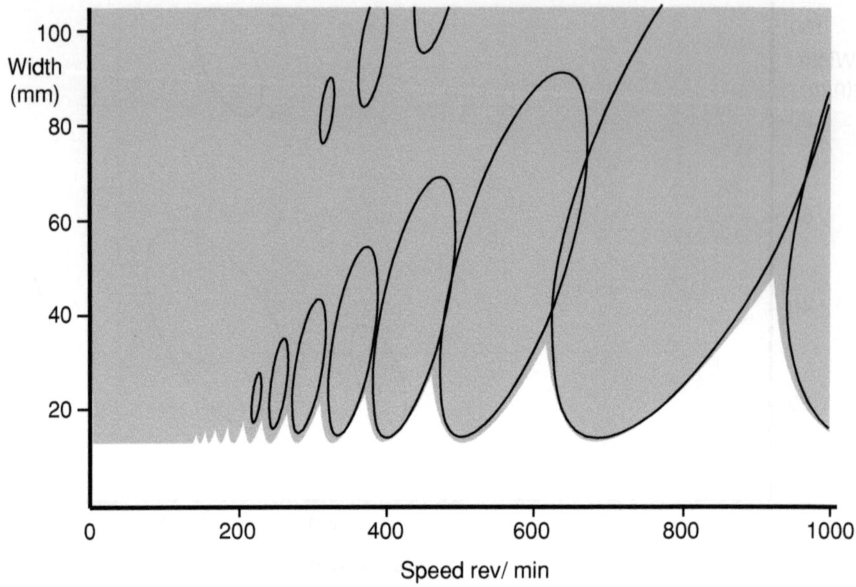

Fig. 3.33 Stability chart for broaching, helix angles 27.5° and 32.5° (Program 3.5)

It is clear from the two stability charts that there is not much improvement at high speeds as the wavelengths will be sufficiently large so that not even one wavelength will be crossed.

At this point, it is very important to review some of the assumptions made and consider their validity. For broaching, the arc of cut is assumed to be much greater than the width. However, in both Figs. 3.32 and 3.33, the length of arc of cut for the reference machining conditions is of the same order of magnitude as the lower unstable widths. As the width is not significantly smaller than the arc, the model that needs to be investigated further is what was called milling in Fig. 3.30b when the width is greater than the arc.

3.5.2 Theoretical Stability Chart for Milling

If reference is made to Fig. 3.30b, it is seen that for a small arc of cut and a relatively large width of the work, the effective width of the work in cut on any single tooth is constant and equal to the arc of cut. Increasing the width of the work will increase the average number of teeth (z_c) in contact with the work. From Fig. 3.30b, the effective width of cut is the arc of cut. Because for the chatter model that is being used, it is the width of cut perpendicular to the vibration direction that is of importance, it is more appropriate to consider the chord of the arc.

This allows the analysis to be applied to much greater depths of cut. From Fig. 2.17, the chord length, which is the effective width b, is given by

$$b = \sqrt{d^2 + R^2 - (R - d^2)} = \sqrt{d^2 + R^2 - R^2 + 2RD - d^2} = \sqrt{2RD}$$

The equation for the force Eq. 3.3 needs to be adjusted for z_c teeth so that

$$F = -Rbz_c\left[X_o \sin \omega t - \frac{2X_o sz}{\omega\tau(m_1 - m_2)b} \sin\left(\omega t - \frac{\omega\tau}{z}\right) \sin\left(\frac{\omega\tau}{z}\frac{(m_1 - m_2)b}{2s}\right)\right] \tag{3.10}$$

However, because the effective direction of motion of the teeth is across the width of the work,

$$s = 2\pi R/z \tan \theta_m \quad \text{where } \theta_m = (\theta_1 + \theta_2)/2 \text{ is the mean helix angle}$$
$$m_1 = \tan(\pi/2 - \theta_1) \quad \text{and} \quad m_2 = \tan(\pi/2 - \theta_2)$$

The graphical solution for chatter is the same as for broaching (see Fig. 3.31) but with different parameter values. The major difference is that b is known and it is z_c, the average number of teeth in contact that needs to be found. The same method may be employed as was used for the other cutters.

Hence,

$$\tan \phi = \frac{-2\xi\omega/\omega_n}{\left(1 - \frac{\omega^2}{\omega_n^2}\right)} \quad \text{where } \phi = \beta - \pi \quad \text{and}$$

$$\frac{kX}{F} = \frac{1}{\left(\left(1 - \frac{\omega^2}{\omega_n^2}\right)^2 + 4\xi^2 \frac{\omega^2}{\omega_n^2}\right)^{1/2}}$$

Noting that $X = X_0$ and that from the geometry of Fig. 3.31, the force

$$F = z_c RbX_o \cos \beta - \frac{z_c R2X_o sz}{\omega\tau(m_1 - m_2)} \sin\left(\frac{\omega\tau}{z}\frac{(m_1 - m_2)b}{2s}\right) \cos\left(\frac{\omega\tau}{z} - \beta\right)$$

Then,

$$\frac{X_o}{F} = \frac{1}{k\left(\left(1 - \frac{\omega^2}{\omega_n^2}\right)^2 + 4\xi^2 \frac{\omega^2}{\omega_n^2}\right)^{1/2}}$$
$$= \frac{X_o}{z_c RbX_o \cos \beta - \frac{z_c R2X_o sz}{\omega\tau(m_1 - m_2)} \sin\left(\frac{\omega\tau}{z}\frac{(m_1 - m_2)b}{2s}\right) \cos\left(\frac{\omega\tau}{z} - \beta\right)}$$

And solving for z_c at the stability boundary gives

$$z_c = \frac{k\left(\left(1 - \frac{\omega^2}{\omega_n^2}\right)^2 + 4\xi^2 \frac{\omega^2}{\omega_n^2}\right)^{1/2}}{Rb\left(\cos \beta - \frac{z}{\omega\tau}\frac{2s}{(m_1 - m_2)b}\sin\left(\frac{\omega\tau}{z}\frac{(m_1 - m_2)b}{2s}\right)\cos\left(\frac{\omega\tau}{z} - \beta\right)\right)} \tag{3.11}$$

All that remains is to find the width of the work $b_{\lim}$ that results in an average of z_c teeth being in cut. For there to be an average of one tooth in cut, $b_{\lim} = s$ where s is as defined for this milling case, i.e. the pitch on the constant pitch line shown in Fig. 3.30b. Thus, when there are z_c teeth in cut, $b_{\lim} = z_c s$.

Figure 3.34 shows a stability chart for helix angles of 27.5° and 32.5° with the reference conditions used. As with broaching, there are significant improvements at lower speeds when the short wavelengths ensure that more than one wave is crossed. At higher speeds, the improvement is not as great. Increasing the difference in the helix angles improves performance. Figure 3.35 shows a stability chart for helix angles of 25° and 35° and for a higher speed range.

3.5.2.1 Effect of One Tooth with a Zero Helix Angle

The mean helix angle was also found to be significant, especially as the helix angle on one tooth approached zero. When this is the case, it was found that the process becomes unconditionally stable. The reason for this may be understood from the diagram in Fig. 3.36. It is seen that, except for entry and exit from the arc of cut, the non-zero helix tooth "sees" a constant chip cross section. As a result, there is no regeneration on this tooth. However, the real situation is complex and involves many different effects. One that is not immediately apparent is the different effective cutting speeds on each tooth as considered in Sect. 3.5.2.2.

In practice, however, the effect of having a tooth with zero helix angle would result in the tooth instantaneously taking up the cut over the full width of the work and producing large intermittent forces. However, it may be possible to have one tooth with a suitably small helix angle. Figure 3.37 indicates the improvement obtained when the angles are changed from 27.5° and 32.5° (Fig. 3.34) to 7.5° and 12.5°. Note that the speed range in Fig. 3.37 has been increased.

3.5.2.2 Effect of Each Tooth Having a Different Speed

The milling situation shown in Fig. 3.30b has a constant width b and the teeth move across the width of the work. However, since the peripheral speed of the teeth is constant (ΩR), the speed across the work (shown by arrows) for tooth 1 is $\Omega R/\tan\theta_1 = \Omega R/m_1$ and for tooth 2 $\Omega R/\tan\theta_2 = \Omega R/m_2$. The consequence is that the effective mean pitch is continually changing. If one of the helix angles is zero, the effective speed across the work is infinite. This is another reason why a

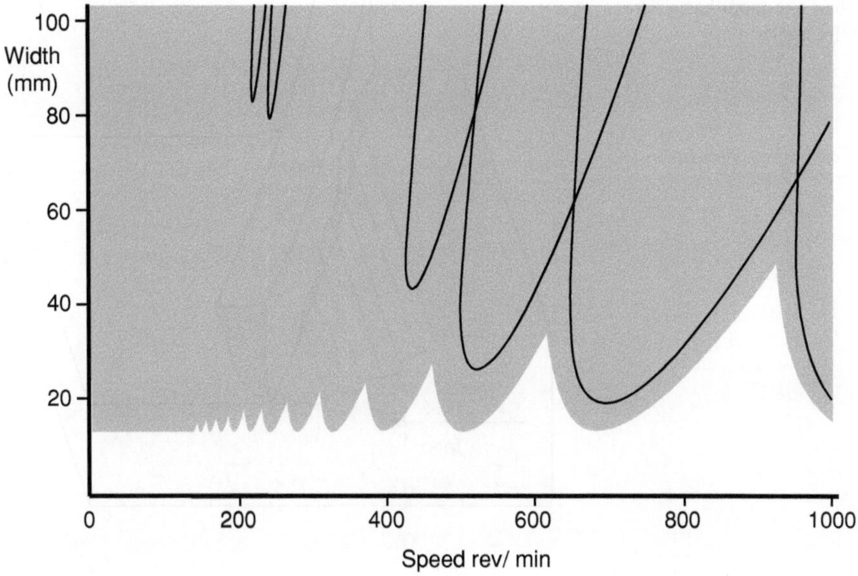

Fig. 3.34 Stability chart for milling, helix angles 27.5° and 32.5° (Program 3.6)

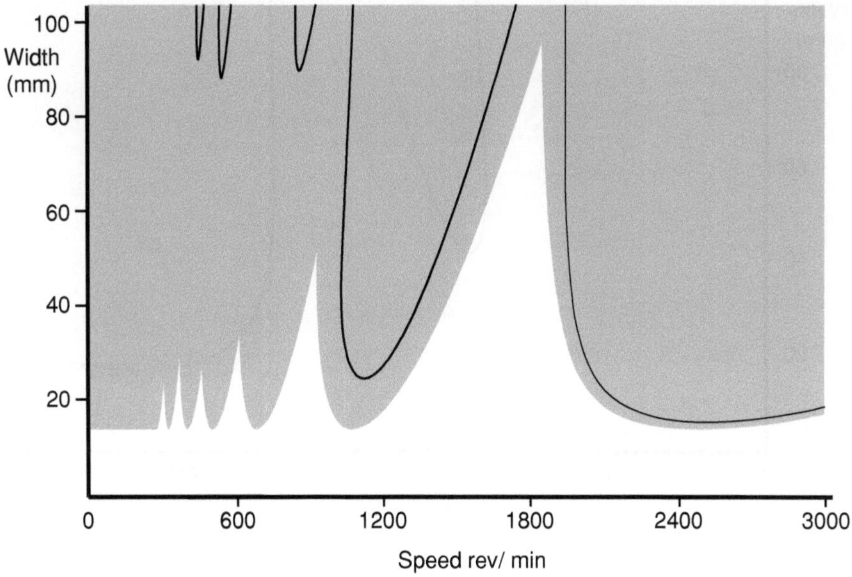

Fig. 3.35 Stability chart for milling, helix angles 25° and 35° (Program 3.6)

Fig. 3.36 Chip cross section when one tooth has a zero helix angle

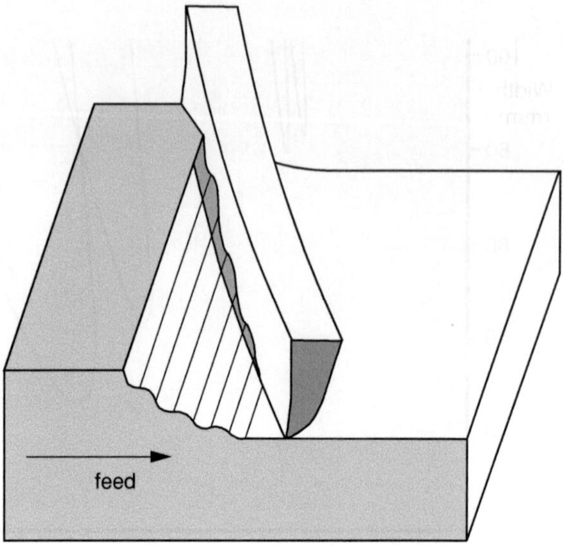

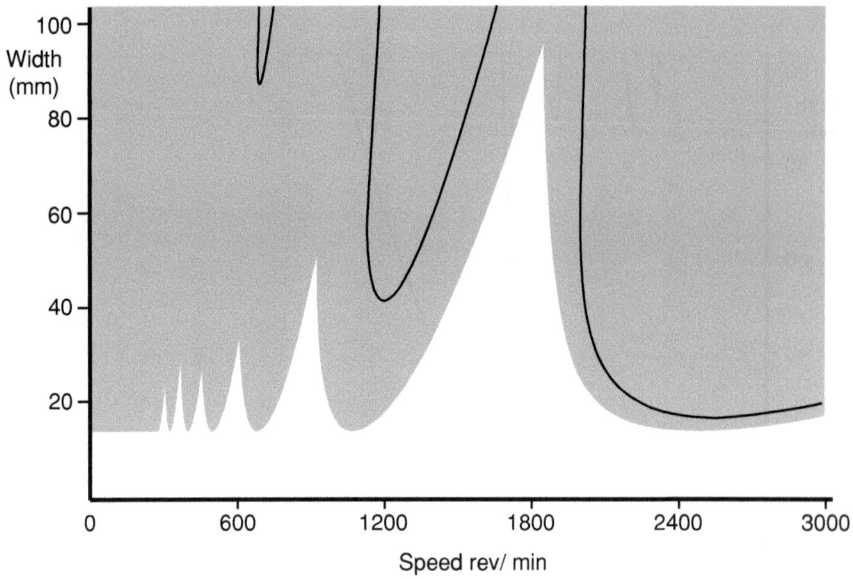

Fig. 3.37 Stability chart for milling, helix angles 7.5° and 12.5° (Program 3.6)

zero angle helix in a bi-helix cutter would be beneficial. The analysis presented above has considered only the effect of crossing waves, and the continuously varying mean pitch effect has been ignored. For small depths of cut, however, this latter effect is the most significant as only a small portion of a wave is crossed and, as will be shown below, the improved performance obtained in practice is far in excess of that anticipated from consideration merely of the number of waves crossed.

3.5.3 Tunable Alternating Pitch Effect

For both broaching and milling, there is one final effect that is beneficial for chatter performance. It is possible to impose a mean alternating pitch by moving the cutter laterally relative to the work so that the constant pitch line is no longer on the centre line of the work. This effect is tunable. As was found for alternating pitch cutters, for a particular speed and frequency combination, only certain pitch variations will produce large improvements. Thus, if chatter does occur, these optimum pitch variations may be found by changing the position of the cutter as shown in Fig. 3.38. The mean pitch difference between teeth increases continuously with lateral movement of the work relative to the cutter. For milling with small depths of cut, the situation is slightly more complicated as the pitch is varying continuously. However, lateral movement of the cutter produces different minimum and maximum pitches.

3.5.4 Experimental Results

There seems to be very little experimental work in the published literature relating to the types of cutter discussed in this chapter. As a consequence, the results published by Stone [7, 8] for milling will be considered here. Slab mills having six teeth were made with alternating helix angles. The helix angles were chosen about a mean of 30° and, to allow for the investigation of the effect of the magnitude of the difference in helix angles, cutters were made with differences in helix angle of 1.5°, 3.0°, 5.0° and 10.0°. The latter cutter had alternating helix angles of 25° and 35° and is shown in Fig. 3.39.

The performance of these cutters was compared with that of a constant 30° helix angle cutter with six teeth. All the cutters were manufactured from the same bar stock with nominally identical rake and clearance angles and each received the same heat treatment. The cutters were 4 in diameter, 4 in long, manufactured from 18/4/1 high-speed steel and heat treated to C62–C63 Rockwell.

Cutting tests were conducted on tapered mild-steel workpieces (see Fig. 3.40) manufactured so that the width of cut increased linearly from 0.25 into 4.0 in. The required speed, feed and depth of cut were set and then the workpiece machined

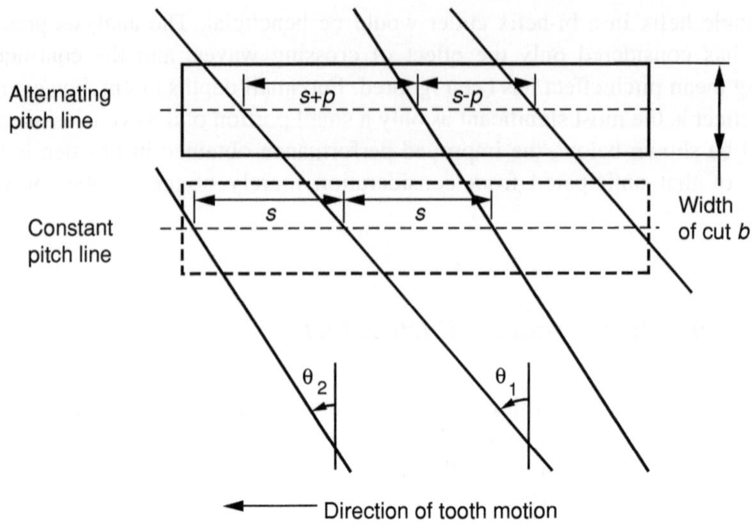

Fig. 3.38 Alternating pitch effect resulting from lateral movement of work relative to cutter

Fig. 3.39 Bi-helix slab mill with alternating 25° and 35° helices

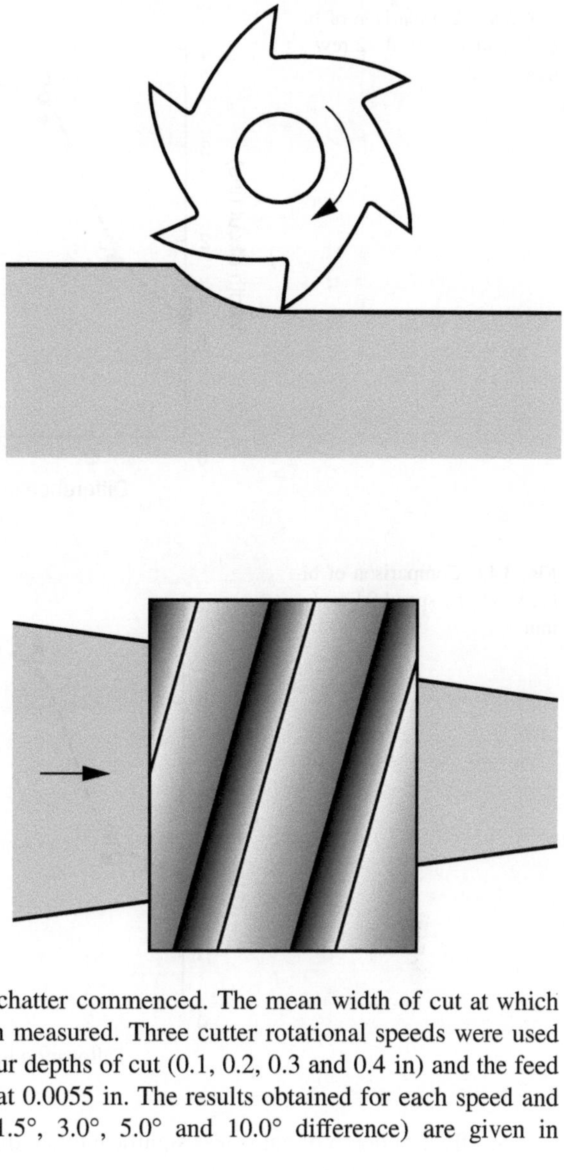

Fig. 3.40 Tapered workpiece cutting test

(conventional milling) until chatter commenced. The mean width of cut at which chatter commenced was then measured. Three cutter rotational speeds were used (72, 94 and 122 rev/min), four depths of cut (0.1, 0.2, 0.3 and 0.4 in) and the feed per tooth was kept constant at 0.0055 in. The results obtained for each speed and the five cutters (standard, 1.5°, 3.0°, 5.0° and 10.0° difference) are given in Figs. 3.41, 3.42, and 3.43.

These results indicate that the major effect is due to the magnitude of the difference in helix angle; for the larger differences, an improvement in performance of a factor of four was generally achieved. It is of interest to note that the improvement in performance does not vary linearly with the difference in helix angles so that smaller differences in helix angles may be used under certain conditions with only a small adverse effect on performance. This is useful in that for wide cutters, there is a limit to the allowable helix angle difference before the minimum pitch between teeth becomes too small.

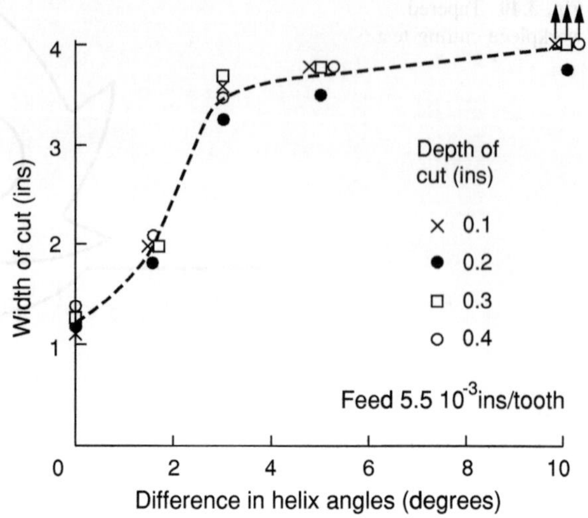

Fig. 3.41 Comparison of bi-helix cutters, speed 72 rev/min

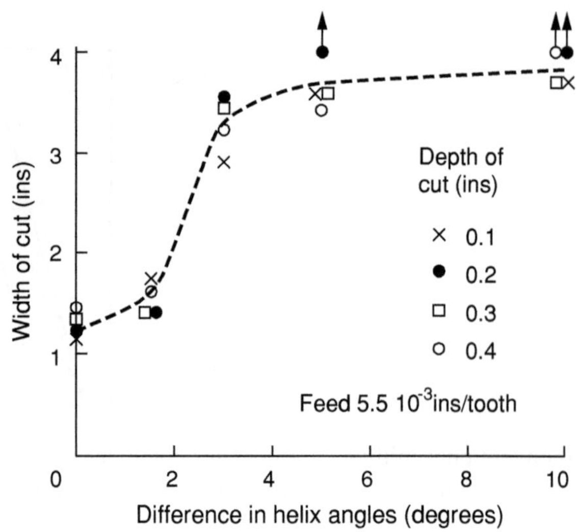

Fig. 3.42 Comparison of bi-helix cutters, speed 94 rev/min

The effect of depth of cut is less significant than that of difference in helix angle but over the range 0.1–0.4 in, up to 50 % variations in performance occurred for particular combinations of speed and difference in helix angle. At the larger depths of cut, it was observed from the chatter marks left on the arc of cut that at least one wave was being crossed; however, at the smallest depth of cut, 0.1 in, only a small part of a wave was crossed. It is notable that, in general, the improvement in performance was of the same order for depths of cut of both 0.1 and 0.4 in. This indicates that the non-constant speed effect is likely to be the major effect for milling. An alternating helix angle slab mill with helix angles of zero and 10.0°

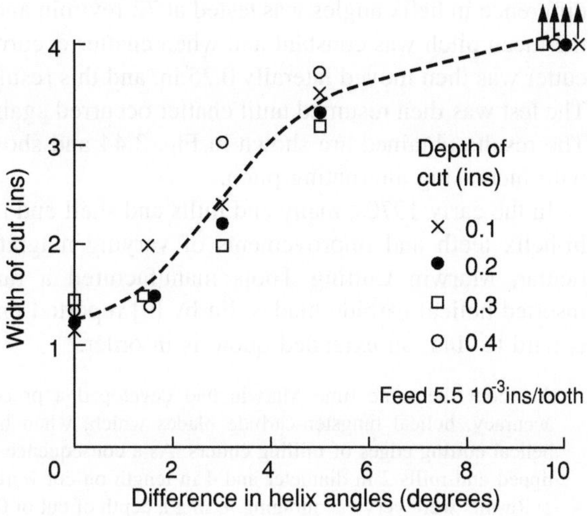

Fig. 3.43 Comparison of bi-helix cutters, speed 122 rev/min

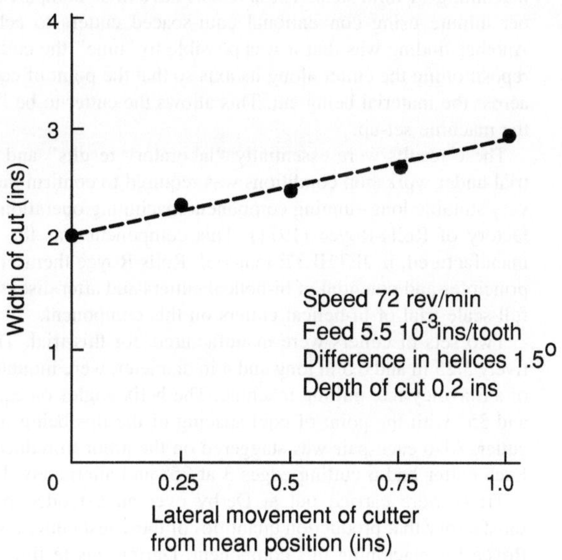

Fig. 3.44 Effect of lateral movement of the bi-helix cutter

was manufactured to extend the non-constant speed effect to its limit. With this cutter, the machine did not chatter but, as previously discussed, machining suffered from the presence of severe forced vibrations resulting from the zero helix teeth taking up the full width of cut instantaneously.

All the tests described so far were conducted with the workpiece symmetrically positioned about the constant pitch plane of the cutters, i.e. the mean pitch was kept constant. To investigate the effect of varying the mean pitch a previous set-up (for an alternating helix cutter that had resulted in a small width of cut at the chatter boundary) was used to investigate varying pitch. The cutter with a 1.5°

difference in helix angles was tested at 72 rev/min and 0.1 in depth of cut. Initially, the mean pitch was constant and when chatter occurred, the width was noted. The cutter was then moved laterally 0.25 in, and this resulted in some alternating pitch. The test was then resumed until chatter occurred again and the procedure repeated. The results obtained are shown in Fig. 3.44 and show a continuous improvement with increasing alternating pitch.

In the early 1970s, many end mills and shell end mills were manufactured with bi-helix teeth and improvements of varying magnitudes were obtained. In particular, Marwin Cutting Tools manufactured a range of milling cutters with inserted helical carbide blades. Furby [9] reported on the results. As reference [9] is hard to find, an extended quote is in order.

> At about the same time Marwin had developed a process for producing, with great accuracy, helical tungsten-carbide blades which, when brazed in position, formed the helical cutting edges of milling cutters. As a consequence of discussions helical carbide-tipped end mills 2 in diameter and 4 in length on cut were manufactured.
>
> Results were very encouraging. Using a depth of cut of 0.2 in and a length of cut of 3.5 in with a feed rate of 19 in per minute, a surface finish of 32 CLA was produced when machining 11 mild steel. These results have to be compared with a feed rate as low as 3 in per minute using conventional equi-spaced cutters to achieve the same surface finish. Another finding was that it was possible to "tune" the cutter to give minimum chatter by repositioning the cutter along its axis so that the point of equi-spacing of the tips is moved across the material being cut. This allows the cutter to be "tuned" to be sympathetic with the machine set-up.
>
> These results were essentially "laboratory results" and it was agreed that a full-scale trial under workshop conditions was required to confirm the benefits suggested. Happily a very suitable long-running component machining operation was commenced, at the Derby factory of Rolls-Royce (1971). This component was the RB211 engine exhaust casing manufactured, in JETHETE material. Rolls-Royce themselves were very interested in the principles and potential of bi-helical cutters and after discussions with Marwin, agreed to a full-scale trial of bi-helical cutters on this component.
>
> Two sets of cutters were manufactured, for this trial. The cutters, which were respectively 3.25 in and 6.5 in long and 4 in diameter, were mounted in pairs on the vertical arbor of a Staveley n.c. milling machine. The helix angles on each cutter were respectively 25° and 35° with the point of equi-spacing of the tips being midway along the length of the cutter. Also each pair was staggered on the arbor to reduce induced harmonic vibrations. Each cutter had 6 cutting edges 3 at 25° and alternately 3 at 35°.
>
> Trials were carried out at Derby over an extended period and the results were so satisfactory that production quantities of these test cutters were ordered and used by Rolls-Royce for machining this component. During this testing, feeds and speeds were altered constantly to get the highest metal-removal rates consistent with acceptable surface finish on the components. Also meticulous attention to such details as coolant flow, steady bearing lubrication, position of the cutter on the arbor, depth of cut etc. enabled components' machining time to be halved—i.e. from 560 min to 270 min,
>
> Further testing was carried out to compare the component machining times when using HSS bi-helical cutters with bi-helical TC tipped cutters in the Coventry factory of Rolls-Royce. The components were front flanges, milled two at a time—stacked one on top of the other—and the important parameters were machining time and surface finish.
>
> The final result, after much testing, was that on the centre-band roughing operation the machining time was reduced from 454 min to 316 min, and on the finish profiling operation from 304 min to 129 min.

3.6 Conclusions

Four high-performance cutters have been described, and their advantages were analysed using many assumptions and a reference set of conditions. It is important to remember that any particular machining operation will involve a far more complex set of conditions than those used in these analyses. However, the major benefit that may be obtained by using such cutters is evident.

When chatter occurs on a milling machine, the simplest solution is to investigate implementing the methods described at the end of Chap. 2. If chatter is still a problem, then the high-performance cutters described in this chapter should be considered. The alternating pitch cutter gives improvements over a restricted speed range that depends on the particular machine and speed. As a result, it is only used when it is worth the effort to find the appropriate cutter with the required pitch variation. The other cutters have a wider range of application.

The Strasman or ripping cutter has the best performance overall, except for the poor surface finish. However, it could be used for a roughing cut and finishing completed with a different cutter. The trapezoidal cutters overcome the surface finish problem but almost without exception those commercially available have a small gap in the profile. Manufacturers do not seem to be aware of the benefits, for chatter performance, of using larger gaps in the profile.

In the 1970s, bi-helix cutters were used for difficult machining operations with some success. In fact, Rolls Royce (1971) acquired the patent from MTIRA and this seems to have limited their use by other tool manufacturers. In recent years, there has been some discussion in the literature of alternating pitch and bi-helix cutters [10–14] but the earlier work described above appears to have been largely unknown to these researchers. Their modelling techniques generally follow the method of representing the cutter by a stack of alternating pitch cutters.

The question that now needs to be considered is why machine tools cannot be designed so that chatter would not be a problem in the first place. The next chapter considers structural modifications as a means of preventing chatter.

References

1. Hahn RS (1952) Metal-cutting chatter and its elimination. Trans ASME J Eng Ind Ser B 74:1073–1080
2. Slavicek J (1965) The effect of irregular tooth pitch on stability of milling. In: Proceedings of 6th MTDR conference, pp 15–22
3. Opitz H, Dregger EU, Roese H (1966) Improvement of the dynamic stability of the milling process by irregular tooth pitch. In: Proceedings of 7th MTDR conference, pp 213–227
4. Stone BJ (1976) Improved-performance milling cutters. In: Invited paper at conference on high-metal-removal rate milling. Machine Tool Industry Research Association, Macclesfield
5. Pye CJ, Stone BJ (1980) A critical comparison of some high performance milling cutters. In: International conference on manufacturing engineering, Melbourne, pp 11–15

6. Vanherck P (1967) Increasing milling machine productivity by use of cutters with non-constant cutting-edge pitch. In: Proceedings of 8th MTDR conference, pp 947–960
7. Stone BJ (1970) The effect on the chatter behaviour of machine tools of cutters with different helix angles on adjacent teeth. In: Proceedings of 11th MTDR conference, pp 169–180
8. Stone BJ (1970) The effect, on the regenerative chatter behaviour of machine tools, of cutters with different approach angles on adjacent teeth. Machine Tool Industry Research Association, Research Report No. 34, pp 1–31
9. Furby J (1976) Practical experience with Strasmann and Bi-helix cutters. In: Conference on high-metal-removal rate milling. Machine Tool Industry Research Association, Macclesfield
10. Altintas Y, Engin S, Budak E (1999) Analytical prediction of chatter stability and design for variable pitch cutters. J Manuf Sci Eng 121(2):173–179
11. Budak E (2003) An analytical design method for milling cutter with nonconstant pitch to increase stability, part 2: application. J Manuf Sci Eng 125(1):35–38
12. Turner S, Merdol D, Altintas Y, Ridgway K (2007) Modelling of the stability of variable helix end mills. Int J Mach Tools Manuf 47(9):1410–1416
13. Sims ND, Mann B, Huyanan S (2008) Analytical prediction of chatter stability for variable pitch and variable helix milling tools. J Sound Vib 317(3–5):664–686
14. Nakagawa H, Ogawa K, Demachi S, Hasegawa H (2009) An experimental study on suppression of chatter vibration with different helix angles end-mill. Key Eng Mater 407(408):37–40

Chapter 4
Structural Modifications

4.1 Introduction

In the previous three chapters, regenerative chatter theory has been developed with the objective of finding solutions to chatter problems. This has involved the use of simplified models that have allowed the most significant parameters to be determined. It has been shown that any existing machining operation may be improved by reducing the width of cut, changing speed and selecting a machining configuration that best utilises the mode directions. A range of high-performance multi-tooth cutters were investigated and the reasons for their improvement determined. In this chapter, we turn our attention to the machine itself. Can machine tools be designed that have improved performance, and/or can existing machines be modified so that their performance is improved?

From the theory developed thus far, it has been shown that the most important machine characteristic is the maximum negative in-phase component of the chatter receptance. It might be considered that it should be possible for machine designers, with current computing power, to predict this characteristic and optimise their designs. In practice, there is a limitation to such predictions. It is possible to predict with some accuracy both the stiffness and mode frequencies of any particular design, but it is not possible to predict the damping as accurately. Most modelling of machine tools involves assuming a damping ratio (or Q factor—see Appendix A.8) for each mode and then superposing the modal responses as described in Chap. 2. The values of the Q factors are largely based on experience with existing machines.

In practice, the Q factors vary, even for very similar structures. Also the modes of vibration of machine tools may not just be superposed as they may be coupled through the damping. This is mainly because the damping in machine tools is not spread uniformly throughout the structure but is located at discrete positions such as bearings and metal-to-metal joints. The damping measured in machine tools is significantly higher than that which is available from the material of the structural elements.

B. Stone, *Chatter and Machine Tools*, DOI: 10.1007/978-3-319-05236-6_4,

To illustrate these issues with damping, the prediction of the chatter receptance of machine tool spindle systems will be described. It is often found that spindle characteristics are the most significant contributor, in the form of flexibility, to the chatter receptance of the complete machine.

4.2 Spindle Design

To predict the chatter receptance of a spindle system, the following characteristics are required:

1. The geometry and material properties of the shaft,
2. The location of the bearings, and
3. The stiffness and damping characteristics of the bearings.

Once these are known, the method of calculating the chatter receptance needs to be chosen. The systems approach to modelling complex systems (see Appendix C) is used here as it allows a combination of measured bearing characteristics and predicted shaft stiffnesses and inertias. As in the previous chapters, some simplifying assumptions will be made so that principal effects can be highlighted.

Consider the typical spindle system shown in cross section in Fig. 4.1a. The major assumption is that the bearings are connected to an immovable frame; that is, the rest of the machine structure is relatively rigid. The spindle may be separated into various subsystems, e.g. shaft elements and bearings, that can subsequently be rejoined to form the complete spindle. These components (tubes and bearings) are chosen because they constitute the basic building blocks of any spindle system. The further assumption that will be made is that the spindle flexural (transverse) vibration is in one plane. The subsystems chosen are as shown in Fig. 4.1b. Their addition involves enforcing equilibrium and compatibility conditions at the joins as described in Appendix C. Figure 4.2 shows, in block diagram form, the seven subsystems that make up the complete spindle. Subsystems 1, 3, 4, 6 and 7 are shaft elements, and subsystems 2 and 5 are bearings. There are two coupling connections between each subsystem as, for the transverse bending vibration of shafts, both the displacement and slope due to bending must be compatible at the coupled ends of each subsystem. Before developing a method for adding the subsystems, the receptances of each subsystem are required.

4.2.1 Receptances of Shafts, Including Shear and Rotary Inertia

Most researchers who use the systems approach model the shaft elements using receptances determined from a Bernoulli–Euler model of the shaft so that shear and rotary inertia effects are not included. These latter effects become significant

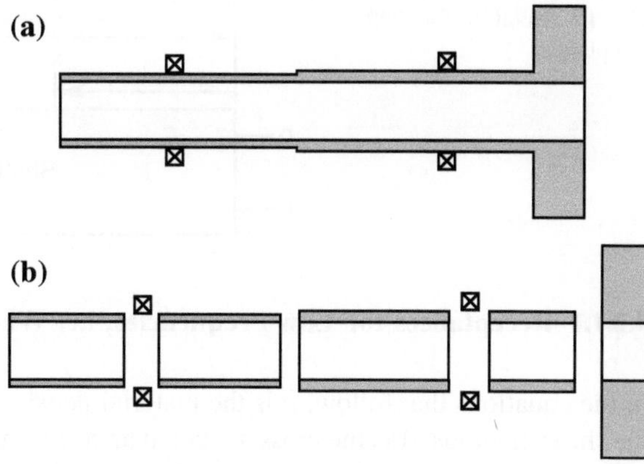

Fig. 4.1 **a** Typical spindle **b** and subsystem components

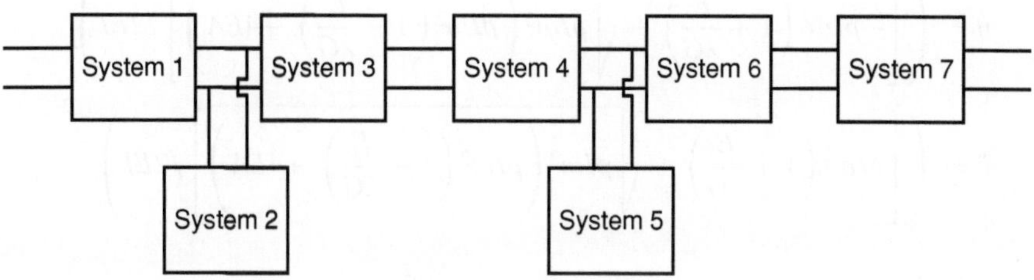

Fig. 4.2 Addition of shaft and bearing subsystems

when the length-to-diameter ratio of the shaft is less than five. An indication that the model has become inaccurate is when the predicted static stiffness (that includes shear effects) differs from the low-frequency (quasi-static) prediction of the vibration model. The receptances of shafts including shear and rotary inertia effects were derived by Potter and Stone [1], Stone [2]. However, the equations presented in the publication that is generally available [2] contain some typographical errors. The correct equations are presented here but without the method used to derive them, which can be found in [1, 2].

The notation used in Fig. 4.3 is that of Bishop and Johnson [3]. They presented the receptances for the Bernoulli–Euler shaft model. As an example of the notation, the receptance

$$\alpha_{xL'} = \frac{\text{transverse displacement at } x}{\text{moment applied at } x = L}$$

A subscript without a prime indicates either a displacement or a force, whereas a prime indicates a slope due to bending or a moment. It is found that receptances that include shear and rotary inertia effects have different forms depending on the frequency.

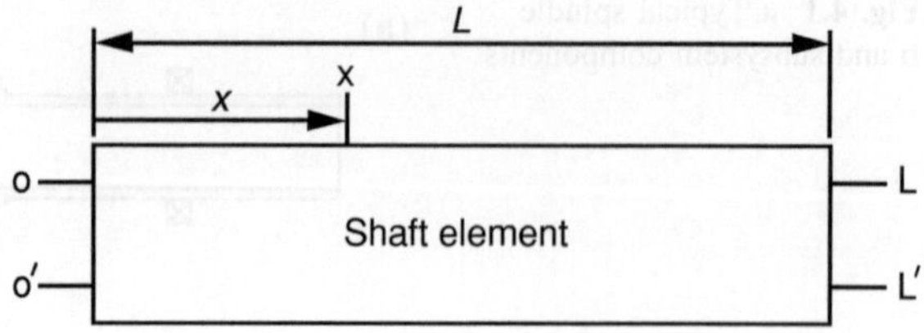

Fig. 4.3 Notation for shaft receptances

4.2.1.1 Receptances for Low Frequencies, i.e. When $\omega^2 < \frac{\lambda GA}{\rho I}$

In the equations that follow, ρ is the material density, E is Young's modulus, G is the shear modulus, A is the cross-sectional area, I is the second moment of area and λ is the shear factor for the cross section. Defining

$$\eta = \left(\left[-\rho I\omega^2\left(1+\frac{E}{\lambda G}\right)+\sqrt{\rho I\omega^2\left(\rho I\omega^2\left(1-\frac{E}{\lambda G}\right)^2+4EA\right)}\right]/2EI\right)^{1/2}$$

$$\xi = \left(\left[\rho I\omega^2\left(1+\frac{E}{\lambda G}\right)+\sqrt{\rho I\omega^2\left(\rho I\omega^2\left(1-\frac{E}{\lambda G}\right)^2+4EA\right)}\right]/2EI\right)^{1/2}$$

$$s = \frac{\rho\omega^2}{\lambda G\xi} - \xi \quad r = \frac{\rho\omega^2}{\lambda G\eta} + \eta w$$

$$F_1 = \sin\xi L\sinh\eta L; \quad F_2 = \cos\xi L\cosh\eta L; \quad F_3 = \cos\xi L\cosh\eta L - 1;$$
$$F_4 = \sin\xi L\cosh\eta L; \quad F_5 = \cos\xi L\sinh\eta L; \quad F_6 = \cos\xi L - \cosh\eta L;$$

and

$$\begin{aligned}\Delta &= \left(2 - 2\cos\xi L\cosh\eta L + \left(\frac{r\eta^2}{s\xi^2} - \frac{s\xi^2}{r\eta^2}\right)\sin\xi L\ \sinh\eta L\right)\\ &= \rho\omega^2 A\left(2 - 2F_2 + \left(\frac{r\eta^2}{s\xi^2} - \frac{s\xi^2}{r\eta^2}\right)F_1\right)\end{aligned}$$

The tip receptances are

$$\alpha_{oo} = \alpha_{LL} = \frac{F_4\left(\frac{s\xi^2}{r\eta} - \xi\right) - F_5\left(\frac{r\eta^2}{s\xi} - \eta\right)}{\Delta}$$

$$\alpha_{Lo} = \alpha_{oL} = \frac{\sinh\eta L\left(\eta - \frac{r\eta^2}{s\xi}\right) + \sin\xi L\left(\frac{s\xi^2}{r\eta} - \xi\right)}{\Delta}$$

$$\alpha_{LL'} = \alpha_{L'L} = -\alpha_{o'o} = -\alpha_{oo'} = \frac{F_1\left(\frac{s\xi^2}{\eta} - \frac{r\eta^2}{\xi}\right) + F_3(s\xi + r\eta)}{\Delta}$$

$$\alpha_{o'L} = \alpha_{Lo'} = -\alpha_{oL'} = \alpha_{L'o} = \frac{F_6(r\eta - s\xi)}{\Delta}$$

$$\alpha_{o'o'} = \alpha_{L'L'} = \frac{F_4\left(sr\eta - \frac{r^2\eta^2}{\xi}\right) + F_5\left(rs\xi - \frac{s^2\xi^2}{\eta}\right)}{\Delta}$$

$$\alpha_{o'L'} = \alpha_{L'o'} = \frac{\sinh\eta L\left(rs\xi - \frac{s^2\xi^2}{\eta}\right) + \sin\xi L\left(sr\eta - \frac{r^2\eta^2}{\xi}\right)}{\Delta}$$

and for some position x along the shaft

$$\alpha_{xo} = \frac{\begin{bmatrix} -\xi\left(\frac{r\eta^2}{s\xi^2}F_5 + F_4\right)\cos\xi x - \xi\left(\frac{r\eta^2}{s\xi^2}F_1 - F_3\right)\sin\xi x \\ +\eta\left(F_5 + \frac{s\xi^2}{r\eta^2}F_4\right)\cosh\eta x - \eta\left(\frac{s\xi^2}{r\eta^2}F_1 + F_3\right)\sinh\eta x \end{bmatrix}}{\Delta}$$

$$\alpha_{xo'} = \frac{\begin{bmatrix} -\xi\left(\frac{s\xi}{\eta}F_1 + \frac{r\eta}{\xi}F_3\right)\cos\xi x + \xi\left(\frac{s\xi}{\eta}F_5 - \frac{r\eta}{\xi}F_4\right)\sin\xi x \\ +\eta\left(\frac{r\eta}{\xi}F_1 - \frac{s\xi}{\eta}F_3\right)\cosh\eta x + \eta\left(\frac{s\xi}{\eta}F_5 - \frac{r\eta}{\xi}F_4\right)\sinh\eta x \end{bmatrix}}{\Delta}$$

$$\alpha_{xL} = \frac{\left(\sinh\eta L + \frac{s\xi^2}{r\eta^2}\sin\xi L\right)\left(\eta\cosh\eta x - \frac{r\eta^2}{s\xi}\cos\xi x\right) + F_6(\xi\sin\xi x + \eta\sinh\eta x)}{\Delta}$$

$$\alpha_{xL'} = \frac{\begin{bmatrix} -r\eta F_6\cos\xi x + \left(\frac{s\xi^2}{\eta}\sinh\eta L - r\eta\sin\xi L\right)\sin\xi x \\ -s\xi F_6\cosh\eta x + \left(s\xi\sinh\eta L - \frac{r\eta^2}{\xi}\sin\xi L\right)\sin\eta x \end{bmatrix}}{\Delta}$$

4.2.1.2 Receptances for High Frequencies, i.e. When $\omega^2 > \frac{\lambda GA}{\rho I}$

In the equations that follow, the variables are changed so that now

$$\eta = \left(\left[\rho I\omega^2\left(1 + \frac{E}{\lambda G}\right) - \sqrt{\rho I\omega^2\left(\rho I\omega^2\left(1 - \frac{E}{\lambda G}\right)^2 + 4EA\right)}\right]/2EI\right)^{1/2}$$

$$\xi = \left(\left[\rho I\omega^2\left(1+\frac{E}{\lambda G}\right)+\sqrt{\rho I\omega^2\left(\rho I\omega^2\left(1-\frac{E}{\lambda G}\right)^2+4EA\right)}\right]/2EI\right)^{1/2}$$

$$s = \frac{\rho\omega^2}{\lambda G\xi} - \xi; \quad r = \frac{\rho\omega^2}{\lambda G\eta} - \eta$$

$$F_1 = \sin\xi L\sin\eta L; \quad F_2 = \cos\xi L\cos\eta L; \quad F_3 = \cos\xi L\cos\eta L - 1;$$
$$F_4 = \sin\xi L\cos\eta L; \quad F_5 = \cos\xi L\sin\eta L; \quad F_6 = \cos\xi L - \cos\eta L;$$

$$\begin{aligned}\Delta &= \left(2 - 2\cos\xi L\cos\eta L - \left(\frac{r\eta^2}{s\xi^2}+\frac{s\xi^2}{r\eta^2}\right)\sin\xi L\sin\eta L\right)\\ &= \rho\omega^2 A\left(2 - 2F_2' - \left(\frac{r\eta^2}{s\xi^2}+\frac{s\xi^2}{r\eta^2}\right)F_1'\right)\end{aligned}$$

The tip receptances are

$$\alpha_{oo} = \alpha_{LL} = \frac{F_5\left(\frac{r\eta^2}{s\xi}-\eta\right)+F_4\left(\frac{s\xi^2}{r\eta}-\xi\right)}{\Delta}$$

$$\alpha_{Lo} = \alpha_{Lo} = \frac{\sin\eta L\left(\frac{r\eta^2}{s\xi}-\eta\right)+\sin\xi L\left(\frac{s\xi^2}{r\eta}-\xi\right)}{\Delta}$$

$$\alpha_{LL'} = \alpha_{L'L'} = -\alpha_{o'o} = \alpha_{oo'} = \frac{F_1\left(\frac{s\xi^2}{\eta}+\frac{r\eta^2}{\xi}\right)+F_3(s\xi+r\eta)}{\Delta}$$

$$\alpha_{o'L} = \alpha_{Lo'} = -\alpha_{oL'} = \alpha_{L'o} = \frac{F_6(r\eta - s\xi)}{\Delta}$$

$$\alpha_{o'o'} = \alpha_{L'L'} = \frac{F_4\left(sr\eta-\frac{r^2\eta^2}{\xi}\right)+F_5\left(rs\xi-\frac{s^2\xi^2}{\eta}\right)}{\Delta}$$

$$\alpha_{o'L'} = \alpha_{L'o'} = \frac{\sinh\eta L\left(rs\xi-\frac{s^2\xi^2}{\eta}\right)+\sin\xi L\left(sr\eta\frac{r^2\eta^2}{\xi}\right)}{\Delta}$$

and for some position x along the shaft

$$\alpha_{xo} = \frac{\left[\begin{array}{l}\xi\left(\frac{r\eta^2}{s\xi^2}F_5-F_4\right)\cos\xi x+\xi\left(\frac{r\eta^2}{s\xi^2}F_1+F_3\right)\sin\xi x\\ +\eta\left(\frac{s\xi^2}{r\eta^2}F_4-F_5\right)\cos\eta x+\eta\left(\frac{s\xi^2}{r\eta^2}F_1+F_3\right)\sin\eta x\end{array}\right]}{\Delta}$$

$$\alpha_{xo'} = \frac{\begin{vmatrix} -\xi\left(\frac{s\xi}{\eta}F_1 + \frac{r\eta}{\xi}F_3\right)\cos\xi x + \xi\left(\frac{s\xi}{\eta}F_5 - \frac{r\eta}{\xi}F_5\right)\sin\xi x \\ -\eta\left(\frac{r\eta}{\xi}F_1 + \frac{s\xi}{\eta}F_3\right)\cos\eta x + \eta\left(\frac{r\eta}{\xi}F_4 - \frac{s\xi}{\eta}F_5\right)\sin\eta x \end{vmatrix}}{\Delta}$$

$$\alpha_{xL} = \frac{\begin{bmatrix} \xi\left(\frac{r\eta^2}{s\xi^2}\sin\eta L - \sin\xi L\right)\cos\xi x + \xi F_6 \sin\xi x \\ -\eta\left(\sin\eta L - \frac{s\xi^2}{r\eta^2}\sin\xi L\right)\cos\eta x - \eta F_6 \sin\eta x \end{bmatrix}}{\Delta}$$

$$\alpha_{xL'} = \frac{\begin{vmatrix} -r\eta F_6 \cos\xi x - r\eta\left(\sin\xi L - \frac{s\xi^2}{r\eta^2}\sin\eta L\right)\sin\xi x \\ +s\xi F_6 \cos\eta x + s\xi\left(\frac{r\eta^2}{s\xi^2}\sin\xi L - \sin\eta L\right)\sin\eta x \end{vmatrix}}{\Delta}$$

The above equations allow all the shaft element receptances to be calculated for any excitation frequency ω. The remaining receptances required to predict the response of a spindle are those of the bearings.

4.2.2 Radial and Tilt Characteristics of Bearings

It is always surprising to find how little is known about the stiffness and damping of bearings under oscillating conditions. Bearings are used so widely that it might be expected that all the characteristics required for machine tool spindle design would be well documented. However, a search of the literature will find very little experimental data relating to the stiffness and damping of bearings under working conditions. For some time, it was falsely assumed that the damping of rolling element bearings would be low and could be ignored. It was reasoned that only the stiffness was required, and most manufacturers give only stiffness values. In practice, even the stiffness values are often theoretical values and not confirmed experimentally. The main reason for the lack of experimental data is the difficulty encountered in making the necessary measurements. It might be thought that measurements on a complete spindle would allow the bearing characteristics to be a determined. However, the simplest spindle has two bearings and these will have both radial and tilt (resisting bending) stiffnesses and associated damping. Thus, for a two-bearing spindle system, there will be, at any one excitation frequency, eight unknowns relating to the bearings (for each bearing the radial stiffness, radial damping, tilt stiffness and tilt damping). These bearing characteristics would have to be calculated by modelling the isolated shaft and removing its contribution from the total system response.

In a well-cited paper [4], the state of the art in measuring bearing characteristics was reviewed. This paper was published in 1982, and not many experimental papers on the topic have been published since, though there are some [5–10]. The major drawback with much of the experimental work is that the characteristics are only determined for the resonant frequency of the research rig. For machine tool spindle design, any variation of stiffness and damping with the vibration frequency may be important. The speed of rotation and the preload on the bearing(s) may also affect the characteristics. Some early work, published in a restricted circulation [11], was made more widely known [12–14]. This showed that the damping in rolling element bearings was not negligible and that there was also damping in the necessary interface between the outer race of the bearing and its housing. It is appropriate to describe the method used for the measurements as they were achieved in a reasonably direct manner. The initial work was that of Walford [15], Walford and Stone [16–18] and extended by Lambert et al. [12–14].

The experimental rig used is shown in Fig. 4.4 and has a short, rigid shaft mounted in two bearings that could be preloaded against each other. Matching and large inertias were mounted on the ends of the shaft. The housing was suspended on a vertical cable so that it could be excited horizontally. The shaft was driven by a motor connected through a very flexible connection. Assuming the housing and shaft plus inertias to be rigid bodies, a schematic diagram of the test rig is as shown in Fig. 4.5 where the value of K_{T} is the total stiffness of the pair of bearings that will each have stiffness and damping components. The results were measured as the amplitude and phase of the bearing stiffness as shown in Fig. 4.6 and as used below $K_{\mathrm{T}} = K_{\mathrm{real}} + iK_{\mathrm{imaginary}}$.

The value of K_{T} is given by the force on the bearings $(-m_s\ddot{x}_s)$ divided by the displacement across them (x_r). Thus,

$$K_{\mathrm{T}} = -m_s\ddot{x}_s/x_r \text{ where } x_{\mathrm{r}} = x_{\mathrm{s}} - x_{\mathrm{h}}.$$

For steady-state sinusoidal excitation, $\ddot{x}_{\mathrm{r}} = -\omega^2 x_{\mathrm{r}}$; thus, $\ddot{x}_{\mathrm{s}} = \ddot{x}_{\mathrm{h}} - \omega^2 x_{\mathrm{r}}$, and hence,

$$K_{\mathrm{T}} = \frac{-m_s(\ddot{x}_{\mathrm{h}} - \omega^2 x_{\mathrm{r}})}{x_{\mathrm{r}}}$$

so that

$$K_{\mathrm{T}} = m_{\mathrm{s}}\left(\omega^2 - \frac{\ddot{x}_{\mathrm{h}}}{x_{\mathrm{r}}}\right)$$

It was a simple matter to measure the acceleration of the housing ($\ddot{x}_{\mathrm{h}}$), and the displacement across the bearings (x_{r}) was measured using a proximity probe. These two signals were not in phase, and thus, K_{T} was described by both amplitude and phase. The final model of the bearing developed by Lambert as an extension of Walford's is shown in Fig. 4.7. Walford showed that there was significant damping originating from the entry region to the hydrodynamic zone. However, the stiffness contribution from this zone was not significant. The stiffness and

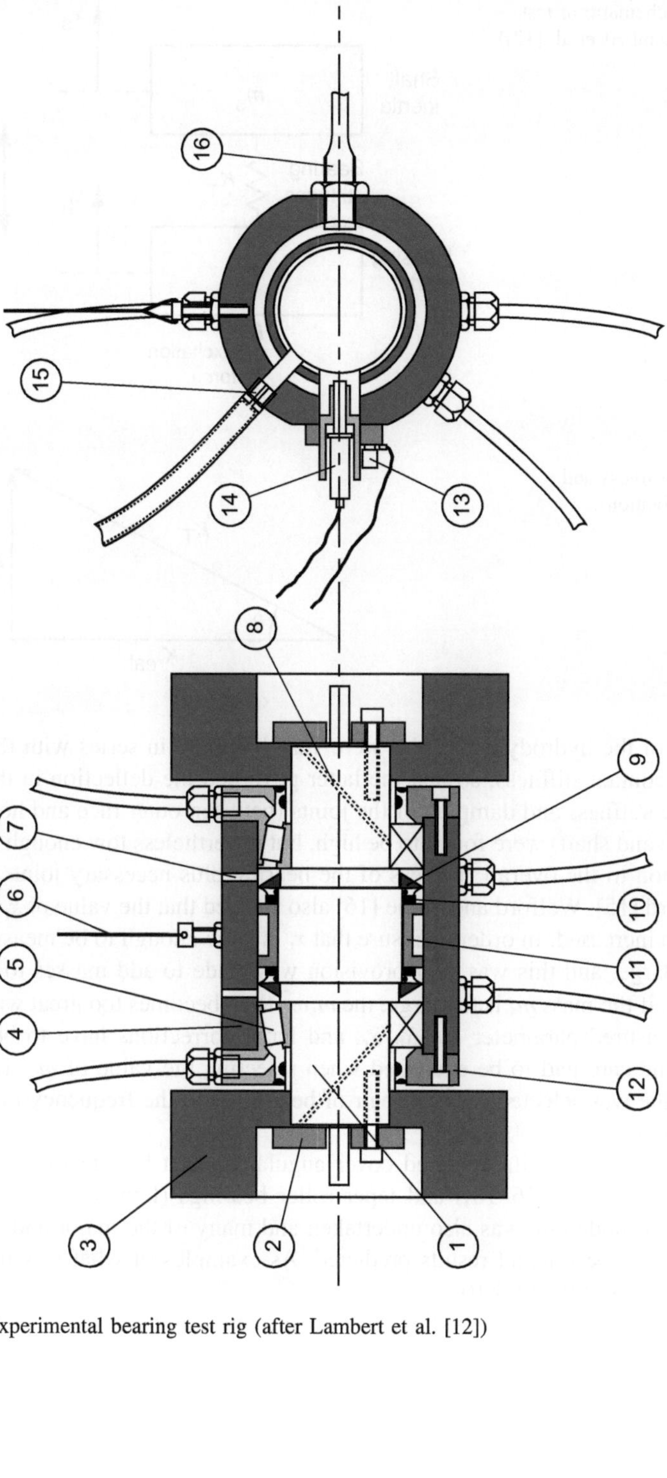

Fig. 4.4 Experimental bearing test rig (after Lambert et al. [12])

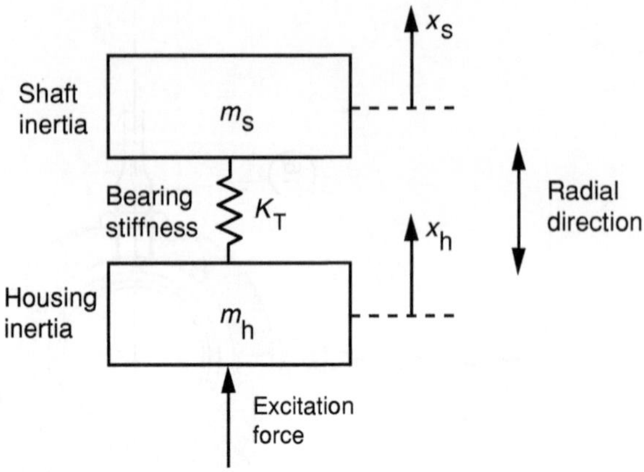

Fig. 4.5 Schematic of test rig (after Lambert et al. [12])

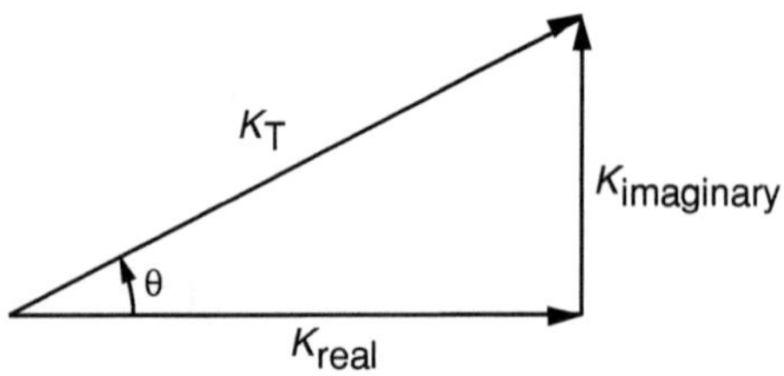

Fig. 4.6 Stiffness and damping notation

damping of the hydrodynamic film were very high and in series with the smaller Hertzian contact stiffness, so that the latter governed the deflection in the contact zone. The stiffness and damping of the joints (between outer race and housing and inner race and shaft) were found to be high, but nevertheless low enough to make a contribution to the overall stiffness of the bearing plus necessary joints.

Walford [15], Walford and Stone [16] also showed that the value of x_r increased as m_s was increased. In order to ensure that x_r is large enough to be measurable, m_s must be large, and this was why provision was made to add masses to the shaft. However, if the mass m_s is too large, the $m_s\omega^2$ term becomes too great with respect to the measured parameter $-m_s\ddot{x}_h/x_r$, and large corrections have to be applied. Thus, some care had to be exercised when selecting the value of m_s. The appropriate value was selected for each pair of bearings and the frequency range being investigated.

The extensive results obtained cover angular contact bearings (Walford [15], Walford and Stone [16–18]) and taper roller bearings (Lambert et al. [11–14]). Theoretical modelling was also undertaken and many of the major and somewhat unexpected experimental results predicted. As examples of what was found consider Figs. 4.8, 4.9 and 4.10.

Summarising the major findings of Walford and Lambert,

1. The level of damping in rolling element bearings is not negligible in the context of the chatter response of machine tool spindles.

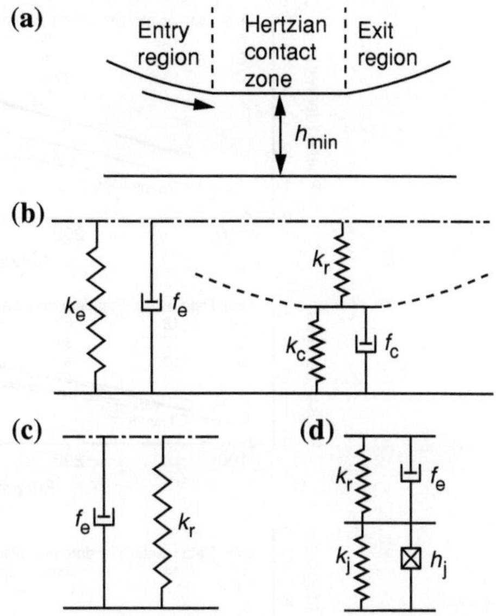

Fig. 4.7 Stiffness and damping contributions to the stiffness and damping of roller bearings (after Lambert et al. [12])

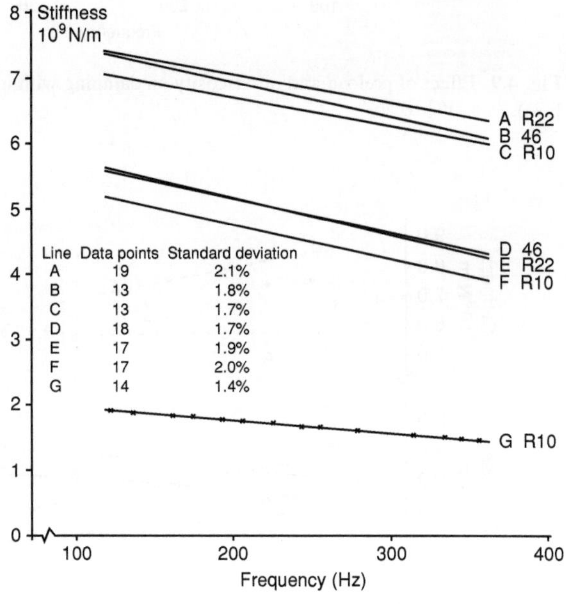

Fig. 4.8 Effect of preload and oil viscosity on stiffness with interference fit (after Lambert et al. [13])

2. Walford found that for angular contact bearings the measured stiffness was far less than that provided by the manufacturer. This was found to be due to the result of the flexibility of the housing to outer race joint.
3. Walford found that every time he dismantled his rig and reassembled it the bearing stiffness was reduced.

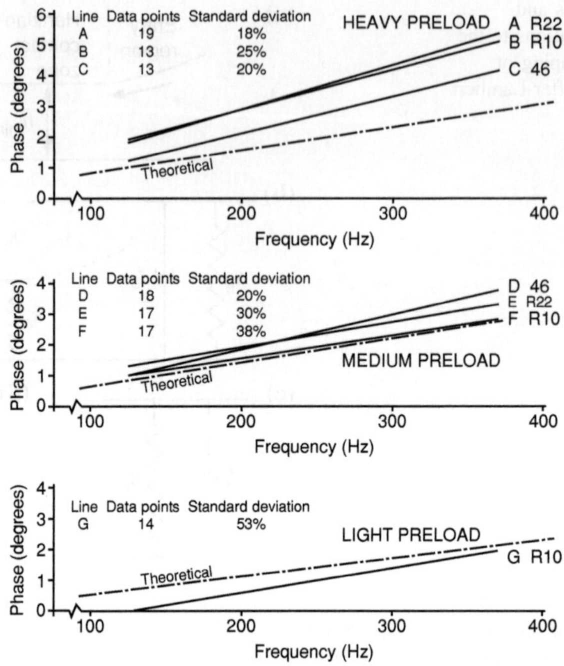

Fig. 4.9 Effect of preload and oil viscosity on damping with interference fit (after Lambert et al. [13])

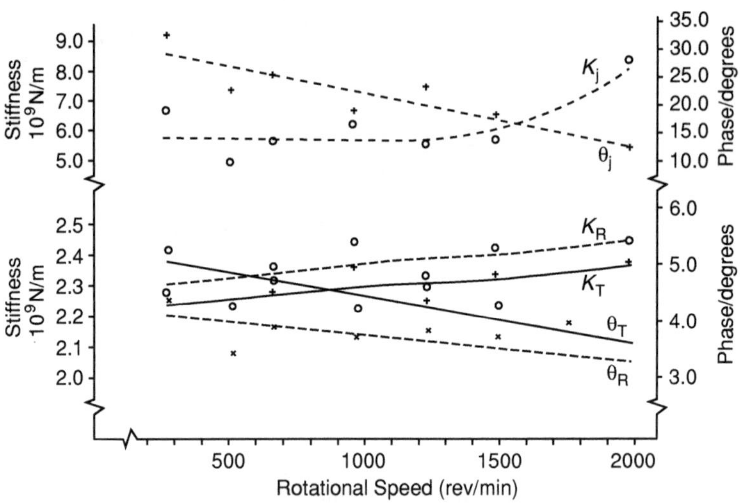

Fig. 4.10 Stiffness and damping components including joints (after Lambert et al. [14])

4. As might be expected, bearing stiffness increased with preload, but the damping also increased.
5. Lambert predicted that the damping would not vary with oil viscosity for typical bearing lubricants, and this was confirmed experimentally.
6. The necessary joints employed when fitting bearings may have a considerable effect on flexibility and, more importantly, on damping.
7. Both Walford and Lambert found that during warm-up, differential expansions occur which changed the preload and hence the characteristics. For repeatable results, it was essential to attain a uniform steady temperature.

It is appropriate to review the limitations of the rig used. The main limitation is that the pair of bearings was used in a back-to-back configuration and all the rolling elements were under similar load. There was no external radial load applied, as would be the case when machining, and this would cause each of the rolling elements to carry a different load—some would have an increased load and others a reduced load compared to the test situation. As preload (and hence contact forces) was found to be very significant, the inability to apply an external radial load was a significant restriction. However, the results obtained allow reasonable values for bearing stiffnesses to be used in the prediction of spindle responses.

The other restriction imposed by the rig was that only the radial characteristics could be measured. In practice, the shaft will tilt in the bearings and experience an applied moment. There has been some discussion of the relative importance of the radial and tilt characteristics. Elsermans et al. [19] measured tilt characteristics and stated: "It is assumed, in this study, that the experimentally determined over-all damping of spindles is almost entirely due to clamping (i.e. tilt) damping occurring in the taper-roller bearings. The radial damping may usually be neglected in this respect. Indeed, due to the usually very high radial static stiffness of the bearing, no significant radial deformation and consequently no measurable damping effect is to be expected from this mode of deformation". This statement will be reexamined when predictions are made of the spindle response.

4.2.3 Subsystem Addition

This section describes the method employed for adding subsystems to form the complete spindle. This method was the basis for a computer program that allowed the response to be predicted for various locations and excitations along the spindle.

Figure 4.11 shows the notation used for the addition of two subsystems. The receptances of a complex built-up system will be found by multiple applications of the equations that will be derived below. For the reader who is not familiar with the systems approach to vibration, Appendix C contains an introduction to the topic. The coordinates P, Q, R and S are necessary if all the cross-receptances of the final system are required. Two coupling coordinates are required for transverse vibration.

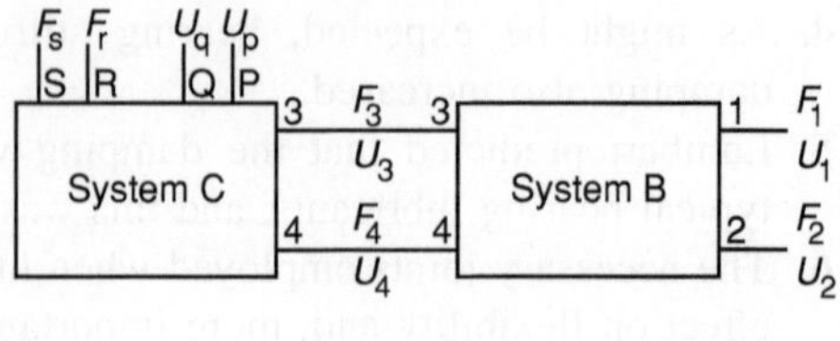

Fig. 4.11 Addition of two subsystems

The systems B and C are separated at the location of coordinates 3 and 4 but are constrained in such a way that they will vibrate in the same manner as when joined. This requires forces and moments to be applied at the joining coordinates so that the separate subsystems have the forces shown in Fig. 4.12.

Note that F may represent a force or a moment. Also U may represent a displacement or a slope. Using conventional (see Appendix C) receptance notation,

For system B,

$$\begin{bmatrix} U_1 \\ U_2 \end{bmatrix} = \begin{bmatrix} \beta_{11} & \beta_{12} \\ \beta_{12} & \beta_{22} \end{bmatrix} \begin{bmatrix} F_1 \\ F_2 \end{bmatrix} + \begin{bmatrix} \beta_{13} & \beta_{14} \\ \beta_{23} & \beta_{24} \end{bmatrix} \begin{bmatrix} F_{b3} \\ F_{b4} \end{bmatrix} \tag{4.1}$$

$$\begin{bmatrix} U_{b3} \\ U_{b4} \end{bmatrix} = \begin{bmatrix} \beta_{13} & \beta_{23} \\ \beta_{14} & \beta_{24} \end{bmatrix} \begin{bmatrix} F_1 \\ F_2 \end{bmatrix} + \begin{bmatrix} \beta_{33} & \beta_{34} \\ \beta_{34} & \beta_{44} \end{bmatrix} \begin{bmatrix} F_{b3} \\ F_{b4} \end{bmatrix} \tag{4.2}$$

For system C,

$$\begin{bmatrix} U_{c3} \\ U_{c4} \end{bmatrix} = \begin{bmatrix} \gamma_{33} & \gamma_{34} \\ \gamma_{34} & \gamma_{44} \end{bmatrix} \begin{bmatrix} F_{c3} \\ F_{c4} \end{bmatrix} + \begin{bmatrix} \gamma_{3r} & \gamma_{3s} \\ \gamma_{4r} & \gamma_{4s} \end{bmatrix} \begin{bmatrix} F_r \\ F_s \end{bmatrix} \tag{4.3}$$

$$\begin{bmatrix} U_p \\ U_q \end{bmatrix} = \begin{bmatrix} \gamma_{3p} & \gamma_{4p} \\ \gamma_{3q} & \gamma_{4q} \end{bmatrix} \begin{bmatrix} F_{c3} \\ F_{c4} \end{bmatrix} + \begin{bmatrix} \gamma_{pr} & \gamma_{ps} \\ \gamma_{qr} & \gamma_{qs} \end{bmatrix} \begin{bmatrix} F_r \\ F_s \end{bmatrix} \tag{4.4}$$

At the join, for compatibility,

$$\begin{bmatrix} U_3 \\ U_4 \end{bmatrix} = \begin{bmatrix} U_{b3} \\ U_{b4} \end{bmatrix} = \begin{bmatrix} U_{c3} \\ U_{c4} \end{bmatrix} \tag{4.5}$$

At the boundary, for force equivalence,

$$\begin{bmatrix} F_{b3} \\ F_{b4} \end{bmatrix} + \begin{bmatrix} F_{c3} \\ F_{c4} \end{bmatrix} = \begin{bmatrix} F_3 \\ F_4 \end{bmatrix} \tag{4.6}$$

Substituting in Eq. 4.5 from Eqs. 4.2 and 4.3,

Fig. 4.12 Separated subsystems, with associated forces and displacements

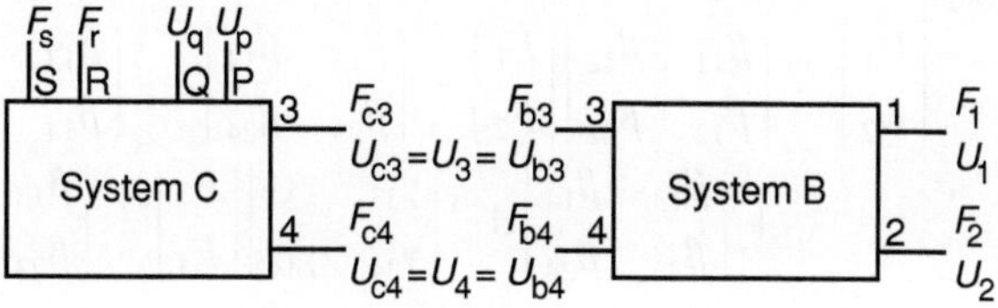

$$\begin{bmatrix} \beta_{13} & \beta_{23} \\ \beta_{14} & \beta_{24} \end{bmatrix} \begin{bmatrix} F_1 \\ F_2 \end{bmatrix} + \begin{bmatrix} \beta_{33} & \beta_{34} \\ \beta_{34} & \beta_{44} \end{bmatrix} \begin{bmatrix} F_{b3} \\ F_{b4} \end{bmatrix} = \begin{bmatrix} \gamma_{33} & \gamma_{34} \\ \gamma_{34} & \gamma_{44} \end{bmatrix} \begin{bmatrix} F_{c3} \\ F_{c4} \end{bmatrix} + \begin{bmatrix} \gamma_{3r} & \gamma_{3s} \\ \gamma_{4r} & \gamma_{4s} \end{bmatrix} \begin{bmatrix} F_r \\ F_s \end{bmatrix} \tag{4.7}$$

Substituting for $\begin{bmatrix} F_{c3} \\ F_{c4} \end{bmatrix}$ from Eq. 4.6 in Eq. 4.7,

$$\begin{bmatrix} \beta_{13} & \beta_{23} \\ \beta_{14} & \beta_{24} \end{bmatrix} \begin{bmatrix} F_1 \\ F_2 \end{bmatrix} + \begin{bmatrix} \beta_{33} & \beta_{34} \\ \beta_{34} & \beta_{44} \end{bmatrix} \begin{bmatrix} F_{b3} \\ F_{b4} \end{bmatrix} = \begin{bmatrix} \gamma_{33} & \gamma_{34} \\ \gamma_{34} & \gamma_{44} \end{bmatrix} \begin{bmatrix} F_3 \\ F_4 \end{bmatrix} - \begin{bmatrix} \gamma_{33} & \gamma_{34} \\ \gamma_{34} & \gamma_{44} \end{bmatrix} \begin{bmatrix} F_{b3} \\ F_{b4} \end{bmatrix} + \begin{bmatrix} \gamma_{3r} & \gamma_{3s} \\ \gamma_{3r} & \gamma_{4s} \end{bmatrix} \begin{bmatrix} F_r \\ F_s \end{bmatrix}$$

and rearranging

$$\begin{bmatrix} \beta_{33}+\gamma_{33} & \beta_{34}+\gamma_{34} \\ \beta_{34}+\gamma_{34} & \beta_{44}+\gamma_{44} \end{bmatrix} \begin{bmatrix} F_{b3} \\ F_{b4} \end{bmatrix} = -\begin{bmatrix} \beta_{13} & \beta_{23} \\ \beta_{14} & \beta_{24} \end{bmatrix} \begin{bmatrix} F_1 \\ F_2 \end{bmatrix} + \begin{bmatrix} \gamma_{33} & \gamma_{34} \\ \gamma_{34} & \gamma_{44} \end{bmatrix} \begin{bmatrix} F_3 \\ F_4 \end{bmatrix} + \begin{bmatrix} \gamma_{3r} & \gamma_{3s} \\ \gamma_{4r} & \gamma_{4s} \end{bmatrix} \begin{bmatrix} F_r \\ F_s \end{bmatrix}$$

so that

$$\begin{bmatrix} F_{b3} \\ F_{b4} \end{bmatrix} = -\Delta^{-1} \begin{bmatrix} \beta_{13} & \beta_{23} \\ \beta_{14} & \beta_{24} \end{bmatrix} \begin{bmatrix} F_1 \\ F_2 \end{bmatrix} + \Delta^{-1} \begin{bmatrix} \gamma_{33} & \gamma_{34} \\ \gamma_{34} & \gamma_{44} \end{bmatrix} \begin{bmatrix} F_3 \\ F_4 \end{bmatrix} + \Delta^{-1} \begin{bmatrix} \gamma_{3r} & \gamma_{3s} \\ \gamma_{4r} & \gamma_{4s} \end{bmatrix} \begin{bmatrix} F_r \\ F_s \end{bmatrix} \tag{4.8}$$

where

$$\Delta = \begin{bmatrix} \beta_{33}+\gamma_{33} & \beta_{34}+\gamma_{34} \\ \beta_{34}+\gamma_{34} & \beta_{44}+\gamma_{44} \end{bmatrix} \tag{4.9}$$

We need all the receptances and cross-receptances of the combined system. Substituting Eq. 4.8 in Eq. 4.1 gives

$$\begin{vmatrix} U_1 \\ U_2 \end{vmatrix} = \begin{vmatrix} \beta_{11} & \beta_{12} \\ \beta_{12} & \beta_{22} \end{vmatrix} \begin{vmatrix} F_1 \\ F_2 \end{vmatrix} - \begin{vmatrix} \beta_{13} & \beta_{14} \\ \beta_{23} & \beta_{24} \end{vmatrix} \Delta^{-1} \begin{vmatrix} \beta_{13} & \beta_{23} \\ \beta_{14} & \beta_{24} \end{vmatrix} \begin{vmatrix} F_1 \\ F_2 \end{vmatrix}$$
$$+ \begin{vmatrix} \beta_{13} & \beta_{14} \\ \beta_{23} & \beta_{24} \end{vmatrix} \Delta^{-1} \begin{vmatrix} \gamma_{33} & \gamma_{34} \\ \gamma_{34} & \gamma_{44} \end{vmatrix} \begin{vmatrix} F_3 \\ F_4 \end{vmatrix} + \begin{vmatrix} \beta_{13} & \beta_{14} \\ \beta_{23} & \beta_{24} \end{vmatrix} \Delta^{-1} \begin{vmatrix} \gamma_{3r} & \gamma_{3s} \\ \gamma_{4r} & \gamma_{4s} \end{vmatrix} \begin{vmatrix} F_r \\ F_s \end{vmatrix},$$

and to find $\begin{bmatrix} \alpha_{11} & \alpha_{12} \\ \alpha_{12} & \alpha_{22} \end{bmatrix}$, we put $\begin{bmatrix} F_3 \\ F_4 \end{bmatrix} = 0$ and $\begin{bmatrix} F_r \\ F_s \end{bmatrix} = 0$ so that

$$\begin{vmatrix} U_1 \\ U_2 \end{vmatrix} = \begin{vmatrix} \beta_{11} & \beta_{12} \\ \beta_{12} & \beta_{22} \end{vmatrix} \begin{vmatrix} F_1 \\ F_2 \end{vmatrix} - \begin{vmatrix} \beta_{13} & \beta_{14} \\ \beta_{23} & \beta_{24} \end{vmatrix} \Delta^{-1} \begin{vmatrix} \beta_{13} & \beta_{23} \\ \beta_{14} & \beta_{24} \end{vmatrix} \begin{vmatrix} F_1 \\ F_2 \end{vmatrix}$$

$$\therefore \begin{vmatrix} U_1 \\ U_2 \end{vmatrix} \begin{vmatrix} \frac{1}{F_1} & \frac{1}{F_2} \end{vmatrix} = \begin{vmatrix} \beta_{11} & \beta_{12} \\ \beta_{12} & \beta_{22} \end{vmatrix} \begin{vmatrix} F_1 \\ F_2 \end{vmatrix} \begin{vmatrix} \frac{1}{F_1} & \frac{1}{F_2} \end{vmatrix}$$
$$- \begin{vmatrix} \beta_{13} & \beta_{14} \\ \beta_{23} & \beta_{24} \end{vmatrix} \Delta^{-1} \begin{vmatrix} \beta_{13} & \beta_{23} \\ \beta_{14} & \beta_{24} \end{vmatrix} \begin{vmatrix} F_1 \\ F_2 \end{vmatrix} \begin{vmatrix} \frac{1}{F_1} & \frac{1}{F_2} \end{vmatrix},$$

and hence,

$$\begin{vmatrix} \frac{U_1}{F_1} & \frac{U_1}{F_2} \\ \frac{U_2}{F_1} & \frac{U_2}{F_2} \end{vmatrix} = \begin{vmatrix} \alpha_{11} & \alpha_{12} \\ \alpha_{21} & \alpha_{22} \end{vmatrix} = \begin{vmatrix} \beta_{11} & \beta_{12} \\ \beta_{12} & \beta_{22} \end{vmatrix} - \begin{vmatrix} \beta_{13} & \beta_{14} \\ \beta_{23} & \beta_{24} \end{vmatrix} \Delta^{-1} \begin{vmatrix} \beta_{13} & \beta_{23} \\ \beta_{14} & \beta_{24} \end{vmatrix} \tag{4.10}$$

Similarly, to find $\begin{bmatrix} \alpha_{13} & \alpha_{14} \\ \alpha_{23} & \alpha_{24} \end{bmatrix}$, we put $\begin{bmatrix} F_1 \\ F_2 \end{bmatrix} = 0$ and $\begin{bmatrix} F_r \\ F_s \end{bmatrix} = 0$ so that

$$\begin{vmatrix} U_1 \\ U_2 \end{vmatrix} = \begin{vmatrix} \beta_{13} & \beta_{14} \\ \beta_{23} & \beta_{24} \end{vmatrix} \Delta^{-1} \begin{vmatrix} \gamma_{33} & \gamma_{34} \\ \gamma_{34} & \gamma_{44} \end{vmatrix} \begin{vmatrix} F_3 \\ F_4 \end{vmatrix},$$

and hence,

$$\begin{vmatrix} \alpha_{13} & \alpha_{14} \\ \alpha_{23} & \alpha_{24} \end{vmatrix} = \begin{vmatrix} \beta_{13} & \beta_{14} \\ \beta_{23} & \beta_{24} \end{vmatrix} \Delta^{-1} \begin{vmatrix} \gamma_{33} & \gamma_{34} \\ \gamma_{34} & \gamma_{44} \end{vmatrix}, \tag{4.11}$$

and to find $\begin{bmatrix} \alpha_{1r} & \alpha_{1s} \\ \alpha_{2r} & \alpha_{2s} \end{bmatrix}$, we just put $\begin{bmatrix} F_1 \\ F_2 \end{bmatrix} = 0$ and $\begin{bmatrix} F_3 \\ F_4 \end{bmatrix} = 0$ so that

$$\begin{vmatrix} U_1 \\ U_2 \end{vmatrix} = \begin{vmatrix} \beta_{13} & \beta_{14} \\ \beta_{23} & \beta_{24} \end{vmatrix} \Delta^{-1} \begin{vmatrix} \gamma_{3r} & \gamma_{3s} \\ \gamma_{4r} & \gamma_{4s} \end{vmatrix} \begin{vmatrix} F_r \\ F_s \end{vmatrix},$$

and hence,

$$\begin{vmatrix} \alpha_{1r} & \alpha_{1s} \\ \alpha_{2r} & \alpha_{2s} \end{vmatrix} = \begin{vmatrix} \beta_{13} & \beta_{14} \\ \beta_{23} & \beta_{24} \end{vmatrix} \Delta^{-1} \left\| \begin{matrix} \gamma_{3r} & \gamma_{3s} \\ \gamma_{4r} & \gamma_{4s} \end{matrix} \right\| \tag{4.12}$$

Now, substituting Eq. 4.8 in Eq. 4.2 and using a similar approach to that used above, it can be shown that

$$\begin{vmatrix} \alpha_{33} & \alpha_{34} \\ \alpha_{34} & \alpha_{44} \end{vmatrix} = \begin{vmatrix} \beta_{33} & \beta_{34} \\ \beta_{34} & \beta_{44} \end{vmatrix} \Delta^{-1} \begin{vmatrix} \gamma_{33} & \gamma_{34} \\ \gamma_{34} & \gamma_{44} \end{vmatrix} \tag{4.13}$$

$$\begin{vmatrix} \alpha_{3r} & \alpha_{3s} \\ \alpha_{4r} & \alpha_{4s} \end{vmatrix} = \begin{vmatrix} \beta_{33} & \beta_{34} \\ \beta_{34} & \beta_{44} \end{vmatrix} \Delta^{-1} \left\| \begin{matrix} \gamma_{3r} & \gamma_{3s} \\ \gamma_{4r} & \gamma_{4s} \end{matrix} \right\| \tag{4.14}$$

Finally, we substitute Eq. 4.8 in Eq. 4.4 and find that

$$\begin{vmatrix} \alpha_{pr} & \alpha_{ps} \\ \alpha_{qr} & \alpha_{qs} \end{vmatrix} = \begin{vmatrix} \gamma_{pr} & \gamma_{ps} \\ \gamma_{qr} & \gamma_{qs} \end{vmatrix} - \begin{vmatrix} \gamma_{3p} & \gamma_{4p} \\ \gamma_{3q} & \gamma_{4q} \end{vmatrix} \Delta^{-1} \begin{vmatrix} \gamma_{3r} & \gamma_{3s} \\ \gamma_{4r} & \gamma_{4s} \end{vmatrix} \tag{4.15}$$

We now have all the required receptances. For the program mentioned, matrix methods were used in the solutions.

For a bearing taking the position of subsystem B in Fig. 4.12 with a complex stiffness, $k(1 + i\eta)$, where η is the hysteretic damping ratio (see Appendix A.9), and also assuming no cross-coupling between displacement and slope,

$$\beta_{33} = \frac{1 - i\eta_r}{k_r(1 + \eta_r^2)}; \quad \beta_{34} = 0; \quad \beta_{43} = 0; \quad \beta_{44} = \frac{1 - i\eta_t}{k_t(1 + \eta_t^2)}$$

These are the only receptances required as a bearing is added in parallel (see Fig. 4.2). The notation uses odd numbers to refer to forces and displacements and even numbers to refer to slopes and moments. A complete spindle system with as many bearings as required may now be assembled by successively adding the subsystems using Eqs. 4.10–4.15.

To find the deflected shape at any particular frequency and for excitation at any shaft joint, we can find the displacement and slope at each end of every shaft element. Then, using the shaft subsection receptances, it is possible to determine the forces and moments at each end required to produce those end responses. Finally, the displacements, at any position x, caused by these moments and forces can be calculated by superposition. As the system has damping, all the displacements, slopes, moments and forces will be complex numbers. A series of programs have been written so that the effect of significant parameters on the spindle response may be demonstrated.

4.2.4 Predicted Responses and Q Factors

Lambert et al. [14] used a reference spindle for comparison of spindles incorporating the various bearings tested using their rig previously described. These involved angular contact, sliding fit taper roller and press-fitted taper roller

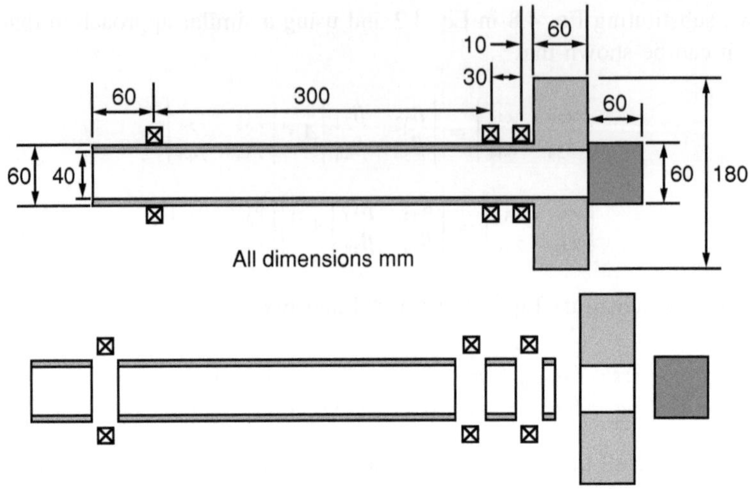

Fig. 4.13 Typical spindle and subsystems

bearings. All the bearings tested had a 60 mm diameter bore, and the spindle system shown in Fig. 4.13 was used as a reference. The main shaft was kept as simple as possible, and a pair of bearings was located at the workpiece end so that the significance of the effects of tilt could be investigated. Two axially separated bearings allowed both radial and tilt stiffness to be varied. From the experimental results, it was known that typical values of stiffness and phase, for the bearing combinations tested, were for a single bearing,

	Stiffness K_T N/m	Phase (degrees)
Angular contact	4.85×10^7	6.2
Taper roller—with joint	7.5×10^8	6.5
Taper roller—no joint	2.55×10^9	1.0

The response at the cutting point for each type of bearing was calculated. The left-hand end bearing was the same for all the spindles and, as is usual practice, was a sliding fit in its housing. Its stiffness and phase were therefore assumed to be the same as for a single angular contact bearing. The three steady-state frequency responses are shown in Fig. 4.14.

For each of the spindles, it was possible to find the first resonant frequency, the associated peak response and Q factor (see Appendix A.8).

	Frequency (Hz)	Peak response (m/N)	Q factor
Angular contact	298	6.18×10^{-7}	12.6
Taper roller—with joint	486	4.06×10^{-7}	17.3
Taper roller—no joint	584	5.0×10^{-7}	31.6

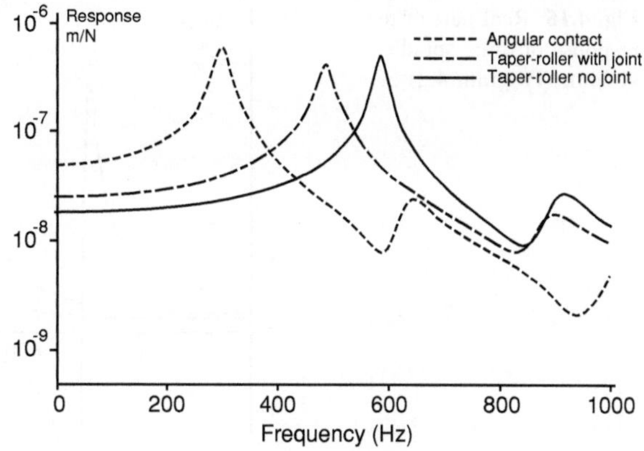

Fig. 4.14 Responses of three spindle systems (Program 4.1)

Figure 4.14 shows that the (quasi) static stiffness of each of the spindles increased significantly as the stiffness of the front bearing increased. However, the peak response does not vary by the same amount and the stiffest pair of bearings (with the small phase) does not have the smallest peak. To cater for the wide range of values, Fig. 4.14 uses a logarithmic scale for the response. If the real part of the response is plotted (Fig. 4.15) for the same conditions, the maximum negative real part (the chatter sensitivity) follows the same trend as the peak response.

The effect of the axial location of the middle bearing was investigated to find an optimum in terms of peak response. For the three bearing types considered, the optimum location relative to the left-hand end, along with the first resonant frequency, the associated peak response and Q factor is shown in the table below.

	Position (mm)	Frequency (Hz)	Peak response (m/N)	Q factor
Angular contact	13	246	6.02×10^{-7}	10.2
Taper roller—with joint	30	581	2.53×10^{-7}	14.5
Taper roller—no joint	34.8	638	4.73×10^{-7}	34.9

The point of some practical application is that for the same bearings, but with a different spatial arrangement, the Q factor changes. This is because the damping is localised. Much modern modelling ignores this fact and assumes a Q factor for a spindle mode that is independent of the bearing locations. Programs 4.1 and 4.2 allow the position of the exciting force and the response to be changed simply by clicking on the diagram. It will be found that for any particular spindle the response changes significantly with the force and displacement positions, but not the resonant frequencies nor the Q factors for the resonant peaks.

An examination of the deflected shape that results from the excitation aids in understanding the results. As the shaft receptances include the contributions from all the modes, the predicted deflected shapes are those that would be measured experimentally for a single-frequency excitation. Most modern modelling presents

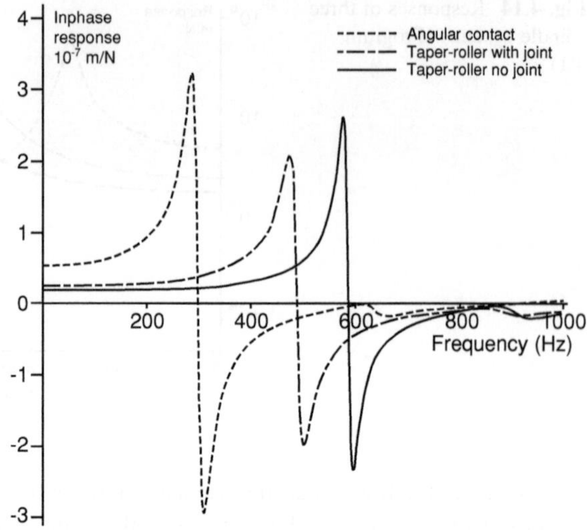

Fig. 4.15 Real part of the response of three spindle systems (Program 4.2)

separate modal responses and superimposes them to find the total response. Figure 4.16 shows the deflected shape for the first mode of each arrangement by depicting the extreme positions half a cycle apart. These plots are scaled to have the same amplitude at the workpiece end. The modes may be animated using Program 4.3. It appears that there are nodes, but this is not entirely true, especially for the arrangement with the angular contact front bearings. When the animation Program 4.3 is run, it is apparent that the point where the deflected shape crosses the axis moves back and forth along the axis. When mode shapes are measured experimentally using a constant excitation frequency, it is not always possible to find a node with zero amplitude for any selected mode.

A general conclusion may be made from the above discussion of spindles. The magnitude of the response peak is governed by the energy dissipation that is occurring. This requires not only a damping source but also motion where the damping is located. Energy dissipation depends on the product of the damping coefficient and the motion. It is possible for a highly damped bearing to have little energy dissipation because there is little motion at its location. It is for this reason that the angular contact bearings, which are relatively flexible, produce peak shaft responses of similar magnitude to the much stiffer bearings. An examination of Fig. 4.16a shows how large the vibration amplitude is at the front pair of bearings.

This concept of energy dissipation explains the earlier comments about the significance of tilt as opposed to radial characteristics. With a single stiff bearing at the front of the spindle, there would be little radial amplitude and little energy dissipation. The tilt across the bearing may then be providing the largest source of energy dissipation.

The complexity of predicting the response of even one component of a machine tool (e.g. the spindle) has indicated how much more difficult it would be to predict the response of a complete machine with several joints. The levels of damping are

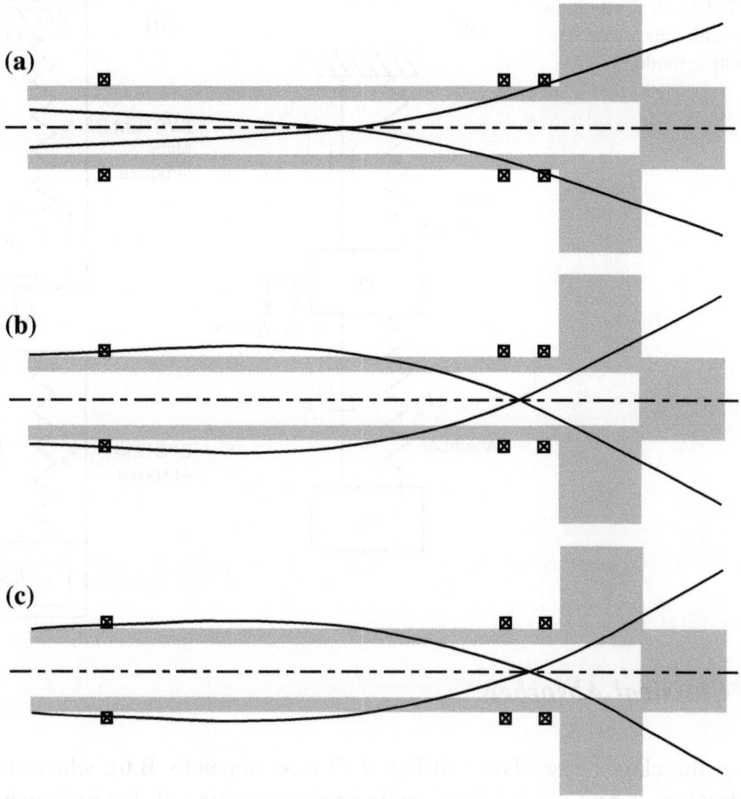

Fig. 4.16 Deflected shapes for **a** angular contact, **b** taper roller with joint and **c** taper roller without joint (Program 4.3)

difficult to predict, and nominally identical machines have been found to have different damping. As a result, it is important to consider ways of adding damping once a machine has been manufactured.

4.3 Vibration Absorbers

The basic concept behind vibration absorbers is well known, and a search of the literature using keywords "vibration" and "absorber" in the title will produce numerous papers. Appendix B.6 includes the analysis of a typical vibration absorber added to a single degree of freedom system. Appendix C.6 gives the example of a vibration absorber added at the end of an undamped cantilever bar. It is important to consider the optimisation of such vibration absorbers with the object of improving chatter performance.

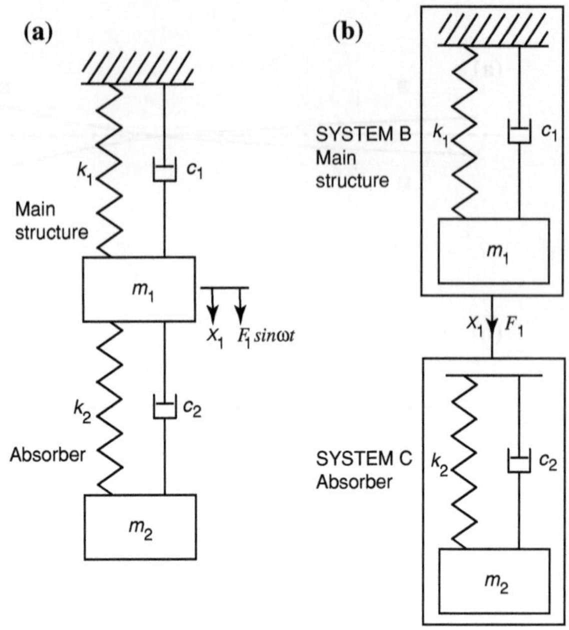

Fig. 4.17 Classical vibration absorber added to a structure with a single mode

4.3.1 Classical Absorber

Consider the classic case shown in Fig. 4.17 (see Appendix B.6), where the original structure is represented by a single mode consisting of a spring, mass and viscous damper with an absorber as an additional system. It is helpful to use the systems approach to find the response as this may also be applied to more complex structures. The analysis can be made by examining two subsystems as illustrated in Fig. 4.17b and adding them in the manner described in Appendix C.3.

The receptance of the main system is

$$\beta_{11} = \frac{1}{k_1 - m_1\omega^2 + \mathrm{i}\omega c_1}$$

and of the absorber with viscous damping

$$\frac{X_1}{F_1} = \gamma_{11} = \frac{1}{k_2 + \mathrm{i}\omega c_2} - \frac{1}{m_2\omega^2}$$

When the two subsystems are joined

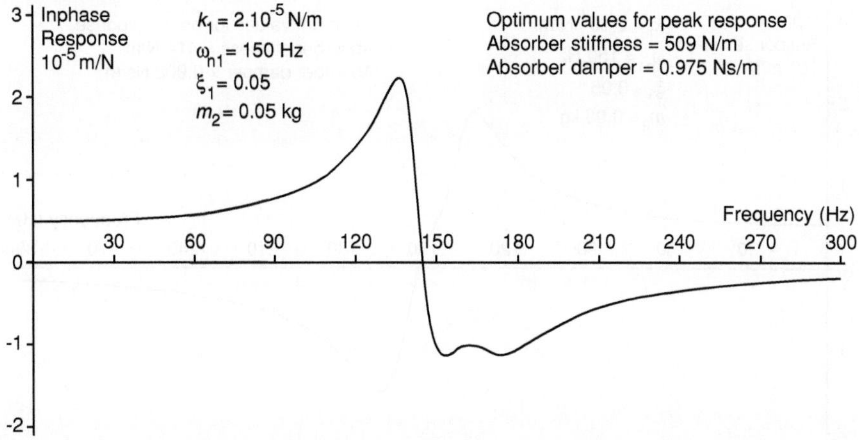

Fig. 4.18 In-phase response for absorber optimised for chatter performance (Program 4.4)

$$\begin{aligned}\frac{1}{\alpha_{11}} &= \frac{1}{\beta_{11}} + \frac{1}{\gamma_{11}} \\ \frac{1}{\alpha_{11}} &= k_1 - m_1\omega^2 + \mathrm{i}\omega c_1 + \frac{1}{\frac{1}{k_2+\mathrm{i}\omega c_2} - \frac{1}{m_2\omega^2}}\end{aligned} \tag{4.16}$$

This equation is best solved using a computer program. In Appendix B.6, the absorber is optimised to minimise the peak response of the main mass, whereas here, it is designed to give the best chatter performance by optimising the maximum negative in-phase component of the response by making it as small as possible. Figure 4.18 shows the optimum in-phase response for a particular main system and absorber mass.

An absorber optimised to minimise the peak response may be quite different from the optimum for chatter performance. Figure 4.19 shows the in-phase response when the peak response is minimised for the same main system as Fig. 4.18.

4.3.2 Clamped/Free Bar with an Absorber

Appendix C.6 contains an example of an absorber attached to a cantilever bar as shown in Fig. 4.20. The model is a reasonable approximation of a boring bar with a large length-to-diameter ratio. The only difference from the analysis of the classical absorber is that the main system receptance is given by

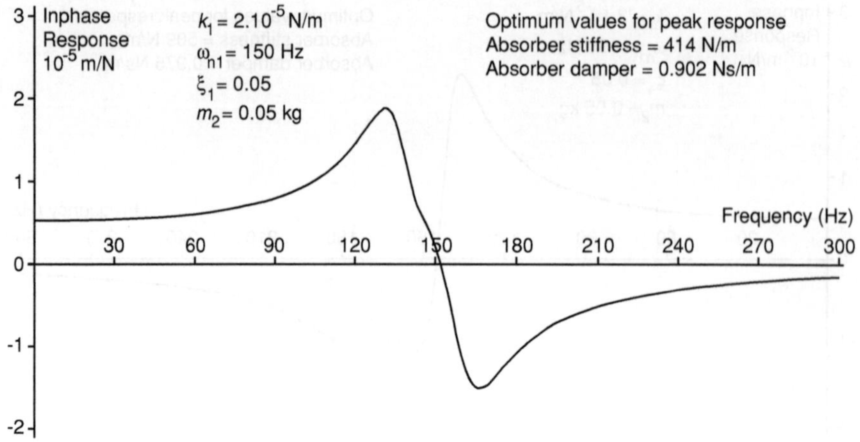

Fig. 4.19 In-phase response for absorber optimised for minimal peak response (Program 4.5)

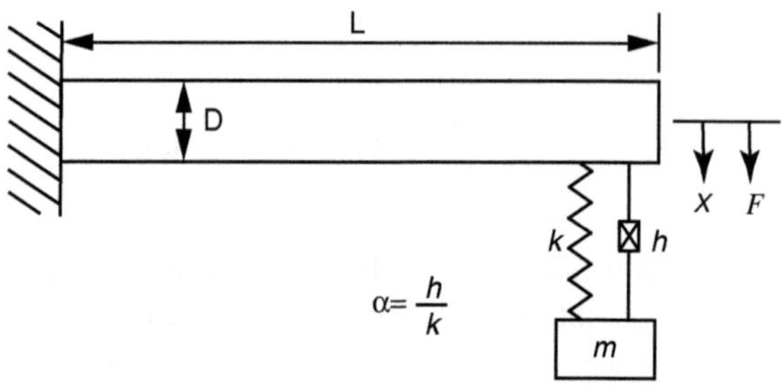

Fig. 4.20 Clamped/free bar with a vibration absorber added

$$\beta_{11} = -\frac{(\cos\lambda L \sinh\lambda L - \sin\lambda L \cosh\lambda L)}{EI\lambda^3(\cos\lambda L \cosh\lambda L + 1)} \quad \text{where } \lambda^4 = \frac{\rho A\omega^2}{EI}$$

This is a beam tip receptance from Bishop and Johnson [3] that does not include shear and rotary inertia effects. For large length-to-diameter ratios, it is adequate. The receptance at the free end of the bar with the absorber added is changed from that of Eq. 4.16 to

$$\frac{1}{\alpha_{11}} = -\frac{EI\lambda^3(\cos\lambda L \cosh\lambda L + 1)}{(\cos\lambda L \sinh\lambda L - \sin\lambda L \cosh\lambda L)} + \frac{1}{\frac{1}{k_2 + \mathrm{i}h_2} - \frac{1}{m_2\omega^2}} \tag{4.17}$$

The absorber damping is assumed to be hysteretic (see Appendix A.9) as this is a better model of the damping in the rubber-like materials used in many absorbers. A computer program is required to optimise the absorber for any given bar. For a bar of length $L = 0.3$ m, diameter $D = 0.02$ m and an absorber mass of 0.05 kg,

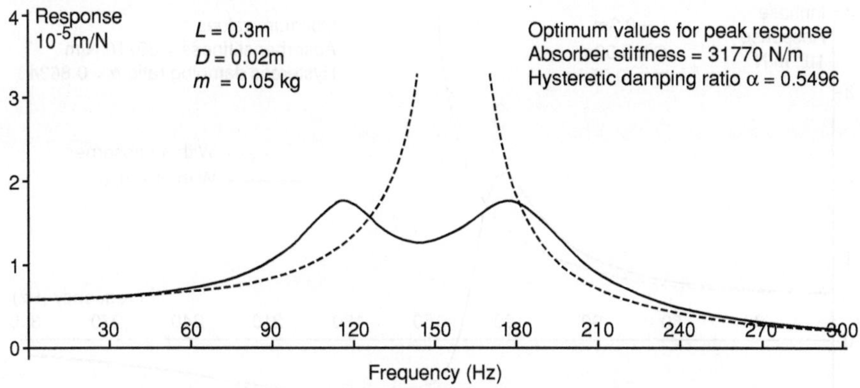

Fig. 4.21 Bar response with and without absorber optimised for peak response (Program 4.6)

the optimum response is shown in Fig. 4.21. The goal of the optimisation was to minimise the peak response.

As the model used is for an undamped bar, it has an infinite response at resonance. In practice, such bars have little damping so that the improvement that may be achieved by using an absorber is significant. To optimise for chatter performance, we need to optimise the maximum negative in-phase component of the response. Figure 4.22 shows the result for the same bar.

In practice, it is not always possible to obtain the required optimal hysteretic damping ratio. For a particular damping material, the damping ratio will be fixed and the optimisation involves adjusting the value of stiffness. There is also a further practical complication for boring bars. The absorber cannot be mounted external to the bar but has to be included within the bar. There can be some difficulty in mounting the absorber mass securely, and the use of a second bar to act in the same manner as an absorber mass can be a feasible solution.

4.3.3 A Second Bar as an Absorber

Figure 4.23 shows a typical free-end receptance of a clamped-free bar and of a rigid mass. It is apparent that above the first natural frequency of the bar, the receptance of the bar behaves somewhat like that of a rigid mass. The conventional absorber mass may therefore be replaced by a second bar of suitable dimensions. This arrangement for a boring bar is as shown in Fig. 4.24.

The receptance at the end of the internal bar is a modification of Eq. 4.17 where the subscript 2 indicates the free tip of the internal bar. Thus,

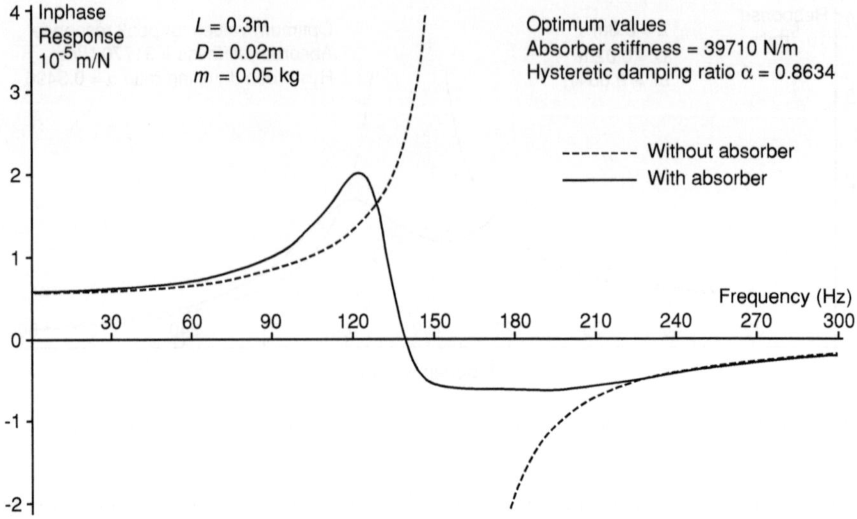

Fig. 4.22 Bar response with and without absorber optimised for chatter performance (Program 4.7)

Fig. 4.23 Comparison of the receptances of a clamped/free bar and a mass

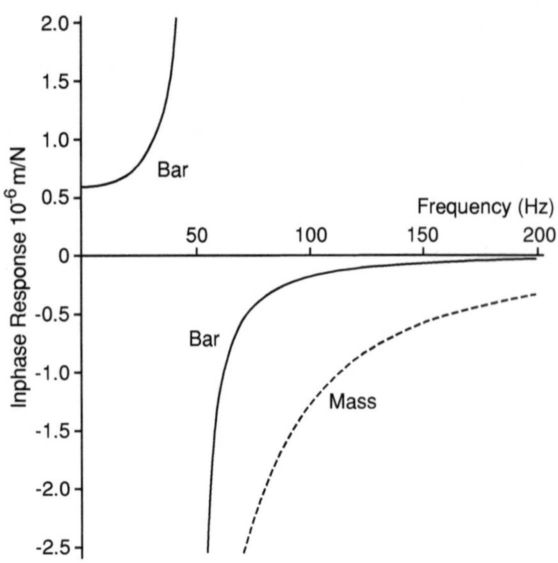

$$-\frac{(\cos\lambda_2 L_2 \sinh\lambda_2 L_2 - \sin\lambda_2 L_2 \cosh\lambda_2 L_2)}{EI_2\lambda_2^3(\cos\lambda_2 L_2 \cosh\lambda_2 L_2 + 1)} \quad \text{where } \lambda_2^4 = \frac{\rho A_2 \omega^2}{EI_2}$$

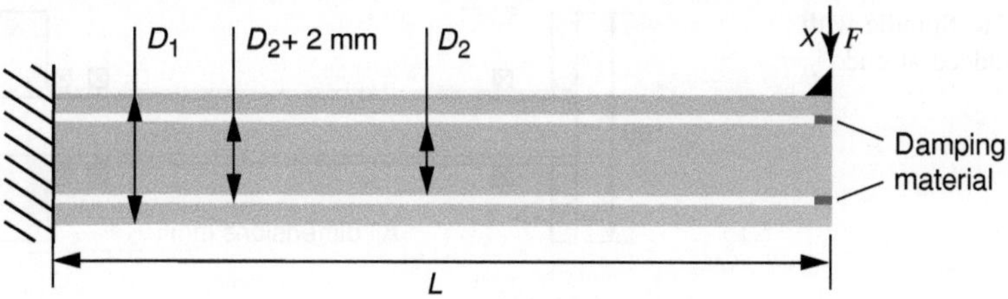

Fig. 4.24 Boring bar with internal bar replacing an absorber mass

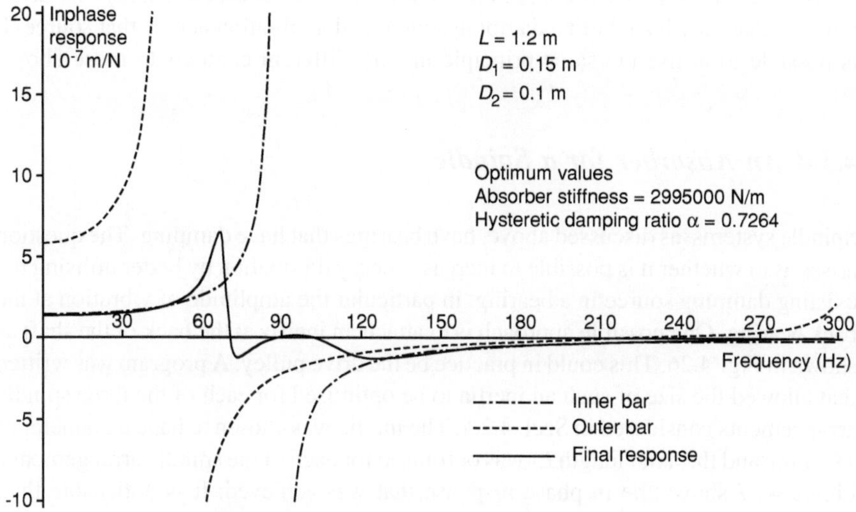

Fig. 4.25 Optimised boring bar with internal bar (Program 4.8)

replaces the receptance of the conventional absorber mass, i.e. $-1/m_2\omega^2$. The receptance at the cutting point, α_{11}, is thus given by

$$\frac{1}{\alpha_{11}} = -\frac{(\cos\lambda_1 L_1 \sinh\lambda_1 L_1 - \sin\lambda_1 L_1 \cosh\lambda_1 L_1)}{EI_1\lambda_1^3(\cos\lambda_1 L_1 \cosh\lambda_1 L_1 + 1)} + \frac{1}{\dfrac{1}{k_2 + ih_2} - \dfrac{EI_2\lambda_2^3(\cos\lambda_2 L_2 \cosh\lambda_2 L_2 + 1)}{(\cos\lambda_2 L_2 \sinh\lambda_2 L_2 - \sin\lambda_2 L_2 \cosh\lambda_2 L_2)}} \tag{4.18}$$

The subscript 1 refers to the outer tube. A program was written to optimise for chatter performance assuming a 1-mm annular clearance with the tube. A typical output is shown in Fig. 4.25.

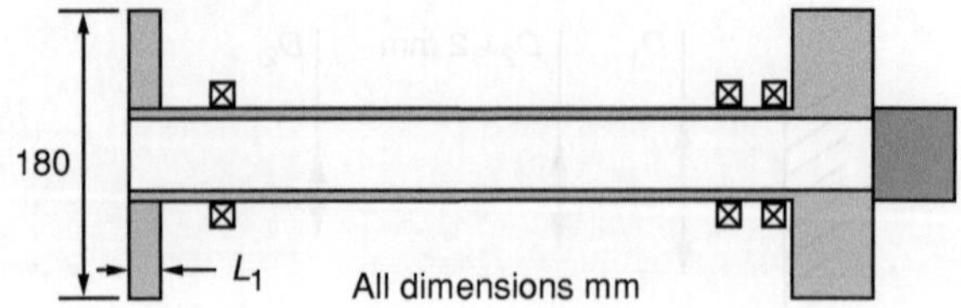

Fig. 4.26 Spindle with inertia added at end

The diameter of the internal bar was fixed for the in-phase response shown Fig. 4.25. Only the stiffness and hysteretic damping ratio of the damping material were variables in the optimisation process.

The main reason absorbers are effective is that they dissipate energy. To do so requires that they have both a damping source and a vibration across that source. It is possible to utilise the same principle in very different contexts to those above.

4.3.4 An Absorber for a Spindle

Spindle systems, as discussed above, have bearings that have damping. The question arises as to whether it is possible to increase energy dissipation by better utilising the existing damping source in a bearing, in particular the amplitude of vibration at the back bearing. One possible approach is to attach an inertia at the back of the shaft as shown in Fig. 4.26. This could in practice be the drive pulley. A program was written that allowed the size of such an inertia to be optimised for each of the three spindle arrangements considered in Sect. 4.2.4. The inertia was chosen to have a diameter of 180 mm, and the axial length L_1 was optimised for each of the spindle arrangements. Figure 4.27 shows the in-phase response that was achieved. It is noticeable that, since the back bearing is the same for each arrangement, the inertia/back bearing combination is "tuned" to the first natural frequency of the existing spindle. Thus, the higher the natural frequency, the smaller the inertia that is required. Improvements of around 100 % are predicted for each optimised arrangement.

4.3.5 Practical Examples

The examples of absorbers given previously do not take into account the fact that boring bars are not connected to a truly rigid abutment and spindle bearings are not in truly fixed housings. In practice, there is also the difficulty of not knowing the exact properties of the damping material. In order to achieve the optimum response, it is useful to have an absorber that may be tuned in situ. Figure 4.28 shows an example of a tunable absorber for a boring bar. The damping material should be chosen to have significant damping, butyl rubber being a good example [20]. Such rubbers have stiffness and damping values that increase with area of contact and preload. If the absorber mass is supported on "O" rings made of the damping material, the stiffness may be tuned by increasing the contact area and/or preload. The example in Fig. 4.28

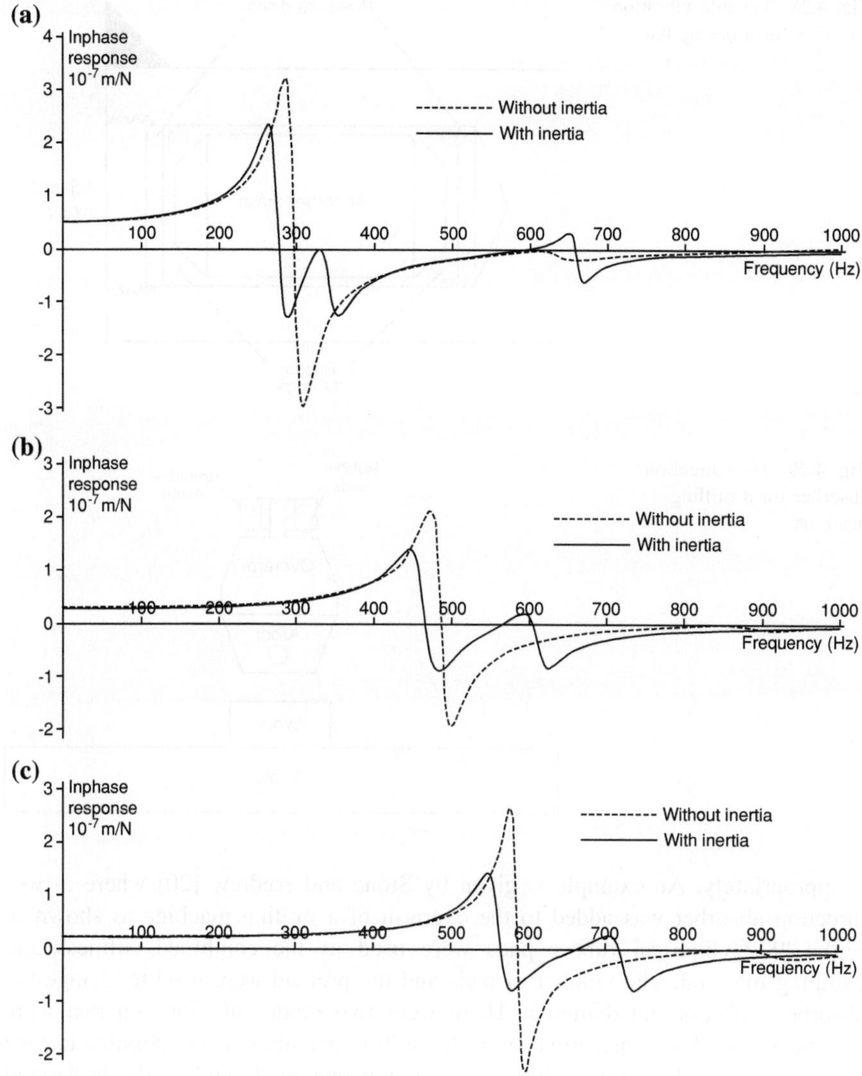

Fig. 4.27 Spindle responses with optimised inertia (Program 4.9)

shows how this may be done in practice for an absorber mass enclosed at the cutting end of a boring bar. The pressure disc moves as the end screw is advanced towards it, and the "O" rings are compressed, increasing the area of contact and preload. It is possible that the "O" rings may be too stiff, even on initial contact. The "O" rings should be made of a rubber that has significant damping and is also suitably "soft". A similar tuning arrangement is possible for the second bar absorber described in Sect. 4.3.3 by again compressing an "O" ring.

All of the examples above have an absorber that is tuned to the frequency of a single mode. It is possible to have an absorber that damps two modes by designing

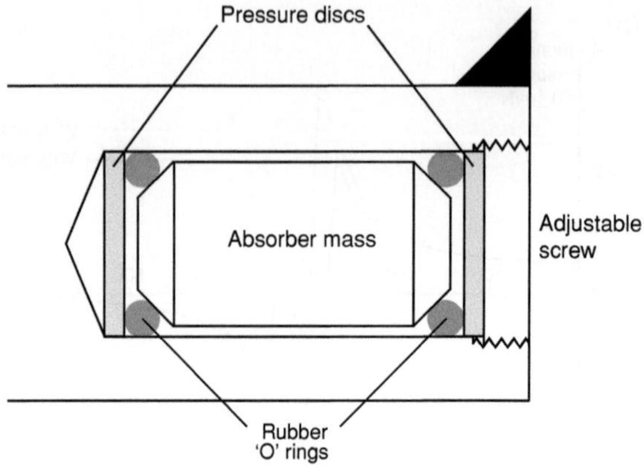

Fig. 4.28 Tunable vibration absorber for a boring bar

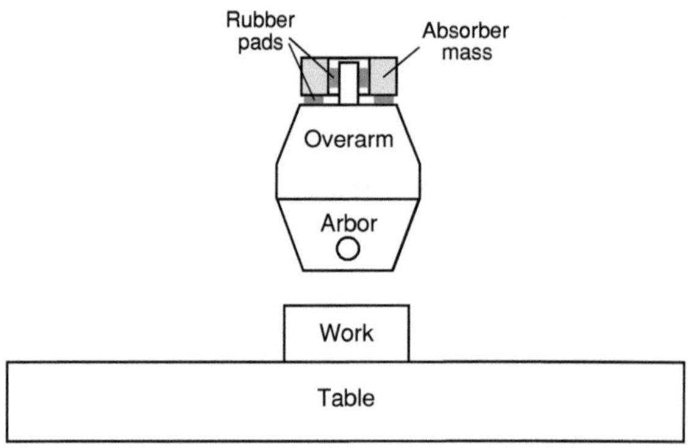

Fig. 4.29 Two-direction absorber for a milling machine

it appropriately. An example is given by Stone and Andrew [20] where a two-direction absorber was added to the overarm of a milling machine as shown in Fig. 4.29. Cylindrical rubber pads were used as the combined stiffness and damping material. The number of pads and the preload were used to change the absorber stiffness and damping. There were two modes of vibration that were significant for chatter performance of the milling machine. It was possible to tune the absorber in the vertical direction to damp one mode and in the horizontal direction to damp the other.

Initially, a one-direction (vertical) absorber was optimised, and subsequently, the two directions (horizontal and vertical) were optimised. The chatter receptances before and after optimising the absorber receptances are shown in Fig. 4.30.

Stone and Andrew [20] measured the machine receptances and used the systems approach to theoretically optimise the absorber. Note that a milling machine requires the absorber to be added remote from the cutting position. The most appropriate location for the absorber is determined by physical constraints and the requirement that the cross-receptance, between the cutting position and the absorber location, allows the chatter receptance to be improved. For the milling

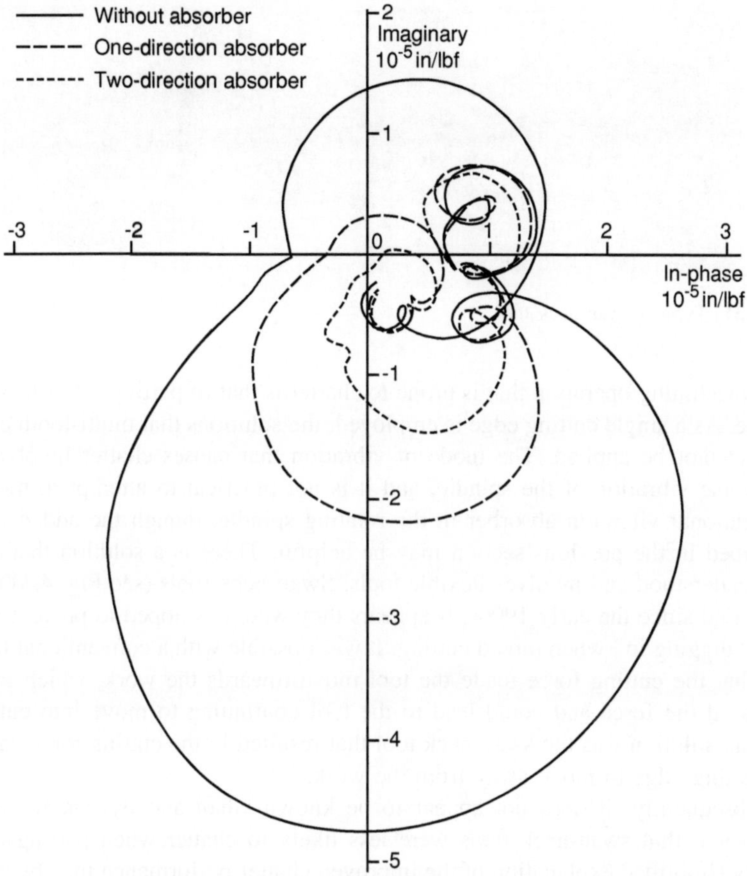

Fig. 4.30 Chatter receptances, with and without absorbers

machine, the chosen location was on the end of the overarm where there were significant amplitudes for both of the two main modes.

Cutting tests on tapered workpieces confirmed a factor of seven improvements in the width of cut when the two-direction absorber was in place.

When a two-direction absorber is tuned experimentally, it must be borne in mind that the tuning in one direction affects the other direction tuning, since the modes of vibration will not necessarily be in the same two directions as the absorber.

4.4 Flexible Tools

There is another and not-so-well-known way of improving the chatter receptance by means of a structural modification. This does not directly involve increasing the energy dissipation of the machine.

Fig. 4.31 Typical swan-neck tool

A machining operation that is prone to chatter is that of parting off or facing on a lathe. As a single cutting edge is employed, the solutions that multi-tooth cutters offer cannot be applied. The mode of vibration that causes chatter involves the transverse vibration of the spindle, and it is not practical to attempt to mount a conventional vibration absorber in the rotating spindle, though the added inertia described in the previous section may be helpful. There is a solution that is not well understood and involves flexible tools. Swan-neck tools (see Fig. 4.31) have been used since the early 1900s. It appears they were developed to prevent a tool from "digging in" when thread cutting. It was possible with a conventional tool to find that the cutting force made the tool move towards the work, which in turn increased the force and could lead to the tool continuing to move into cut. The brilliant solution was the swan-neck tool that resulted in the cutting force causing the cutting edge to move away from the work.

Subsequently (it does not appear to be known when and by whom), it was discovered that swan-neck tools were less likely to chatter when parting off. A greatly simplified explanation of the improved chatter performance may be gained by considering Fig. 4.32. This represents the existing machine as system C. If a flexible spring is inserted between the machine and the cutting edge, the new chatter receptance is given by

$\alpha_{11} = \frac{1}{k} + \gamma_{22}$ as shown in Appendix C.4, where k is the stiffness of the flexible tool.

As may be seen from Fig. 4.32, the result of adding the flexibility is to move the response locus of the tool/machine to the right by $1/k$. If k is chosen correctly, the locus is moved sufficiently to ensure there is no negative in-phase component, and thus, regenerative chatter will not be possible.

However, a real flexible tool will have mass and damping in addition to the flexibility. Considering a simplified model of the flexible tool and machine, the model shown in Fig. 4.33 is appropriate. The machine is represented by a single mode and the flexible tool by a spring/mass system with damping inserted between the cutting edge and the structure. We need to determine the response at the cutting edge, i.e. on the mass m_2.

Using the systems approach (Appendix C), divide the system into two subsystems for which the receptances are known.

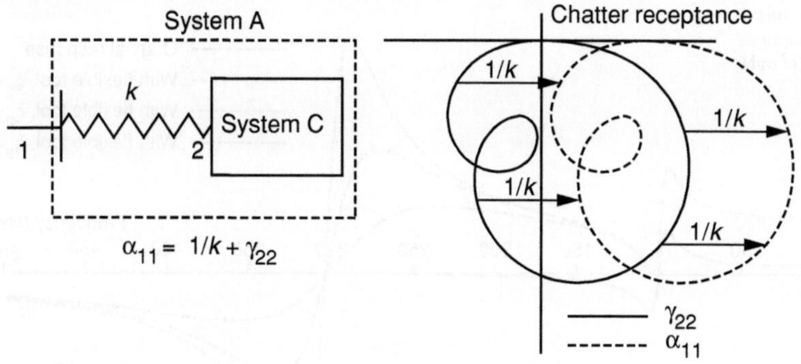

Fig. 4.32 Model of machine with a flexible tool represented by a spring

Fig. 4.33 Model of machine with a flexible tool, both represented by spring/mass systems with damping

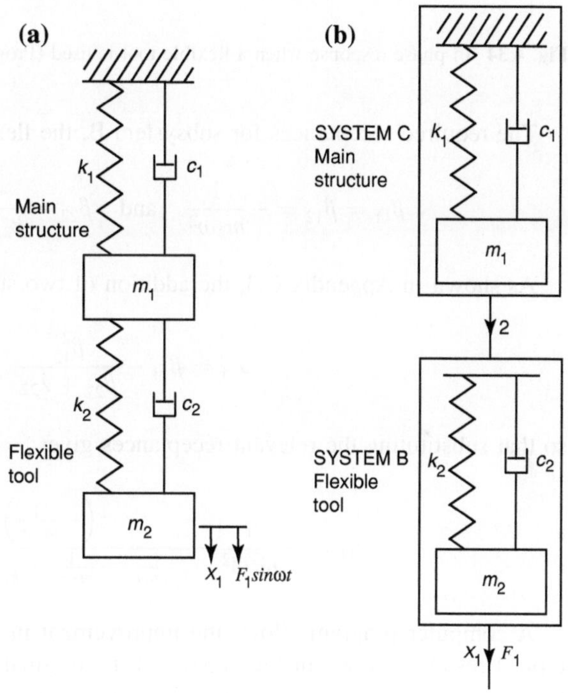

The relevant receptance for subsystem C, the existing structure, is

$$\gamma_{22} = \frac{1}{k_1 - m_1\omega^2 + i\omega c_1}$$

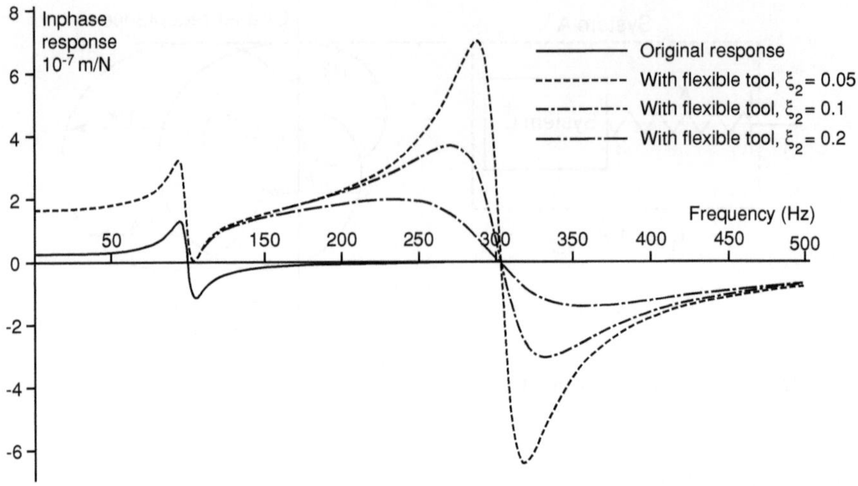

Fig. 4.34 In-phase response when a flexible tool is used (Program 4.10)

The required receptances for subsystem B, the flexible tool, are

$$\beta_{11} = \beta_{12} = -\frac{1}{m_2\omega^2} \quad \text{and} \quad \beta_{22} = \frac{1}{k_2 i\omega c^2} - \frac{1}{m_2\omega^2}$$

As shown in Appendix C.3, the addition of two subsystems in series gives

$$\alpha_{11} = \beta_{11} - \frac{\beta_{12}^2}{\beta_{22} + \lambda_{22}} \tag{C.16}$$

so that substituting the relevant receptances gives

$$\alpha_{11} = -\frac{1}{m^2\omega^2} - \frac{\left(-\frac{1}{m_2\omega^2}\right)^2}{\frac{1}{k_2+i\omega c_2} - \frac{1}{m_2\omega^2} + \frac{1}{k_1 - m_1\omega^2 + i\omega c_1}} \tag{4.19}$$

A computer program allows the improvement in response to be predicted. A typical result is shown in Fig. 4.34, and the original in-phase response has been moved in the positive real direction so that chatter will not occur at the machine-mode frequency as it did originally. However, there is a new negative peak introduced in the in-phase response at a high frequency. This negative peak may be reduced by adding damping to the flexible tool by inserting rubber at the top of the neck. Also it should be remembered that penetration rate damping is present at high frequencies. If the new negative peak is at a very high frequency, chatter will not occur at that frequency.

It is important that any flexibility be introduced close to the cutting edge so that the mass between the flexibility and the cutting edge is minimised and the new in-phase peak occurs at a high frequency. It is also important that this peak occurs at a resonant frequency above any already present so that the existing negative peak(s) will be moved in the positive in-phase direction.

Inevitably, the quasi-static stiffness is reduced and could affect dimensional accuracy. However, in parting off, this is not an issue. Also it is possible to include the flexibility in the direction that affects chip thickness, but not necessarily the finished geometry, e.g. in facing cuts.

The use of flexible tools does not appear to have received much attention. However, flexible grinding wheels have been used with great benefit, and so their use to prevent chatter in grinding will be considered in the next chapter.

4.5 Conclusions

The major emphasis in this chapter has been a consideration of structural modifications that improve chatter performance. The design of a spindle system including the associated bearings showed how difficult it is to predict the levels of damping accurately. Thus, methods of increasing damping in structures have been considered. The main danger is to think that just increasing damping will always be beneficial. Rather, it is increasing energy dissipation that is required and this involves damping and motion.

The use of vibration absorbers was explored, and it was found that optimum levels of absorber stiffness and damping were required to maximise chatter performance. Increasing the damping above the optimum level would reduce performance because, though the damping increases, the amplitude across the damper would be reduced so that the net energy dissipation falls.

Finally, flexible tools were considered and the requirements for their successful implementation established.

References

1. Potter SJ, Stone BJ (1974) The calculation of the response of spindle-bearing systems to oscillating forces MTIRA research report no. 55
2. Stone BJ (1992) The receptances of beams, in closed form, including the effects of shear and rotary inertia. Proc IMechE 206(2):87–94
3. Bishop RED, Johnson DC (1960) The mechanics of vibration. Cambridge University Press, Cambridge
4. Stone BJ (1982) The state of the art in the measurement of the stiffness and damping of rolling element bearings. Annals of CIRP 31(2):529–538
5. Kraus J, Blech JJ, Braun SG (1987) In situ determination of rolling bearing stiffness and damping by modal analysis. J Vib Acoust Stress Reliab 109(3):235–240

6. Lee CW, Hong SW (1989) Identification of bearing dynamic coefficients by unbalance response measurements. Proc IMechE Part C J Mech Eng Sci 203(2):93–101
7. Tiwara R, Vyas NS (1995) Estimation of non-linear stiffness parameters of rolling element bearings from random response of rotor-bearing systems. J Sound Vib 187(2):229–239
8. Marsh ER, Yantek DS (1997) Experimental measurement of precision bearing dynamic stiffness. J Sound Vib 202(1):55–66
9. Chen JH, Lee AC (1997) Identification of linearized dynamic characteristics of rolling element bearings'. Trans ASME J Vib Acoust 119(1):60–69
10. Tiwara R, Lees AW, Friswell MI (2001) Identification of speed-dependant bearing parameters. J Sound Vib 254(5):967–986
11. Lambert RJ, Pollard A, Stone BJ (1981) Some characteristics of rolling element bearings under oscillating conditions (and their relevance to the design of spindle systems with high dynamic stiffness). MTIRA limited circulation report, vol VI, pp 1–71
12. Lambert RJ, Pollard A, Stone BJ (2006) Some characteristics of rolling-element bearings under oscillating conditions: Part 1 theory and rig design. Proc IMechE Part K J Multi-body Dyn 220(3):157–170
13. Lambert RJ, Pollard A, Stone BJ (2006) Some characteristics of rolling-element bearings under oscillating conditions: Part 2 experimental results for interference fitted taper-roller bearings. Proc IMechE Part K J Multi-body Dyn 220(3):171–179
14. Lambert RJ, Pollard A, Stone BJ (2006) Some characteristics of rolling-element bearings under oscillating conditions: Part 3 experimental results for clearance fitted taper-roller bearings and their relevance to the design of spindles with high dynamic stiffness. Proc IMechE Part K J Multi-body Dyn 220(3):181–190
15. Walford TLH (1981) Some stiffness and damping characteristics of angular contact ball bearings. Ph.D. dissertation, University of Bristol
16. Walford TLH, Stone BJ (1980) The measurement of the radial stiffness of rolling element bearings under oscillating conditions. J Mech Eng Sci 22(4):175–181
17. Walford TLH, Stone BJ (1980) Some stiffness and damping characteristics of angular contact bearings under oscillating conditions. In: 2nd International conference on vibrations in rotating machinery, Cambridge C274/80, pp 157–162
18. Walford TLH, Stone BJ (1983) The sources of damping in rolling element bearings under oscillating conditions. Proc IMechE Part C J Mech Eng Sci 197(4):225–232
19. Elsermans M, Hongerloot H, Snoeys R (1976) Damping in taper roller bearings. In: Proceedings of 16th MTDR conference, pp 223–229
20. Stone BJ, Andrew C (1969) Optimisation of vibration absorbers: application to complex structures. J Mech Eng Sci 11(3):221–233

Chapter 5
Grinding

5.1 Introduction

In order to determine the significant parameters that affect chatter in metal cutting, it was necessary to make major assumptions. When chatter in grinding is considered, the process is far more complex than metal cutting. To determine which parameters are significant, it is necessary to make similar but even more assumptions than were necessary for metal cutting. Consider the three types of grinding shown in Fig. 5.1.

When pendulum grinding the main cause of chatter is the regeneration effect that comes from waves being formed on the grinding wheel. Unlike metal cutting, the "wear" of the tool (the grinding wheel) is significant. If the surface left on the work does not cause any regeneration, then this is akin to turning, but where the work "machines" the wheel. However, the wheel is not the same as a metal workpiece in that it is composed of grits embedded in a non-rigid bond material. In addition, there is not a line contact between the "tool" and the work as there is a contact area zone. Grinding is clearly more complex than metal cutting.

For plunge grinding, it is still possible to have surface waves on the wheel, but in addition, waves can be left on the work. This is called double regeneration and adds complexity since the rotational speeds of the work and wheel are different. This type of grinding also includes the complexity of the contact zone.

For roll (or cylindrical) grinding, the wheel traverses the work and the extent of regeneration from waves on the work depends on the overlap factor (the traverse feed as a proportion of the wheel width). Another modelling complexity arises on the returning traverse where any waves left on the work will then be ground by the wheel. Like plunge grinding, waves may again be generated on the wheel and there remains the question of the contact zone.

There are other grinding operations, but these three will suffice to illustrate the complexity of chatter in grinding quite apart from the complexity of the machine dynamics. This extra complexity, compared to metal cutting, was the main reason why the development of the theory of grinding chatter lagged behind that of metal cutting. This is still the case and presents a major problem as grinding is the

B. Stone, *Chatter and Machine Tools*, DOI: 10.1007/978-3-319-05236-6_5,

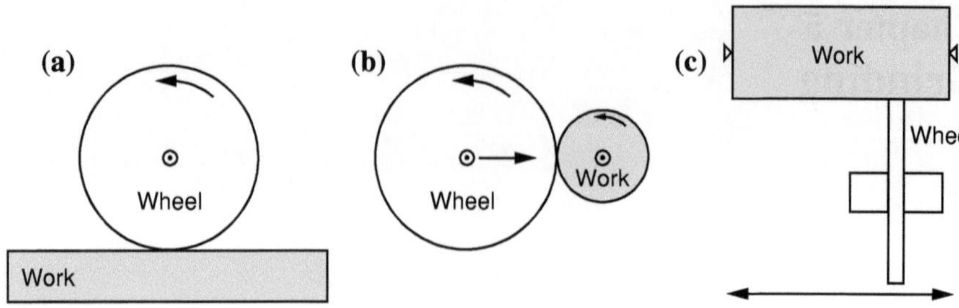

Fig. 5.1 Three examples of grinding: **a** pendulum grinding; **b** plunge grinding and **c** roll grinding

preferred and often only machining process available to the manufacturer for the attainment of tight tolerances coupled with fine surface finishes. The presence of vibrations limits the production rate, reduces surface finish quality, and results in the time between dressings of the wheel being shortened.

The major modelling problem that was encountered was that of determining an appropriate grinding contact force model. Any chatter theory for grinding is completely reliant on the force model adopted. The steady-state grinding forces, which can be measured without too much difficulty, are inadequate for a chatter model, where the forces under oscillating conditions are needed. While such force models were developed for single-point tools, the same was not the case for grinding. Currently, force models for steady-state grinding exist, but none has sufficiently wide applicability to be universally used. The development of dynamic force models for grinding presents an even greater challenge and has received little attention. As a consequence, grinding chatter theories all suffered from the lack of accurate input information in the form of the dynamic grinding force model. The following historical review is based largely on that of Entwistle [1].

While in general, the development of chatter theories for grinding lagged those for other machining processes, one of the earliest contributions was made by Hahn [2], which followed shortly after his work on turning [3]. He considered regeneration on the work and assumed that the grinding wheel did not wear. In addition, the contact between the work and the wheel was considered to be a line contact. Thus, in principle, the analysis method adopted was equally applicable to turning and the application to grinding may have been arbitrarily chosen. As has been common since, Hahn adopted a single-degree-of-freedom model for the vibratory response of the machine tool excited at the cutting zone and constructed a second-order differential–difference equation model. The regeneration was delayed by the constant rotational period of the work. The cutting force was modelled as a linear stiffness that related the force to the cross section of the chip.

The equation was recognised as being similar to those studied by Minorsky [4] and also Ansoff [5]. The solution method of Minorsky [4] was adopted by Hahn and the Nyquist criterion invoked to find positive roots. Using system damping as the adjustable quantity, Hahn described the conditions under which the system was "unconditionally stable" for all work speeds and "conditionally stable" for some combinations of damping and work speed. The result was a

> theory capable of explaining the frequently observed phenomenon of chatter. Furthermore, the theory predicts a spectrum of values of cycles of vibration per revolution of work for which stability should exist.

At this early stage in the development of the theory, correspondent Hunter believed that hard spots and non-uniformities were primary causes of chatter rather than regeneration.

Landberg [6] attempted to better understand the factors influencing chatter in grinding. The waviness of the grinding wheel after grinding compared to that after dressing was documented and the vibration frequencies correlated with the wavelength of the waves on both the work and the wheel. A number of plots showing vibration amplitude against grinding time were given. These revealed the exponential nature of the amplitude growth. It was stated that

> … the design is attached to the spindle and the properties of the wheel itself are very important to the behaviour of the wheel in the grinding process. That is the reason that the researches of the laboratory go in the direction of finding the natural frequencies of parts of the machine especially the spindles, of the work and of the wheel itself. The problem is further whether forced or self-induced vibrations play a part. But it is too early to give results of these experiments: as yet they are not conclusive and must be continued before it is possible to give justified explanations of the observed phenomena.

The state of understanding at the time was given here. The dynamic response of the machine structure and spindle was now understood to be central. The link between the vibrations and the machined surface topology of the work and the wheel was established.

Four papers published towards the end of the 1950s show that the research community was still attempting to gain a clearer understanding of the sources of vibration in grinding. Doi [7] experimentally studied various types and sources of vibration but did not specifically mention regenerative chatter. Lurie [8] reports some observations of trends in grinding chatter, but no theory or possible explanations were presented. Hahn [9] undertook a study of the fundamental mechanics of a flexible internal grinding spindle by setting up the equations of motion and showed that the tangential and radial motions of the grinding wheel were coupled. He concluded that a disturbance was sufficient to cause uneven wheel wear and/or wheel loading that could lead subsequently to regeneration. Puzanov [10] also undertook an experimental survey with some interesting measurements of work torsional vibrations.

The algebraic solution method of Sweeney and Tobias [11] was applied to grinding by Sweeney [12] in his doctoral thesis. The algebraic theory assumes a constant and nonzero vibration amplitude. This research appears to be the first comprehensive study of grinding instability from both theoretical and experimental viewpoints. In addition, the study developed a "transient theory", which considers self-excited vibrations as a transient process commencing with zero amplitude. The cutting force constants required for the algebraic solution method were evaluated and experimentally determined. The experimental stability study showed "partial agreement" with the theory developed. The common practice of assuming the cutting force to be proportional to the chip thickness was followed. As a consequence, the incremental cutting forces were assumed to be in the same direction as the mean force. The wear of the grinding wheel was generally ignored and hence only work regeneration was considered in detail. Sweeney concluded that all grinding instability was regenerative but that the parameters are time dependent. Thus, phenomena such as grinding wheel loading play a significant role. The major contribution of this work was the measurement of the dynamic grinding force coefficients.

Gurney [13] extended his graphical method [14] to cover grinding wheel regenerative chatter, which he terms "surface wave instability". This is one of the earliest analyses of grinding wheel regenerative chatter, and it was concluded that the

> self-excited vibration…is not due to inhomogeneities in the wheel.

He showed that grinding wheel regenerative chatter was independent of the work speed and observed that the lobes slowly precessed around the grinding wheel in time. The conclusion was that the vibration was due primarily to the machine work characteristics.

The fact that some of the assumptions made to this point in analysing grinding chatter were in many cases over simplifications manifested itself in the deviations of experience from the predictions. Snoeys and Vanherck [15] observed from experiments that

- No "limit of stability" can be found implying that all grinding was unstable
- The number of lobes on the grinding wheel altered with vibration amplitude
- No correlation between chatter frequency and some system parameter, such as wheel hardness, had been found.
- The amplitude of the horizontal vibration (between work and wheel) could be twice the amplitude of the grinding wheel lobes and larger than the chip thickness.

Snoeys and Vanherck sought to explain these observations in relation to grinding wheel regenerative chatter by setting up a feedback system (similar to Merritt [16]) that included the machine dynamic transfer function and the effect of the contact deformation between the work and the wheel. This was taken to be nonlinear as per the Hertzian model. The system was modelled on an analogue computer. The results of the simulation showed that the

> necessary elements to explain the practical chatter phenomenon (had been included).

They further concluded that

> the evolution (of amplitude growth during unstable conditions) in the instability is chiefly due to the non-linear characteristics of the grinding wheel contact.

This paper correctly identified the contact zone deformations as having a determining role in the development of regenerative chatter in grinding.

Snoeys and Brown [17] greatly expanded this work into a paper now recognised as being pivotal in the development of chatter theories in grinding. It achieved the following:

- Developed a block diagram for the grinding process (and hence the characteristic equation) that included both work and wheel regenerative paths, machine dynamic transfer function, contact stiffness and wave filtering—all of which are now considered to be of fundamental importance to grinding chatter.
- The stability boundary was determined by the location of the most negative real part of the machine response locus, the same as developed for metal cutting by Tlusty [18].
- A detailed discussion was provided concerning the important system parameters: cutting stiffness, wheel wear resistance, contact stiffness, machine stiffness, contact length filtering. It was shown that, …*the finite length of contact is one of the most important features of the proposed model leading to a significant reduction of the upcome of instability in low workspeed grinding work.*
- Suggestions were offered for the reduction or prevention of the onset of chatter.
- Provided test results which showed both wheel and work regeneration.
- Tabulated machine stiffnesses and cutting stiffnesses for various materials.
- The range of numerical values of the important stability parameters was tabulated.

From the comprehensive nature of the model and the detailed tabulation of measurements of the parameters involved, Snoeys and Brown were able to show that most grinding appeared to take place under unstable conditions and hence the growth rate became important.

> …a great deal of grinding work is performed under unstable work conditions because most grinding wheel widths are larger than the 5 mm critical wheel width.

Precession of the wheel lobes was noted (0.1–0.2 rpm).

The cutting force model that was utilised was very simple, viz. the cutting force was taken to be proportional to the instantaneous depth of cut. It is also noted that the rotational speeds of the work and the wheel were assumed constant. This work was expanded a little further in Brown's doctoral thesis [19] and which claimed it was

> the most comprehensive approach to self-excited and forced chatter in the grinding process.

A detailed study of the growth rate was added by solving for the complex roots of the characteristic equation via a computer algorithm. One addition to Snoeys' approach was the addition of an "overlap factor" in the model that provided for the partial work regeneration, that occurs when roll grinding.

By about the year 1970, the fundamental parameters causing and influencing work and wheel regenerative chatter had been identified and adequate analysis techniques had been devised to model the essential dynamics of the system. Over the following decades, many further developments were achieved, but none discredited the fundamental understandings that had been achieved. This is confirmed if more recent publications on grinding chatter are examined. Inasaki et al. [20] reviewed the current state of research into chatter in grinding for the CIRP annual assembly in 2001. They have 129 references that include some of those noted above and the fundamental parameters were apparently considered not to have changed.

In order to find solutions to chatter in grinding in this book, the same approach will be used as when looking for solutions to chatter in metal cutting; that is, the simplest model that allows real solutions to be anticipated will be used. As far as possible, each significant parameter will be investigated in isolation. It is acknowledged that the real situation will involve all the parameters and be far more complex. However, if the solutions predicted from these simple models work in practice, then a good result has been achieved. As noted above, the grinding force model is crucial and will be considered first.

5.2 Grinding Force

As already noted, most of the early research on chatter in grinding used a force model that was essentially the same as for metal cutting. The force was assumed to be proportional to the chip thickness. However, in time, it was found [21] from steady-state tests that the mean force could be more accurately represented by the following equation

$$P_t = \frac{u_{ch} V_w \delta_w b}{V_g} \quad \text{and} \quad P_n = k_1 P_t \tag{5.1}$$

where referring to Fig. 5.2

P_t is the tangential component of force,
u_{ch} is called the specific chip formation energy,
V_w is the surface speed of the work,
δ_w is the undeformed chip thickness of the work,
b the width of cut is the width of the contact between wheel and work,
V_g is the grinding wheel surface speed,
P_n is the normal/radial component of force and
k_1 is a constant.

To be used in a model of chatter, it needs to be assumed that the oscillating force would act in the same direction as the mean force, i.e. k_1 applies to the oscillating case and $P_n = k_1 P_t$. Also if the equation applies to the time-varying situation, it may be differentiated to give,

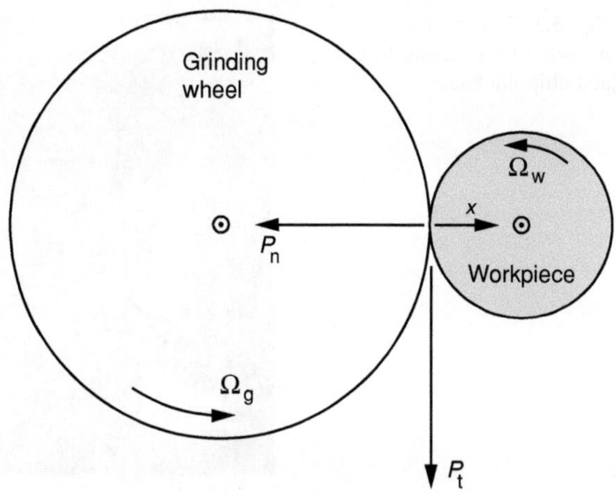

Fig. 5.2 Grinding force terminology

$$\frac{dP_t}{dt} = u_{ch}b\left(\frac{V_w}{V_g}\frac{d\delta_w}{dt} + \frac{\delta_w}{V_g}\frac{dV_w}{dt} - \frac{\delta_w V_w}{V_g^2}\frac{dV_g}{dt}\right) \tag{5.2}$$

Very little work has been published on the measurement of data to confirm Eq. 5.2. However, as a result of developing a torsional vibrator [22], Drew and Stone were able to collaborate with Ong and Mannan to experimentally determine u_{ch} for the dV_w term. Initially, it was hoped to use in-process measurement of the ground surface [23] using lasers. However, because of the small amplitudes and the sensitivity of the signals to the reflectivity of the surface that was continuously changing, this was found to be not possible. As a result, an involved process requiring steady-state torsional excitation of the work and its subsequent sudden removal from cut for profile measurement was adopted. The results are presented in [24], and it was found that for the conditions tested u_{ch} was 1.63×10^{10} N/m^2. It is of interest to note that the value obtained when there was no oscillation present was 3.91×10^{10} N/m^2.

The same authors also investigated oscillating the chip thickness on the work by using an out-of-balance mass on the grinding wheel as shown in Fig. 5.3. The detailed results have not been published, but it was found [25] that the average value of u_{ch} was 1.7×10^{10} N/m^2. This compares with the average value of 1.63×10^{10} N/m^2 found when the work speed was varied [24]. Apparently no attempt has been made to investigate the effect of the variation of oscillating forces with oscillating wheel speed.

Thus, the force model for grinding under oscillating conditions is uncertain. The current state of knowledge suggests that to use more complex force models than described by Eq. 5.2 is difficult to justify. The experimental data that are available suggest that Eq. 5.2 may be used but with reservations. Chatter in grinding will be analysed with this force model, starting from the simplest model possible and then adding more realistic effects.

Fig. 5.3 Test rig for measuring oscillating forces and chip thickness

5.3 Chatter Models

The aim of this book is to describe methods of improving chatter performance by adopting the simplest model and mathematical approach available. For metal cutting, the graphical approach of finding the stability boundary was used because of its simplicity and the aid it gives in both suggesting and understanding solutions. When chatter in grinding is considered, the graphical approach is not always possible. The method to be adopted now will involve assuming any vibration will have the form $Ae^{(\sigma+\mathrm{i}\omega)t}$. This implies a chatter frequency of ω and an exponential growth rate of the vibration controlled by σ. If $\sigma > 0$, then instability occurs as the vibration grows in amplitude, whereas if $\sigma < 0$, the grinding is stable. In most cases, the stability boundary condition (neutral stability) will be investigated, i.e. when $\sigma = 0$ and the vibration will then be of the form $Ae^{(\mathrm{i}\omega t)}$.

To allow some simple conclusions to be drawn for grinding chatter, it is instructive to consider the controlling equation for metal cutting using a solution of the form $Ae^{(\mathrm{i}\omega t)}$. For a machine represented by a single mode, the equation of motion with regenerative chatter was shown in Chap. 1 to be

$$m\frac{\mathrm{d}^2x(t)}{\mathrm{d}t^2} + c\frac{\mathrm{d}x(t)}{\mathrm{d}t^2} + kx(t) = Rb[h - x(t) + x(t-\tau)] \tag{1.22}$$

If only the oscillating contributions are considered and $x(t) = Ae^{(\mathrm{i}wt)}$ the equation becomes,

$$-m\omega^2Ae^{\mathrm{i}\omega t} + \mathrm{i}\omega cAe^{\mathrm{i}\omega t} + kAe^{\mathrm{i}\omega t} = Rb\left[-Ae^{\mathrm{i}\omega t} + Ae^{\mathrm{i}\omega(t-\tau)}\right]$$

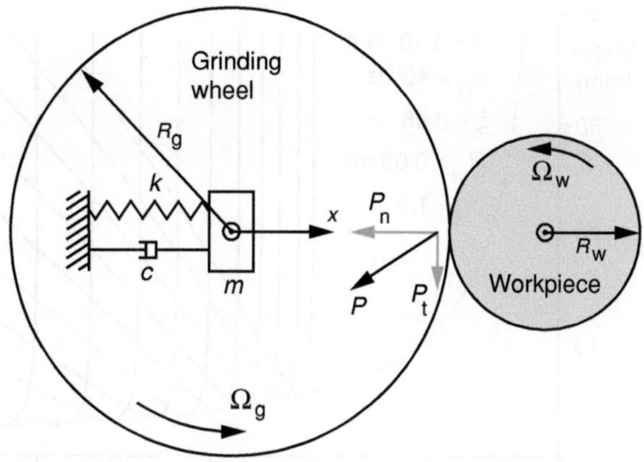

Fig. 5.4 Machine model for simplest case

which reduces to

$$k - m\omega^2 + \mathrm{i}\omega c = -Rb\left(1 - e^{\mathrm{i}\omega\tau}\right) \tag{5.3}$$

It will be found useful to compare the results in the following sections with this metal cutting equation, for which a great deal is now known.

5.3.1 The Simplest Grinding Model

The dynamic characteristics of machines are complex and involve many modes. The use of a single mode to represent the machine allows a simplified model to be developed. This in turn can give an insight into the situation where there are multiple modes. The machine response will be represented by a spring/mass system with viscous damping that vibrates in the direction affecting chip thickness as shown in Fig. 5.4. For plunge grinding, the response will be taken to be the ratio of displacement in the chip thickness direction to a force in the grinding force direction. The grinding wheel is shown supported on the spring/mass model in Fig. 5.4. However, it would be found that the following equations would not change if the work instead of the wheel was considered to be mounted on the spring/mass system. This is because it is the relative deflection between the wheel and work that is important.

Now if Eq. 5.2 is expanded and the work and wheel speeds assumed to be constant so that V_w and V_g are constants

$$\frac{\mathrm{d}P_t}{\mathrm{d}t} = u_{ch} b \frac{V_w}{V_g} \frac{\mathrm{d}\delta_w}{\mathrm{d}t}$$

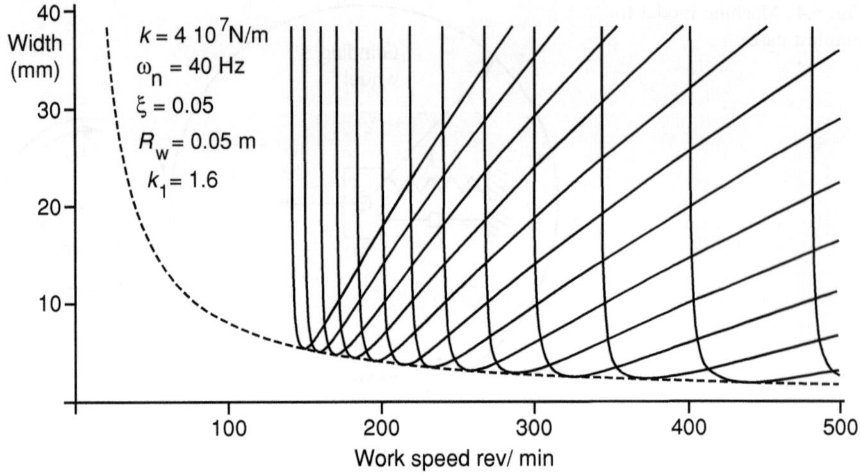

Fig. 5.5 Stability chart against work speed, for the simplest case (Program 5.1)

and hence,

$$\frac{dP_n}{dt} = k_1 u_{ch} b \frac{V_w}{V_g} \frac{d\delta_w}{dt} = Rb \frac{d\delta_w}{dt} \tag{5.4}$$

The force variation in the mode direction is proportional to the chip thickness variation. This equation needs to be compared with the simplest force equation used for the cutting force models $F(t) = Rbx(t)$. Noting the change in notation, Eq. 5.4 is the same as the cutting force model, but with $R = k_1 u_{ch} V_w / V_g$. The equation of motion is the same as Eq. 5.3, but with a modified value of R.

$$k - m\omega^2 + i\omega c = \frac{k_1 u_{ch} V_w}{V_g} b\left(1 - e^{i\omega\tau}\right) \tag{5.5}$$

For plunge grinding and with no regeneration on the wheel, chatter in grinding is equivalent to chatter in turning, as τ in Eq. 5.5 becomes τ_w, the time for one revolution of the work. The cutting force coefficient R is $k_1 u_{ch} V_w / V_g$ and a typical stability chart is shown in Fig. 5.5. If the wheel speed is constant, the cutting force coefficient increases linearly with the work surface speed. As reported by Snoeys and Brown [17], it also follows that the structural characteristic that determines the unconditional stability boundary is the most negative real part of the machine response locus (as was found for metal cutting). Thus, all the solutions that were proposed for single-point metal cutting and presented in the previous chapters are applicable. These include optimising spindle design, adding vibration absorbers and the use of flexible tools. Flexible grinding wheels are considered in Sect. 5.3.4.

There are some important differences from metal cutting that are predicted by this simple model. The limiting width increases as the work diameter is reduced and also as the work speed is reduced. As $R = k_1 u_{ch} V_w / V_g$, increasing the wheel speed reduces R and the limiting width will increase. Finally, this model predicts the mean chip thickness (in-feed) to have no effect as it does not affect R.

The model in this section can also be applied to plunge grinding when only regeneration on the wheel is significant and the work is regarded as a cutting tool machining the wheel. In Sect. 5.3.3, where contact stiffness is discussed, this assumption will be shown to include a significant error. However, some investigators have made this error and it is necessary to show the consequences. With this erroneous model, the difficulty arises in determining the depth of cut on the wheel. The grinding force will be the same on the wheel as on the work but opposite in direction. It is possible to make an estimate of the effective cutting force coefficient for the wheel as follows. We have shown that for the work $R = k_1 u_{ch} V_w / V_g$. If we assume that the grinding ratio is the same for oscillating conditions as for steady state, then the grinding ratio G is

$$G = \frac{\text{Volume removed from work}}{\text{Volume removed from wheel}} = \frac{b\delta_w V_w}{b\delta_g V_g} = \frac{\delta_w V_w}{\delta_g V_g}$$

Thus, the "chip thickness" on the wheel is given by

$$\delta_g = \frac{V_w}{V_g G} \delta_w$$

and Eq. 5.4 becomes, for constant wheel and work speeds,

$$\frac{dP_n}{dt} = k_1 u_{ch} b \frac{V_w}{V_g} \frac{d\delta_g}{dt} \frac{V_g G}{V_w} = k_1 u_{ch} b G \frac{d\delta_g}{dt} \tag{5.6}$$

The effective value of R is $k_1 u_{ch} G$ if the grinding ratio G is the same for all speeds and radii. A stability chart for regeneration on the wheel alone is shown in Fig. 5.6. The limiting width of cut is very small for realistic G values. For super-abrasive wheels when G may exceed 1,000, the width is extremely small. This model may explain why many of the early workers considered that grinding was never stable but is now known to be possible. The presence of very low growth rates was used to explain why acceptable grinding could be conducted for a time. It is therefore important to consider the growth rates of chatter in grinding.

5.3.2 Growth Rates

The equation of motion for the simple case of work regeneration alone is as for metal cutting,

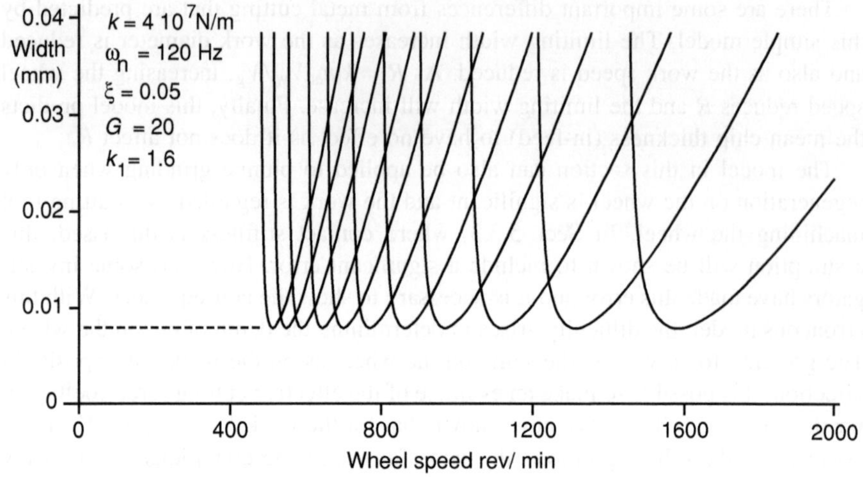

Fig. 5.6 Stability chart against wheel speed, for regeneration on the wheel alone, the work represented by a metal cutting tool (Program 5.2)

$$m\frac{d^2x(t)}{dt^2} + c\frac{dx(t)}{dt^2} + kx(t) = -Rb(x(t) - x(t - \tau_w)) \tag{5.7}$$

but for grinding, $R = k_1 u_{ch} \frac{V_w}{V_g}$

Assuming a solution of the form $x(t) = Ae^{(\sigma + i\omega)t}$ implies that the motion is oscillatory with frequency ω and growing exponentially if σ is positive. Substituting in Eq. 5.5 gives,

$$\begin{aligned} &m(\sigma + i\omega)^2 Ae^{(\sigma+i\omega)t} + c(\sigma + i\omega)Ae^{(\sigma+i\omega)t} + kAe^{(\sigma+i\omega)t} \\ &= -Rb\left(Ae^{(\sigma+i\omega)t} - Ae^{(\sigma+i\omega)(t-\tau_w)}\right) \end{aligned}$$

The term $Ae^{(\sigma+i\omega)t}$ cancels leaving,

$$m(\sigma + i\omega)^2 + c(\sigma + i\omega) + k = -Rb\left(1 - e^{(\sigma+i\omega)(-\tau_w)}\right)$$

Expanding

$$\begin{aligned} &m(\sigma^2 + 2i\sigma\omega - \omega^2) + c(\sigma + i\omega) + k \\ &= -Rb(1 - e^{-\sigma\tau_w}(\cos(\omega\tau_w) - i\sin(\omega\tau_w))) \end{aligned}$$

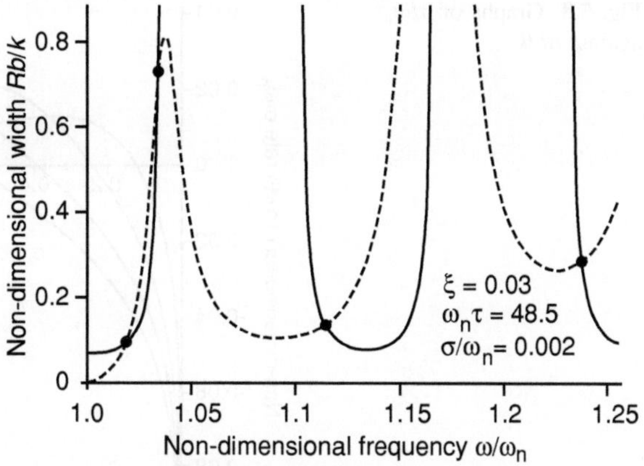

Fig. 5.7 Graph of *Rb*/*k* against ω/ω_n for two conditions to be satisfied

Grouping real and imaginary terms yields two equations,

$$m(\sigma^2 - \omega^2) + c\sigma + k = -Rb(1 - e^{-\sigma\tau_w}(\cos \omega\tau_w)) \tag{5.8}$$

$$m2\sigma\omega + c\omega = -Rbe^{-\sigma\tau_w}\sin(\omega\tau_w) \tag{5.9}$$

To solve for particular values of *m*, *k*, *c*, *R* and τ_w, choose a value of σ and then search for the value of ω that gives the same value of *b* from Eqs. 5.8 and 5.9, so that

$$b = \frac{m(\sigma^2 - \omega^2) + c\sigma + k}{-R(1 - e^{-\sigma\tau_w}\cos \omega\tau_w)} = \frac{m2\sigma\omega + c\omega}{-Re^{-\sigma\tau w}\sin \omega\tau_w} \tag{5.10}$$

If a solution exists, then both the width of cut (*b*) and the growth rate (σ) are known. If σ is negative, then any vibration will decay. However, if σ is positive, the vibration will grow exponentially and chatter will occur. It is useful to non-dimensionalise Eq. 5.10 using the conventional vibration parameters,

$$\omega_n = \sqrt{\frac{k}{m}} \quad \text{and} \quad \xi = \frac{c}{2\sqrt{mk}}$$

Dividing Eq. 5.10 throughout by *k* and rearranging to have all the terms non-dimensional:

$$\frac{Rb}{k} = \frac{\left(\frac{\sigma}{\omega_n}\right)^2 - \left(\frac{\omega}{\omega_n}\right)^2 + 2\xi\frac{\sigma}{\omega_n} + 1}{-\left(1 - e^{-\frac{\sigma}{\omega_n}\omega_n\tau_w}\cos\left(\frac{\omega}{\omega_n}\omega_n\tau_w\right)\right)} = \frac{2\frac{\sigma}{\omega_n}\frac{\omega}{\omega_n} + 2\xi\frac{\omega}{\omega_n}}{-e^{-\frac{\sigma}{\omega_n}\omega_n\tau_w}\sin\left(\frac{\omega}{\omega_n}\omega_n\tau_w\right)} \tag{5.11}$$

As an example, Fig. 5.7 shows the variation of the two values of *Rb*/*k* against ω/ω_n for $\xi = 0.03$, $\sigma/\omega_n = 0.002$ and $\omega_n\tau_w = 48.5$. Over the frequency range

Fig. 5.8 Graphs of σ/ω_n against *Rb/k*

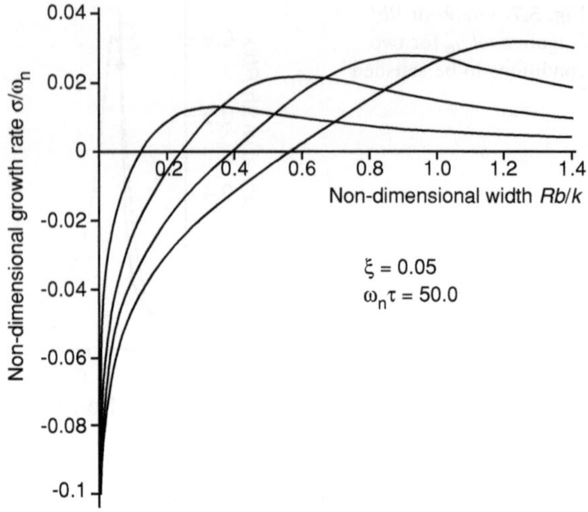

plotted, there are four solutions that satisfy Eqs. 5.11. It is clear that the functions are complicated and are not easily interpreted. Some of the solutions predict negative values of *Rb/k* and so great care is needed to avoid spurious solutions.

The significant variables contained in Eqs. 5.11 are found by plotting solutions of σ/ωn against *Rb/k*. This requires choosing a value for σ/ω_n and then searching to find a value of ω/ω_n that satisfies Eqs. 5.11. It is found that in general, there are multiple values of ω/ω_n for any value of σ/ω_n. Thus, there are many values of *Rb/k* for a given σ/ω_n. Some typical solutions are shown in Fig. 5.8. Each of the curves corresponds to a particular stability lobe and the intersection with the *Rb/k* axis indicates the stability boundary value of *Rb/k* as the growth rate is zero. Below this value of *Rb/k,* any vibration decays as σ/ω_n is negative. Above the stability boundary, the vibration grows in amplitude. It is noticeable that there is a maximum growth rate for each curve and if the width is increased well beyond the stability boundary, the growth rate becomes smaller. This is of some significance when regeneration is only on the wheel. In this case, the value of *Rb/k* can be very large as $R = k_1 u_{ch} G$. Figure 5.9 shows a sample curve with the number of waves on the wheel close to four. When the width greatly exceeds the stability boundary, the growth rate becomes very small.

Figure 5.9 also shows the chatter frequency as a function of the non-dimensional width. The frequency is close to that which would produce an integer number of waves on the wheel. As a consequence, the graph shows the difference in chatter frequency to that which would produce an integer number of waves. When the stability boundary is greatly exceeded, the difference is very small and this would result in the waves on the wheel circumference changing position very slowly. As has been found experimentally, the waves would precess around the circumference quite slowly.

The simple model presented in this section and the previous one has predicted many of the effects observed in experiments. However, the model is very simple

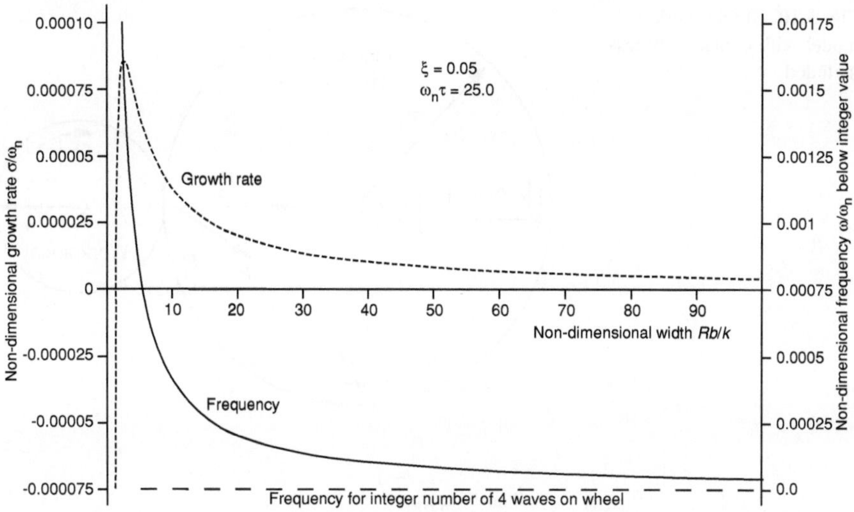

Fig. 5.9 Variation of growth rate and chatter frequency for large values of *Rb*/*k*

and some major effects can only be found by increasing the complexity of the model. The next section introduces a model that includes the stiffness of the contact zone.

5.3.3 Contact Stiffness

Thus far, we have treated the grinding wheel as a cutting edge with a line contact. Grinding wheels are composed of grits mounted in a bond material. They are not rigidly held but displace under load resulting in distortion of the contact zone. It is possible to model the contact zone as in Sect. 5.3.1, but including a contact spring so that the effective cutting edge is mounted on a spring whose stiffness increases with the width of cut and may be represented as $k_c b$.

It has been shown for metal cutting (Sect. 4.4) that a flexible tool can greatly improve the chatter receptance and that is also the case for grinding. To allow just the one parameter to be investigated, consider the simplest case again (the mode direction is on the line between centres) with the contact stiffness included as shown in Fig. 5.10.

To minimise the mathematics, only the envelope of the stability lobes will be considered, i.e. the unconditional stability boundary. It is appropriate to use the metal cutting stability criterion that has been developed as the model for grinding is equivalent to metal cutting but with a non-constant cutting force coefficient. The unconditional stability boundary for single-point cutting is given by

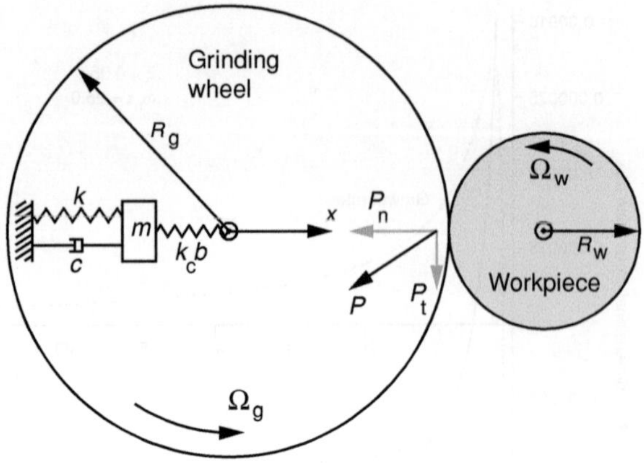

Fig. 5.10 Single-mode model with contact stiffness included

$$b_{\text{lim}} = \frac{-1}{2RG_{R,\max(\omega)}} \tag{1.15}$$

where $G_{\text{R,max}}(\omega)$ is the most negative in-phase component of the structure's response. In Appendix A.10, it is shown that the maximum negative in-phase component of the response of a single-mode system is given by $-m/\left[c\left(2\sqrt{mk}+c\right)\right]$, and in Appendix C.4, it is shown that a spring (k) in series with a structure results in $1/k$ being added to the real part of the response. The maximum negative in-phase component for the single mode is reduced when there is a contact stiffness so that

$$G_{R,\max}(\omega) = \frac{1}{k_c b} - \frac{m}{c\left(2\sqrt{mk}+c\right)}$$

If regeneration on the work alone is considered, the effective cutting force coefficient is $R = k_1 u_u \text{ch} V_{\text{W}}/V_{\text{g}}$. Thus, substituting in Eq. 1.15,

$$b = \frac{-V_g}{2k_1 u_{\text{ch}} V_{\text{w}} \left[\frac{1}{k_c b} - \frac{m}{c\left(2\sqrt{mk}+c\right)}\right]}$$

and expanding and rearranging

$$b_{\text{lim}} = \frac{c\left(2\sqrt{mk}+c\right)}{m}\left[\frac{1}{k_c} + \frac{V_g}{2k_1 u_{\text{ch}} V_{\text{w}}}\right] \tag{5.12}$$

Figure 5.11 shows the envelope for some typical values, and for comparison, the envelope with no contact stiffness is also shown. Clearly, there are possible benefits if grinding wheels with lower contact stiffness are used. The analysis above was for regeneration only on the work. If only regeneration on the wheel is

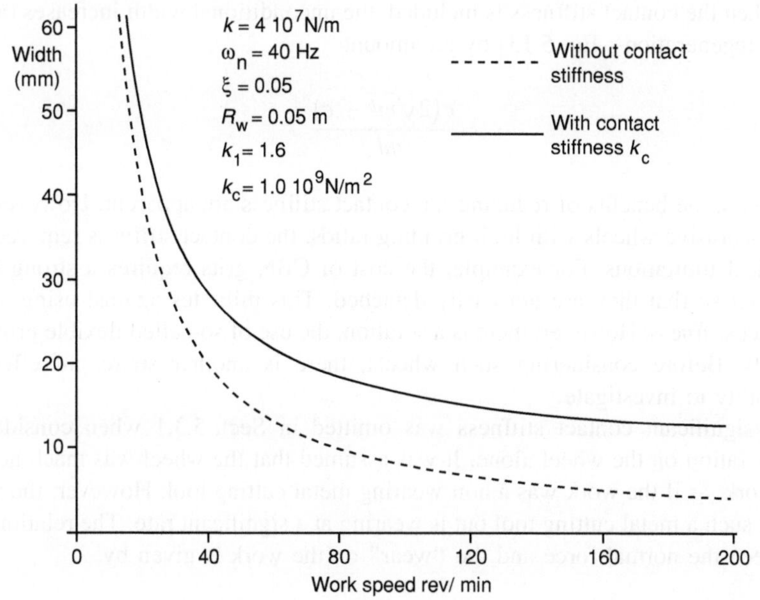

Fig. 5.11 Unconditional stability boundary, with and without contact stiffness (Program 5.3)

considered, the effective cutting force coefficient is $R = k_1 u_{ch} G$ so that the unconditional stability boundary is given by

$$\therefore b = \frac{-1}{2k_1 u_{ch} G\left[\frac{1}{k_c b} - \frac{m}{c\left(2\sqrt{mk}+c\right)}\right]}$$

and rearranging

$$b_{\lim} = \frac{c\left(2\sqrt{mk}+c\right)}{m}\left[\frac{1}{k_c} + \frac{1}{2k_1 u_{ch} G}\right] \tag{5.13}$$

Equation 5.13 indicates that the unconditional stability boundary is independent of the diameters of the wheel and work and also their rotational speeds. This is because it has been assumed that the grinding ratio G does not depend on these variables. In Sect. 5.3.1, it was found that without a contact stiffness and grinding with regeneration on the wheel alone, the unconditional width could be very small. This is particularly true for super-abrasive wheels that provide large values of G. If the contact stiffness in Eq. 5.13 is made infinite, then

$$b = \frac{c\left(2\sqrt{mk}+c\right)}{2mk_1 u_{ch} G} \tag{5.14}$$

When the contact stiffness is included, the unconditional width increases (as for work regeneration—Eq. 5.13) by an amount

$$\frac{c\left(2\sqrt{mk}+c\right)}{mk_c}$$

Again, the benefits of reducing the contact stiffness are apparent. However, for super-abrasive wheels with high grinding ratios, the contact stiffness required has practical limitations. For example, the cost of CBN grits requires a strong-bond material so that they are not easily detached. This militates against using a low contact stiffness. However, there is a solution, the use of so-called flexible grinding wheels. Before considering such wheels, there is another source of effective flexibility to investigate.

A significant contact stiffness was omitted in Sect. 5.3.1 when considering regeneration on the wheel alone. It was assumed that the wheel was machined by the work, as if the work was a non-wearing metal cutting tool. However, the work is not such a metal cutting tool but is wearing at a significant rate. The relationship between the normal force and the "wear" on the work is given by,

$$P_{\mathrm{n}}=\frac{k_1 u_{\mathrm{ch}} V_{\mathrm{w}} \delta_{\mathrm{w}} b}{V_g} \text{ so that } \frac{P_{\mathrm{n}}}{\delta_{\mathrm{w}}}=\frac{k_1 u_{\mathrm{ch}} V_{\mathrm{w}} b}{V_g}$$

This is significant because the work (representing a cutting tool) is not in the position assumed in Sect. 5.3.1. It has moved out of cut because of the effective contact stiffness for a width b of $\frac{P_{\mathrm{n}}}{\delta_{\mathrm{w}}}=\frac{k_1 u_{\mathrm{ch}} V_{\mathrm{w}} b}{V_g}$ so that the contact stiffness (stiffness/unit width) is

$$k_c=\frac{P_{\mathrm{n}}}{\delta_{\mathrm{w}} b}=\frac{k_1 u_{\mathrm{ch}} V_{\mathrm{w}}}{V_g}$$

Equation 5.13 may thus be modified to become

$$b=\frac{c\left(2\sqrt{mk}+c\right)}{m}\left[\frac{V_g}{k_1 u_{\mathrm{ch}} V_{\mathrm{w}}}+\frac{1}{2k_1 u_{\mathrm{ch}} G}\right]$$

The second term in the square brackets was the one that predicted very small values of b, but the contact stiffness effect from the wear of the work has a larger effect. If that term alone is considered,

$$b=\frac{c\left(2\sqrt{mk}+c\right)}{m}\frac{V_g}{k_1 u_{\mathrm{ch}} V_{\mathrm{w}}}$$

As this value of b is usually much greater than that found from assuming a metal cutting model, great care must be taken to ensure that it is always included.

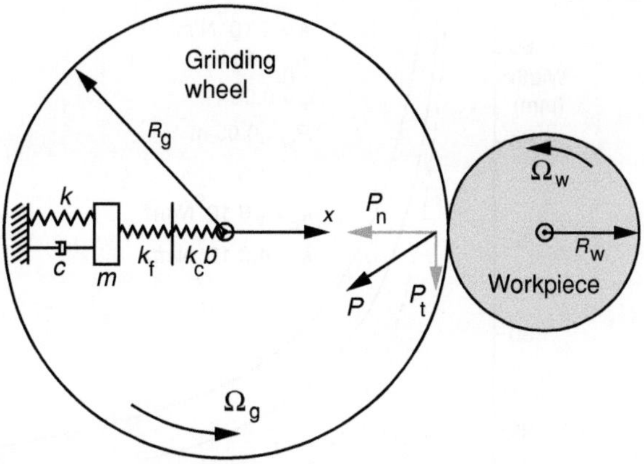

Fig. 5.12 Single-mode model with contact stiffness and a flexible wheel

The contribution from the work is significant even when there is no regeneration on it.

A similar analysis for the wheel, when it is considered to be machining the work, as in Sect. 5.3.1 yields an effective contact stiffness of $k_c = P_n/\delta_g = k_1 u_{\rm ch} Gb$, and this is much greater than that caused by the work. It is thus of less significance for practical values of G.

5.3.4 Flexible Grinding Wheels

The improvements possible from the effect of contact stiffness suggest that the use of a flexible grinding wheel would be useful. Such a wheel is designed to include additional flexibility close to the contact zone without introducing additional local mass. Sexton et al. [28, 29] describe such wheels, and their experimental results will be reviewed in Sect. 5.4. As in the previous section, the simplest structural model will be used and the effect of both a flexible wheel and a contact stiffness will be modelled as shown in Fig. 5.12. The unconditional stability boundary will again be found.

The flexible wheel provides an additional and fixed stiffness in series with the contact stiffness so that the effective flexibility at the contact point is

$$\frac{1}{k_{\rm f}} + \frac{1}{k_{\rm c} b}$$

The analysis of the previous section may be repeated replacing

$$\frac{1}{k_{\rm c} b} \text{with} \frac{1}{k_{\rm f}} + \frac{1}{k_{\rm c} b}$$

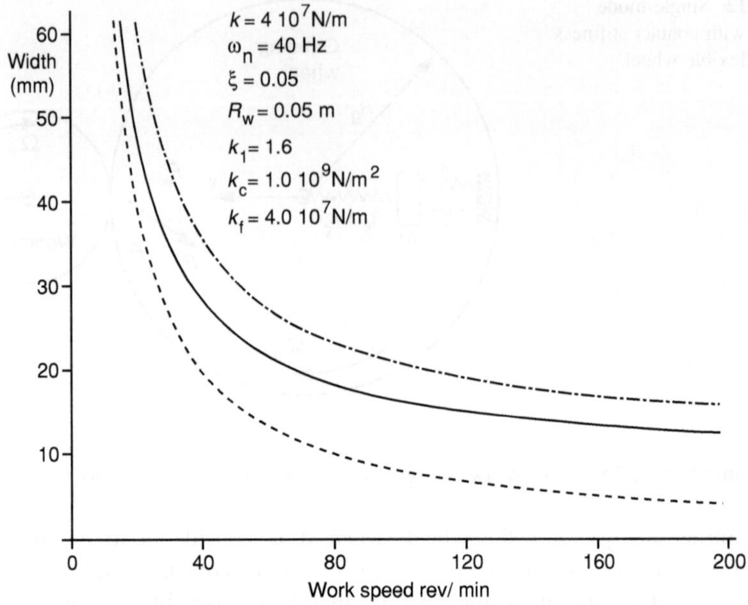

Fig. 5.13 Unconditional stability boundary with a flexible wheel (Program 5.4)

The maximum negative in-phase component of the response becomes

$$G_{\mathrm{R,max}}(\omega) = \frac{1}{k_{\mathrm{f}}} + \frac{1}{k_{\mathrm{c}}b} - \frac{m}{c\left(2\sqrt{mk} + c\right)}$$

and substituting in Eq. 1.15 as above with $R = k_1 u_{\mathrm{ch}} V_{\mathrm{w}}/V_{\mathrm{g}}$, for regeneration on the work alone, gives

$$b = \frac{-V_g}{2k_1 u_{\mathrm{ch}} V_{\mathrm{w}} \left[\frac{1}{k_{\mathrm{f}}} + \frac{1}{k_c b} - \frac{m}{c\left(2\sqrt{mk}+c\right)}\right]}$$

which upon rearranging yields

$$b_{\mathrm{lim}} = \frac{\frac{1}{k_c} + \frac{V_g}{2k_1 u_{\mathrm{ch}} V_{\mathrm{w}}}}{\left[\frac{m}{c\left(2\sqrt{mk}+c\right)} - \frac{1}{k_{\mathrm{f}}}\right]} \tag{5.15}$$

Figure 5.13 shows the envelope for some typical values and for comparison, the envelopes with and without contact stiffness and wheel flexibility (compare Fig. 5.11) are also shown. There are possible benefits if flexible grinding wheels can be realised and used. The analysis above was for regeneration only on the work.

If only regeneration on the wheel is considered, the effective cutting force coefficient is $R = k_1 u_{ch} G$, so that the unconditional stability boundary is given by

$$b_{\lim} = \frac{\frac{1}{k_c} + \frac{1}{2k_1 u_{ch} G}}{\left[\frac{m}{c(2\sqrt{mk}+c) - \frac{1}{k_f}}\right]} \tag{5.16}$$

The value of the unconditional $b_{\lim}$ as in the previous section is independent of the diameters of the wheel and work and also their rotational speeds. Also Eqs. 5.15 and 5.16 predict that there will be no chatter, i.e. no positive values of $b_{\lim}$ if

$$\frac{m}{c(2\sqrt{mk}+c)} - \frac{1}{k_f}\text{i.e. when}\frac{1}{k_f} > \frac{m}{c(2\sqrt{mk}+c)}$$

This is the condition when the chatter receptance no longer has a negative in-phase component. Whether this can be achieved practically is discussed in Sect. 5.4.

Thus far, the structural response has been kept as simple as possible, a single mode with the direction along the line joining the centres of the wheel and work. In practice, the mode direction will normally be at some inclination.

5.3.5 Inclined Mode

The next stage in the development of the chatter model is to incline the mode of vibration at some angle ϕ to the line joining the centres of the wheel and the work as shown in Fig. 5.14. In keeping with the approach of examining one parameter at a time, the model does not include the contact stiffness or a flexible wheel. The equation of motion is

$$m\frac{d^2u(t)}{dt^2} + c\frac{du(t)}{dt} + ku(t) = -P_n \cos\phi - P_t \sin\phi \tag{5.17}$$

where $P_t = \frac{u_{ch} V_w \delta_w b}{V_g}$ and $P_n = k_1 P_t$
so that substituting

$$\begin{aligned} &m\frac{d^2u(t)}{2} + c\frac{du(t)}{dt} + ku(t) \\ &= -\frac{u_{ch} V_w \delta_w b}{V_g}(k_1 \cos\phi + \sin\phi) = -\frac{u_{ch} V_w \delta_w b}{V_g}\varepsilon \end{aligned} \tag{5.18}$$

Fig. 5.14 Machine model for an inclined mode

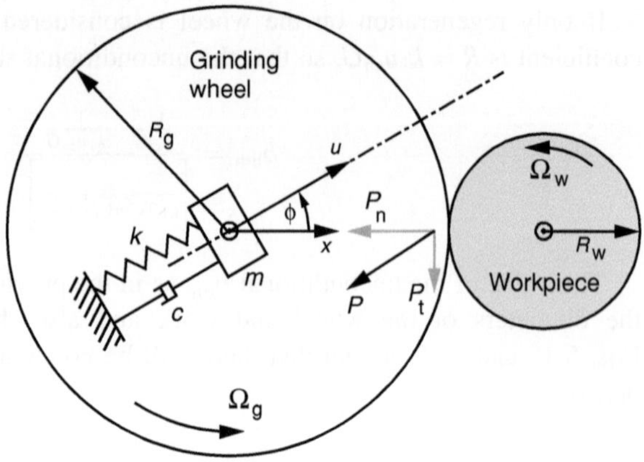

where

$$\varepsilon = k_1 \cos\phi + \sin\phi \tag{5.19}$$

When an oscillation in the u direction occurs, the surface speeds of the work and wheel will be affected. Making the assumption that it is possible to operate at the stability boundary with no growth or decay, we may assume

$$\delta_{\mathrm{w}}(t) = \bar{\delta}_{\mathrm{w}} + \delta_{\mathrm{w}} e^{\mathrm{i}\omega t} V_{\mathrm{w}}(t) = \bar{V}_w + V_{\mathrm{w}} e^{\mathrm{i}\omega t} V_{\mathrm{g}}(t) = \bar{V}_g + V_g e^{\mathrm{i}\omega t}$$

The over-bars represent the mean values, and for the chip thickness, $\bar{\delta}_{\mathrm{w}}$ is the feed per revolution of the wheel into the work.

It is assumed that all three variables will be vibrating in-phase, as the oscillation in each case (in this section) will come from the same source, i.e. the oscillation of the mass. Considering only vibration about the mean values

$$\mathrm{d}\left(\frac{V_{\mathrm{w}}\delta_{\mathrm{w}}}{V_{\mathrm{g}}}\right) = \frac{\bar{V}_{\mathrm{w}}}{\bar{V}_{\mathrm{g}}}\mathrm{d}\delta_{\mathrm{w}} + \frac{\bar{\delta}_{\mathrm{w}}}{\bar{V}_{\mathrm{g}}}\mathrm{d}V_{\mathrm{w}} - \frac{\bar{\delta}_{\mathrm{w}}\bar{V}_{\mathrm{w}}}{\bar{V}_{\mathrm{g}}^2}\mathrm{d}V_g \tag{5.20}$$

The varying components of the three parameters are $\mathrm{d}V_{\mathrm{w}} = V_{\mathrm{w}}e^{\mathrm{i}\omega t}$, $\mathrm{d}V_{\mathrm{g}} = V_{\mathrm{g}}e^{\mathrm{i}\omega t}$ and for regeneration only on the work $\mathrm{d}\delta_{\mathrm{w}} = \delta_{\mathrm{w}}e^{\mathrm{i}\omega t} - \delta_{\mathrm{w}}e^{\mathrm{i}\omega(t-\tau_{\mathrm{w}})}$, so that substituting in Eq. 5.20

$$\mathrm{d}\left(\frac{V_{\mathrm{w}}\delta_{\mathrm{w}}}{V_{\mathrm{g}}}\right) = \frac{\bar{V}_{\mathrm{w}}}{\bar{V}_{\mathrm{g}}}\delta_{\mathrm{w}}e^{\mathrm{i}\omega t} + \frac{\bar{\delta}_{\mathrm{w}}}{\bar{V}_{\mathrm{g}}}V_{\mathrm{w}}e^{\mathrm{i}\omega t} - \frac{\bar{\delta}_{\mathrm{w}}\bar{V}_{\mathrm{w}}}{\bar{V}_{\mathrm{g}}^2}V_g e^{\mathrm{i}\omega t}$$

For the model shown in Fig. 5.14 from geometry, $\delta_{\mathrm{w}}(t) = u(t)\cos\phi$. For small oscillations about the mean $u(t) = Ue^{\mathrm{i}\omega t}$ so that $\delta_{\mathrm{w}}e^{\mathrm{i}\omega t} = Ue^{\mathrm{i}\omega t}\cos\phi$ and hence, $\delta_{\mathrm{w}} = U\cos\phi$. Finally, substituting all the above in Eq. 5.18

$$m\frac{\mathrm{d}^2(Ue^{\mathrm{i}\omega t})}{\mathrm{d}t^2}+c\frac{\mathrm{d}(Ue^{\mathrm{i}\omega t})}{\mathrm{d}t}+k(Ue^{\mathrm{i}\omega t})$$
$$=-u_{\mathrm{ch}}b\varepsilon\left(\frac{\bar{V}_{\mathrm{w}}}{\bar{V}_{\mathrm{g}}}U\left(e^{\mathrm{i}\omega t}-e^{\mathrm{i}\omega(t-\tau_{\mathrm{w}})}\right)\cos\phi+\frac{\bar{\delta}}{\bar{V}_{\mathrm{g}}}V_{\mathrm{w}}e^{\mathrm{i}\omega t}-\frac{\bar{\delta}\bar{V}_{\mathrm{w}}}{\bar{V}_{\mathrm{g}}^2}V_g e^{\mathrm{i}\omega t}\right)$$

so that

$$-m\omega^2U+\mathrm{i}\omega cU+kU$$
$$=-u_{\mathrm{ch}}b\varepsilon\left(\frac{\bar{V}_{\mathrm{w}}}{\bar{V}_{\mathrm{g}}}U\cos\phi\left(1-e^{-\mathrm{i}\omega\tau_{\mathrm{w}}}\right)+\frac{\bar{\delta}}{\bar{V}_{\mathrm{g}}}V_{\mathrm{w}}-\frac{\bar{\delta}\bar{V}_{\mathrm{w}}}{\bar{V}_{\mathrm{g}}^2}V_g\right) \tag{5.21}$$

At this stage, it is helpful to remember that U, V_{w} and V_{g} are, respectively, the amplitudes of the vibration of the mass, the work surface speed and the wheel speed. The variables $\bar{\delta}$, $\overline{V_{\mathrm{w}}}$ and $\overline{V_{\mathrm{g}}}$ are, respectively, the in-feed of the wheel into the work per revolution, the mean work surface speed and the mean wheel speed.

For the particular example that is being considered in this section, i.e. a single inclined mode, there is a connection between U, V_{w} and V_{g} as they are not independent. The vibration $Ue^{\mathrm{i}\omega t}$ produces a velocity of the mass of $\mathrm{i}\omega Ue^{\mathrm{i}\omega t}$, and this results in an increase in the amplitude of the speed of approach of the work of $\mathrm{i}\omega U\sin\phi$. If the small effect of the change in the wheel speed caused by the vibration of the mass is ignored, the wheel speed may be considered constant, Eq. 5.21 becomes

$$\left(k-m\omega^2+\mathrm{i}\omega c\right)U=-u_{\mathrm{ch}}b\varepsilon\left(\frac{\bar{V}_{\mathrm{w}}}{\bar{V}_{\mathrm{g}}}U\cos\phi\left(1-e^{-\mathrm{i}\omega\tau_{\mathrm{w}}}\right)+\mathrm{i}\omega U\sin\phi\frac{\bar{\delta}}{\bar{V}_{\mathrm{g}}}\right)$$

and rearranging

$$\left(k-m\omega^2+\mathrm{i}\omega\left(c+u_{\mathrm{ch}}b\varepsilon\frac{\bar{\delta}}{\bar{V}_{\mathrm{g}}}\sin\phi\right)\right)=-\cos\phi\left(1-e^{-\mathrm{i}\omega\tau_{\mathrm{w}}}\right)u_{\mathrm{ch}}b\varepsilon\frac{\bar{V}_{\mathrm{w}}}{\bar{V}_{\mathrm{g}}} \tag{5.22}$$

If the mode direction is along the line, joining the centres of the work and wheel ϕ is zero and Eq. 5.22 becomes,

$$\left(k-m\omega^2+\mathrm{i}\omega c\right)=\left(1-e^{-\mathrm{i}\omega\tau_{\mathrm{w}}}\right)u_{\mathrm{ch}}b\varepsilon\frac{\bar{V}_{\mathrm{w}}}{\bar{V}_{\mathrm{g}}}=-Rb\varepsilon\left(1-e^{-\mathrm{i}\omega\tau_{\mathrm{w}}}\right) \tag{5.23}$$

This is the same equation derived for the simplest case considered in Sect. 5.3.1. If Eq. 5.22 is compared with Eq. 5.23, it is clear that the effective structural damping has been increased and the regenerative term has been decreased in magnitude. The effective damping is now $c+u_{\mathrm{ch}}b\varepsilon\sin\overline{\phi\delta}/\bar{V}_{\mathrm{g}}$ and the effective cutting force coefficient is now $u_{\mathrm{ch}}\varepsilon\cos\phi\bar{V}_{\mathrm{w}}/\bar{V}_{\mathrm{g}}$. Both of these changes will result in an increase in stability.

The solution of Eq. 5.22 is not as straight forward as for the simple case because the width of cut (*b*) occurs on the left-hand side as well as the right. However, it is important to note that the damping is increased when the width of cut is increased and when the feed per revolution is increased. It is decreased when the grinding wheel speed is increased.

A solution to Eq. 5.22 may be found using the same method as that used for growth rate in Sect. 5.3.2. Two equations for *b* can be found utilising the real and imaginary components of Eq. 5.22. A computer program is then required to search for solutions that satisfy both equations. As with the contact stiffness, we can examine the unconditional stability boundary, i.e. not consider stability lobes but just their envelope.

The condition for unconditional stability is

$$b_{\mathrm{lim}} = \frac{-1}{2RG_{\mathrm{R,max}}(\omega)} \tag{1.15}$$

As used for the previous two sections and as shown in Appendix A.10, the maximum negative in-phase component of the response of a single-mode system is given by $-m/c\left[\left(2\sqrt{mk}+c\right)\right]$. For the inclined mode, the effective damping is $c + u_{\mathrm{ch}}b\,\varepsilon\sin\phi\bar{\delta}/\bar{V}_{\mathrm{g}} = c + \alpha b$ where $\alpha = u_{\mathrm{ch}}\varepsilon\sin\phi\bar{\delta}/\bar{V}_{\mathrm{g}}$. Thus, substituting in Eq. 1.15,

$$b = \frac{(c+\alpha b)\left(2\sqrt{mk}+c+\alpha b\right)}{2Rm}$$

Rearranging to solve for *b*,

$$\alpha^2 b^2 + 2b\left(\alpha[c+\sqrt{mk}] - Rm\right) + c\left(2\sqrt{mk}+c\right) = 0 \tag{5.24}$$

noting that for the inclined mode $R = u_{\mathrm{ch}}\varepsilon\cos\phi\bar{V}_{\mathrm{w}}/\bar{V}_{\mathrm{g}}$

Figure 5.15 shows the envelope for some values of the parameters chosen to illustrate some significant effects. For comparison purposes, three cases are shown, the simple case, i.e. with no inclination, the inclined case without the extra damping and then with the extra damping. As might be expected from the importance of the mode direction in metal cutting (Sect. 2.3), it is also important in grinding. If the mode is perpendicular to either the force or the chip thickness direction, the process is very stable.

The extra damping has little effect even when α is increased by making ϕ as large as 75° and increasing $\bar{\delta}$ and decreasing $\bar{V}_{\mathrm{g}}$. However, the extra damping term may become important if there is a significant torsional mode of the work and its drive system. This can cause large variations in the work surface speed, and this is discussed in the next section.

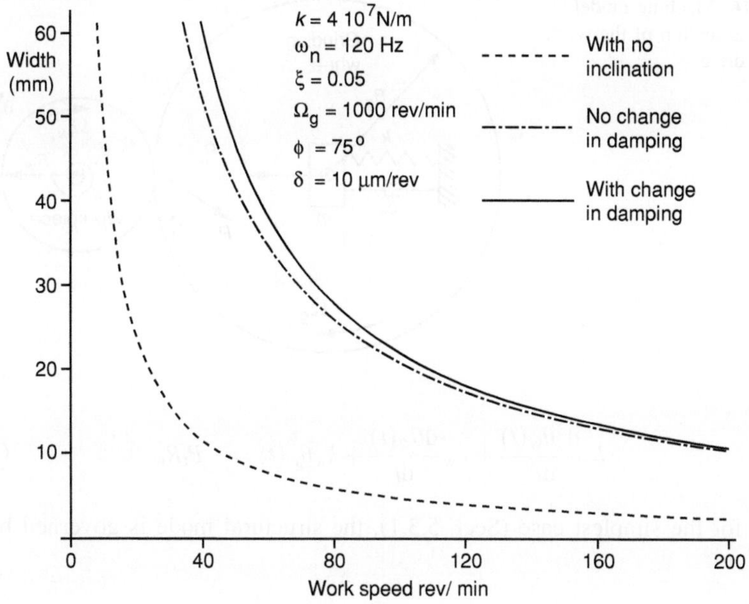

Fig. 5.15 Unconditional stability boundary with an inclined mode (Program 5.5)

In passing, it should be noted that if ϕ is negative, any "extra" damping is negative. This is considered further in Sect. 5.3.8.3. Also the mode direction may be so inclined so as to produce a negative mode, and the analysis above needs to be repeated to find the new most negative in-phase component of the negative mode response (Program 5.5 includes this effect).

The above analysis assumed no regeneration on the wheel. If no regeneration occurs on the work but rather on the wheel, a similar analysis may be conducted with similar effects.

5.3.6 Torsional Effects

The simplest model that allows the possible effects of torsional vibration to be considered is shown in Fig. 5.16. The drive for the work is assumed to have a single torsional mode with a torsional stiffness k_w, torsional natural frequency ω_{nw} and a torsional viscous damping ratio ξ_w. These parameters will then define the inertia of the torsional mode I_w and the viscous damping coefficient c_w. The equation of motion for the torsional mode is therefore

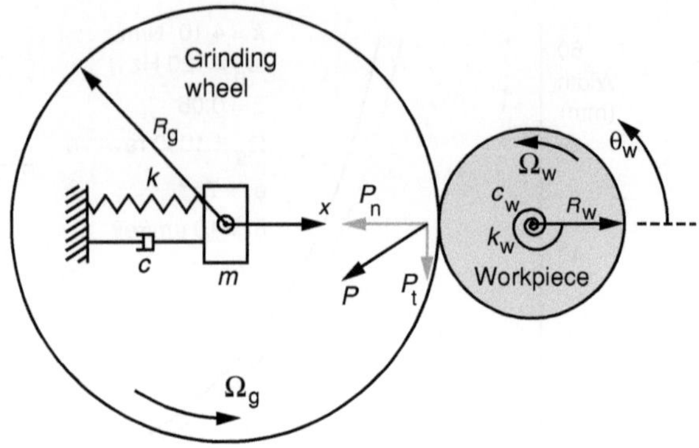

Fig. 5.16 Machine model including torsion of the work and its drive

$$I_w \frac{d^2\theta_w(t)}{dt^2} + c_w \frac{d\theta_w(t)}{dt} + k_w\theta_w(t) = -P_t R_w \tag{5.25}$$

As for the simplest case (Sect. 5.3.1), the structural mode is governed by the equation

$$m\frac{d^2x(t)}{dt^2} + c\frac{dx(t)}{dt} + kx(t) = -P_n \tag{5.26}$$

If conditions at the stability boundary are considered, then the oscillating components of x and θ_w are $Xe^{i\omega t}$ and $\Theta_w e^{i\omega t}$. These may be phase-shifted meaning that if X is defined as real, then Θ_w may be complex.

The force variation, for regeneration only on the work when $\delta_w = x$, and assuming the mean wheel speed is constant, is given by

$$P_t = b_{\lim} u_{ch}\left(\frac{\bar{V}_w}{\bar{V}_g}\left(Xe^{i\omega t} - Xe^{i\omega(t-\tau_w)}\right) + \frac{\bar{\delta}_w}{\bar{V}_g} V_w e^{i\omega t}\right) \text{ and } P_n = k_1 P_t$$

Now the variation in the work surface speed will be caused by the torsional vibration so that $V_w e^{i\omega t} = R_w \Theta_w e^{i\omega t}$. Substituting in Eqs. 5.25 and 5.26 and dividing throughout by $e^{i\omega t}$,

$$\begin{aligned} &-I_w\omega^2\Theta_w + i\omega c_w\Theta_w + k_w\Theta_w \\ &= -u_{ch} b_{\lim} R_w\left(\frac{\bar{V}_w}{\bar{V}_g} X\left(1 - e^{-i\omega\tau_w}\right) + \frac{\bar{\delta}_w}{\bar{V}_g} R_w\Theta_w\right) \end{aligned} \tag{5.27}$$

$$-m\omega^2 X + i\omega cX + kX = -k_1 u_{ch} b_{\lim}\left(\frac{\bar{V}_w}{\bar{V}_g} X\left(1 - e^{-i\omega\tau_w}\right) + \frac{\bar{\delta}_w}{\bar{V}_g} R_w\Theta_w\right) \tag{5.28}$$

From Eq. 5.27,

$$\Theta_{\mathrm{w}}\left(-I_{\mathrm{w}}\omega^2 + \mathrm{i}\omega c_{\mathrm{w}} + k_{\mathrm{w}} + u_{ch}b_{\mathrm{lim}}R_{\mathrm{w}}^2\bar{\delta}_{\mathrm{w}}/\bar{V}_{\mathrm{g}}\right) = -u_{ch}b_{\mathrm{lim}}R_{\mathrm{w}}X\bar{V}_{\mathrm{w}}\left(1 - e^{-\mathrm{i}\omega\tau_w}\right)\bar{V}_g$$

and rearranging

$$\Theta_{\mathrm{w}} = \frac{-u_{ch}b_{\mathrm{lim}}R_{\mathrm{w}}\frac{\bar{V}_{\mathrm{w}}}{\bar{V}_{\mathrm{g}}}X(1 - e^{-\mathrm{i}\omega\tau_w})}{\left(k_w - I_w\omega^2 + \mathrm{i}\omega c_{\mathrm{w}} + u_{ch}b_{\mathrm{lim}}R_{\mathrm{w}}^2\bar{\delta}_{\mathrm{w}}/\bar{V}_{\mathrm{g}}\right)}X$$

where as noted above Θ_{w} may be complex if X is real. We may now substitute for Θ_{w} in Eq. 5.28 to obtain, after some simplification,

$$\begin{aligned}&\left(k - m\omega^2 + \mathrm{i}\omega c\right)\\ &= -k_1u_{ch}b_{\mathrm{lim}}\frac{\bar{V}_{\mathrm{w}}}{\bar{V}_{\mathrm{g}}}\left(1 - e^{-\mathrm{i}\omega\tau_{\mathrm{w}}}\right)\left(1 - \frac{u_{ch}b_{\mathrm{lim}}R_{\mathrm{w}}^2\bar{\delta}_{\mathrm{w}}/\bar{V}_{\mathrm{g}}}{\left(k_{\mathrm{w}} - I_{\mathrm{w}}\omega^2 + u_{\mathrm{ch}}b_{\mathrm{lim}}R_{\mathrm{w}}^2\bar{\delta}_{\mathrm{w}}/\bar{V}_{\mathrm{g}} + \mathrm{i}\omega c_{\mathrm{w}}\right)}\right)\end{aligned} \tag{5.29}$$

It should be noted that X has cancelled. This is to be expected since the chatter boundary is independent of the amplitude for the linear case.

The solution of Eq. 5.29 is taxing as there is no simple way of finding an envelope. First make some substitutions. Let $A = u_{ch}\frac{\bar{\delta}_{\mathrm{w}}}{\bar{V}_{\mathrm{g}}}R_{\mathrm{w}}^2$ and $B = k_1u_{ch}\frac{\bar{V}_{\mathrm{w}}}{\bar{V}_{\mathrm{g}}}$ so that Eq. 5.29 becomes

$$(k - m\omega^2 + \mathrm{i}\omega c) = -Bb_{\mathrm{lim}}(1 - e^{-\mathrm{i}\omega\tau_w})\left(1 - \frac{Ab_{\mathrm{lim}}}{(k_{\mathrm{w}} - I_{\mathrm{w}}\omega^2 + \mathrm{i}\omega c_{\mathrm{w}} + Ab_{\mathrm{lim}})}\right)$$

Now multiply throughout by $(k_{\mathrm{w}} - I_{\mathrm{w}}\omega^2 + Ab_{\mathrm{lim}} + \mathrm{i}\omega c_{\mathrm{w}})$ to obtain

$$\begin{aligned}&\left(k - m\omega^2 + \mathrm{i}\omega c\right)\left(k_{\mathrm{w}} - I_{\mathrm{w}}\omega^2 + Ab_{\mathrm{lim}} + \mathrm{i}\omega c_{\mathrm{w}}\right)\\ &= -Bb_{\mathrm{lim}}\left(1 - e^{-\mathrm{i}\omega\tau_w}\right)\left(k_{\mathrm{w}} - I_{\mathrm{w}}\omega^2 + \mathrm{i}\omega c_{\mathrm{w}}\right)\end{aligned}$$

Now substitute $e^{-\mathrm{i}\omega\tau_w} = \cos\omega\tau_w - i\sin\omega\tau_w$, expand and group real and imaginary terms

$$\begin{aligned}&\left(k - m\omega^2\right)\left(k_{\mathrm{w}} - I_{\mathrm{w}}\omega^2\right) - cc_{\mathrm{w}}\omega^2 + i\left[\omega c(k_{\mathrm{w}} - I_{\mathrm{w}}\omega^2) + \omega c_{\mathrm{w}}(k - m\omega^2)\right]\\ &= -b_{\mathrm{lim}}A(k - m\omega^2) - b_{\mathrm{lim}}\left(B\left[(k_{\mathrm{w}} - I_{\mathrm{w}}\omega^2)(1 - \cos\omega\tau_w) - \omega c_{\mathrm{w}}\sin\omega\tau_w\right]\right)\\ &- ib_{\mathrm{lim}}\left(\omega cA + B\left[\omega c_{\mathrm{w}}(1 - \cos\omega\tau_w) + \sin\omega\tau_w(k_{\mathrm{w}} - I_{\mathrm{w}}\omega^2)\right]\right)\end{aligned}$$

Equating the real parts gives,

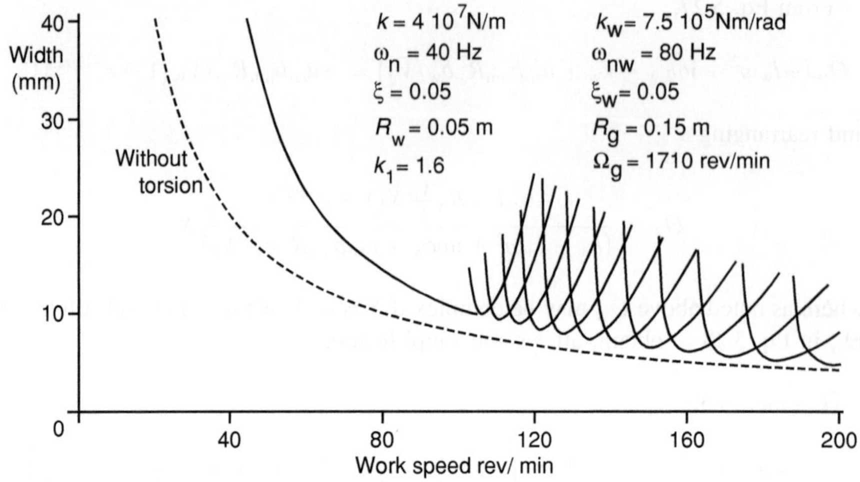

Fig. 5.17 Stability chart with torsion of the work included (Program 5.6)

$$b_{\text{lim}} = \frac{(k - m\omega^2)(k_{\text{w}} - I_{\text{w}}\omega^2) - cc_{\text{w}}\omega^2}{-A(k - m\omega^2) - B((k_{\text{w}} - I_{\text{w}}\omega^2)(1 - \cos\omega\tau_w) - \omega c_{\text{w}}\sin\omega\tau_w)} \tag{5.30}$$

and the imaginary parts give,

$$b_{\text{lim}} = \frac{\omega c(k_{\text{w}} - I_{\text{w}}\omega^2) + \omega c_{\text{w}}(k - m\omega^2)}{-A\omega c - B(\omega c_{\text{w}}(1 - \cos\omega\tau_w) + \sin\omega\tau_w(k_{\text{w}} - I_{\text{w}}\omega^2))} \tag{5.31}$$

A solution may be found in the same manner as described in Sect. 5.3.2 for growth rates. For given values of all the parameters, b_{lim} is plotted against ω for both Eqs. 5.30 and 5.31. The intercepts give solutions for both ω and b_{lim}. The curves are similar to those shown in Fig. 5.7, and care has to be taken not to include spurious solutions. If a stability chart is plotted, it is found that several parameters are significant when it comes to improving stability. Figure 5.17 shows that stability can be improved by the torsional characteristics of the work and its drive. The example shows that the improvement is greater at low work speeds. If the torsional stiffness is decreased while keeping constant the natural frequency and damping ratio for torsion, even greater improvements are achieved. If the torsional natural frequency approaches the transverse natural frequency, chatter is often eliminated. However, if the torsional natural frequency is less than the transverse natural frequency, the stability may be decreased by the torsional effects.

A final important point is to note that if the direction of the work speed was reversed the effect of torsion would be different. The grinding force direction would still be determined by the rotation of the wheel, but the torque on the work would be in the opposite sense to that considered above.

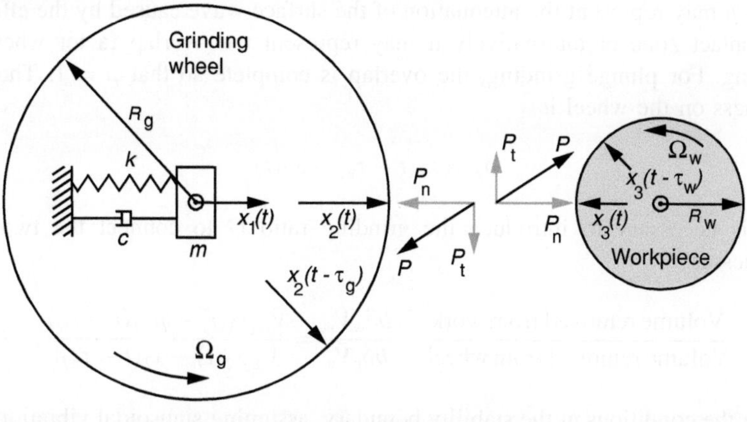

Fig. 5.18 Machine model and parameters for double regeneration

It is not possible to summarise all the interactions in a simple manner. However, the conclusion is clear and torsional effects can be significant.

At this stage, as we investigate one major effect at a time, it is apparent that chatter in grinding is very complex to model, and there is still the outstanding question of what happens when regeneration occurs on both the wheel and the work.

5.3.7 Double Regeneration

The simplest model that includes double regeneration is shown in Fig. 5.18. When regeneration was considered to be only on the wheel or work, the displacement of the structure was the same as that of the surface with the regeneration on it. For double regeneration, we need to define the parameters carefully.

$x_1(t)$ is the relative displacement of the centres of the wheel and work.
$x_2(t)$ is the increase in radius of the wheel.
$x_3(t)$ is the increase in radius of the work.

It follows that the relative motion of the wheel and work centres, $x_1(t)$, causes a reduction in both the radii of the wheel and work so that

$$x_1(t) = -x_2(t) - x_3(t) \tag{5.32}$$

In order to model chatter, we will only investigate the variation from the mean values. Thus, the oscillating chip thickness on the work is

$$\delta_w = \mu x_3(t - \tau_w) - x_3(t)$$

where μ may represent the attenuation of the surface wave caused by the effect of the contact zone or alternatively it may represent the overlap factor when roll grinding. For plunge grinding, the overlap is complete so that $\mu = 1$. The chip thickness on the wheel is

$$\delta_g = x_2(t - \tau_g) - x_2(t)$$

It is necessary to introduce the grinding ratio G to connect the two chip thicknesses.

$$G = \frac{\text{Volume removed from work}}{\text{Volume removed from wheel}} = \frac{b\delta_w \bar{V}_w}{b\delta_g \bar{V}_g} = \frac{\bar{V}_w[x_3(t) - \mu x_3(t - \tau_w)]}{\bar{V}_g[x_2(t) - x_2(t - \tau_g)]}. \tag{5.33}$$

For the conditions at the stability boundary, assuming sinusoidal vibration with a constant amplitude $x_1(t) = X_1 e^{i\omega t}$; $x_2(t) = X_3 e^{i\omega t}$ and $x_3(t) = X_3 e^{i\omega t}$. As all three will not be in-phase, X_2 and X_3 may be complex if X_1 is defined as real. Also if the mean speed of both work and wheel are assumed constant, the oscillating grinding force of interest is

$$P_n = -k_1 u_{ch} b_{lim} \frac{\bar{V}_w}{\bar{V}_g}[x_3(t) - \mu x_3(t - \tau_w)]$$

The equation of motion for the vibration is therefore,

$$m\frac{d^2x_1(t)}{dt^2} + c\frac{dx_1(t)}{dt} + kx_1(t) = -P_n = k_1 u_{ch} b_{lim} \frac{\bar{V}_w}{\bar{V}_g}[x_3(t) - \mu x_3(t - \tau_w)]$$

And substituting

$$x_1(t) = X_1 e^{i\omega t};\ x_2(t) = X_3 e^{i\omega t} \text{ and } x_3(t) = X_3 e^{i\omega t},$$

$$-m\omega^2 X_1 e^{i\omega t} + i\omega c X_1 e^{i\omega t} + kX_1 e^{i\omega t} = k_1 u_{ch} b_{lim} \frac{\bar{V}_w}{\bar{V}_g}\left[X_3 e^{i\omega t} - \mu X_3 e^{i\omega t} e^{-i\omega\tau_w}\right]$$

which simplifies to

$$X_1(k - m\omega^2 + i\omega c) = -k_1 u_{ch} b_{lim} \frac{\bar{V}_w}{\bar{V}_g} X_3\left[1 - \mu e^{-i\omega\tau_w}\right] \tag{5.34}$$

From Eq. 5.32,

$$X_1 e^{i\omega t} = -X_2 e^{i\omega t} - X_3 e^{i\omega t} \quad \text{so that } X_1 = -X_2 - X_3 \tag{5.35}$$

and the grinding ratio, Eq. 5.33, becomes

$$G = \frac{\bar{V}_w[x_3(t) - \mu x_3(t-\tau_w)]}{\bar{V}_g[x_2(t) - x_2(t-\tau_g)]} = \frac{\bar{V}_w X_3[1 - \mu e^{-i\omega\tau_w}]}{\bar{V}_g X_2[1 - e^{-i\omega\tau_g}]} \tag{5.36}$$

To solve Eqs. 5.34–5.36, first rearrange Eq. 5.34

$$X_1 = \frac{k_1 u_{ch} b_{lim} \frac{\bar{V}_w}{\bar{V}_g}[1 - \mu e^{-i\omega\tau_w}]}{(k - m\omega^2 + i\omega c)} X_3 \tag{5.37}$$

And from Eq. 5.36,

$$X_2 = \frac{\bar{V}_w(1 - \mu e^{-i\omega\tau_w})}{G\bar{V}_g(1 - e^{-i\omega\tau_g})} X_3 \tag{5.38}$$

Now substitute Eqs. 5.37 and 5.38 in Eq. 5.35,

$$\frac{k_1 u_{ch} b_{lim} \frac{\bar{V}_w}{\bar{V}_g}(1 - \mu e^{-i\omega\tau_w})}{(k - m\omega^2 + i\omega c)} X_3 = -\frac{\bar{V}_w(1 - \mu e^{-i\omega\tau_w})}{G\bar{V}_g(1 - e^{-i\omega\tau_g})} X_3 - X_3$$

Rearranging

$$\frac{-k_1 u_{ch} b_{lim} \frac{\bar{V}_w}{\bar{V}_g}(1 - \mu e^{-i\omega\tau_w})}{(k - m\omega^2 + i\omega c)} = \frac{\bar{V}_w(1 - \mu e^{-i\omega\tau_w}) + G\bar{V}_g(1 - e^{-i\omega\tau_g})}{G\bar{V}_g(1 - e^{-i\omega\tau_g})} \tag{5.39}$$

Note that X_3 cancels.

Now substitute $e^{-i\omega\tau_w} = \cos\omega\tau_w - i\sin\omega\tau_w$ and

$$e^{-i\omega\tau_g} = \cos\omega\tau_g - i\sin\omega\tau_g,$$

$$\frac{-k_1 u_{ch} b_{lim} \frac{\bar{V}_w}{\bar{V}_g}(1 - \mu\cos\omega\tau_w + i\mu\sin\omega\tau_w)}{(k - m\omega^2 + i\omega c)}$$

$$= \frac{\bar{V}_w(1 - \mu\cos\omega\tau_w + i\mu\sin\omega\tau_w) + G\bar{V}_g(1 - \cos\omega\tau_g + i\sin\omega\tau_g)}{G\bar{V}_g(1 - \cos\omega\tau_g + i\sin\omega\tau_g)}$$

And rearrange to get an explicit equation for b

$$b_{lim} = \frac{\bar{V}_w\left(\begin{matrix}[(1 - \mu\cos\omega\tau_w) + i\mu\sin\omega\tau_w] \\ +G\bar{V}[(1 - \cos\omega\tau_g) + i\sin\omega\tau_g]\end{matrix}\right)[(k - m\omega^2) + i\omega c]}{-k_1 u_{ch} G\bar{V}_w[(1 - \cos\omega\tau_g) + i\sin\omega\tau_g][(1 - \mu\cos\omega\tau_w) + i\mu\sin\omega\tau_w]} \tag{5.40}$$

The method of solution is to note that b_{lim} is real. So if the imaginary part of the right-hand side of Eq. 5.40 is equated to zero, it is possible to search for values of ω that satisfy the equation. If these values are substituted in Eq. 5.40, real solutions for b_{lim} may be obtained. This may seem trivial, but it takes much computation, especially to find the solutions that involve a near-integer number of waves on the wheel. The latter are often termed "wheel chatter" and the other solutions "work chatter".

It will be found that there may be complex interactions between waves on the wheel and the work. To avoid these, for reasons of clarity, some specific conditions have been chosen that result in relatively "clean" results. Figure 5.19 shows the stability chart for work speed for the parameter values shown on the graph. The overlap factor is 0.5 and some lobes for "work chatter" are shown. The envelope of these lobes should be compared with the simplest case result that was calculated for the same parameters but with only work regeneration while plunge grinding ($\mu = 1.0$). As might be expected, because the regenerative effect has been reduced ($\mu = 0.5$), the stability boundary has increased.

There are several envelopes for wheel chatter that correspond to different numbers of waves around the periphery of the wheel. These solutions are difficult to find when searching for the values of ω that give real solutions to Eq. 5.40. As noted previously in Sect. 5.3.2, the waves precess around the wheel circumference and the value of ω is just less than that required for an integer number of waves. For the results shown in Fig. 5.19, the search frequency step required was 10^{-8} times the natural frequency. When searching for solutions for "work chatter", the frequency step was 0.0005 times the natural frequency. It is therefore not possible to have an online program that can be used for both wheel and work chatter, rather two programs are provided. If the programs are used, it will be found that they run slowly and not all results are found.

A stability chart against wheel speed is shown in Fig. 5.20. For clarity, the envelopes for work chatter are not shown in this figure but in a separate one, Fig. 5.21. Figure 5.20 presents results for overlap factors of 1.0 (plunge grinding) and zero (e.g. roll grinding). It is important to note, that for realistic wheel speeds, the stability boundary is much higher than that predicted from the simple model where the wheel was "machined" by a metal cutting tool (Sect. 5.3.1 and Fig. 5.6). This is because the work is not a metal cutting tool and provides a contact stiffness (Sect. 5.3.3).

Figure 5.21 shows a stability chart against wheel speed but with only the envelopes for work chatter shown. These envelopes all tend to zero for low (and unrealistic) wheel speeds. This is because the cutting force model is inversely proportional to the surface speed of the wheel.

For the purposes of eliminating chatter, it is important to note that all the stability envelopes are crucially dependent on the maximum negative in-phase component of the machine response. The question arises as to whether this is generally true, and therefore, it is necessary to re-examine the assumptions made in the models used. However, before considering the assumptions, it is important

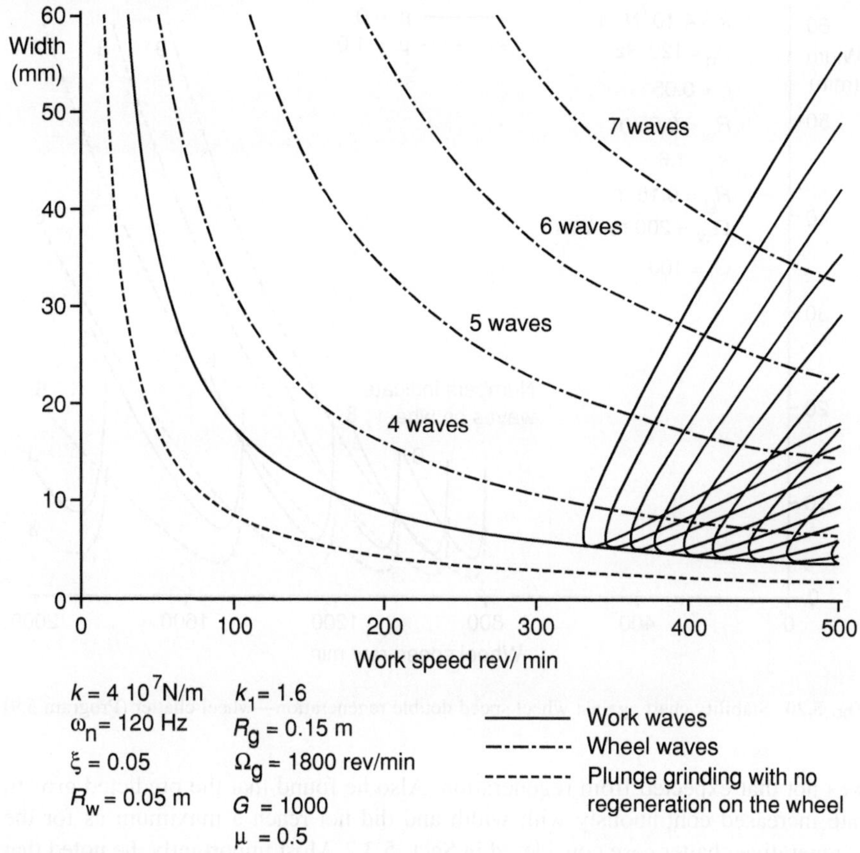

Fig. 5.19 Stability chart against work speed with double regeneration (Program 5.7—work chatter and Program 5.8—wheel chatter)

to discuss a possible source of chatter that appears not to have received much attention. Entwistle and Stone [33] have suggested that non-regenerative chatter may occur in grinding.

5.3.8 Non-regenerative Chatter?

In his doctoral thesis, Entwistle [1] obtained some solutions to his equations that were unexpected. These he labelled "Arnold Chatter" (i.e. non-regenerative chatter) and to quote, *In all cases the stability boundary is formed by a chatter mode which occurs at the torsional natural frequency of the grinding wheel degree of freedom. The type of chatter involved is that described by Arnold, labelled 'Type B' by Tobias and Fishwick.* He drew this conclusion because the frequency

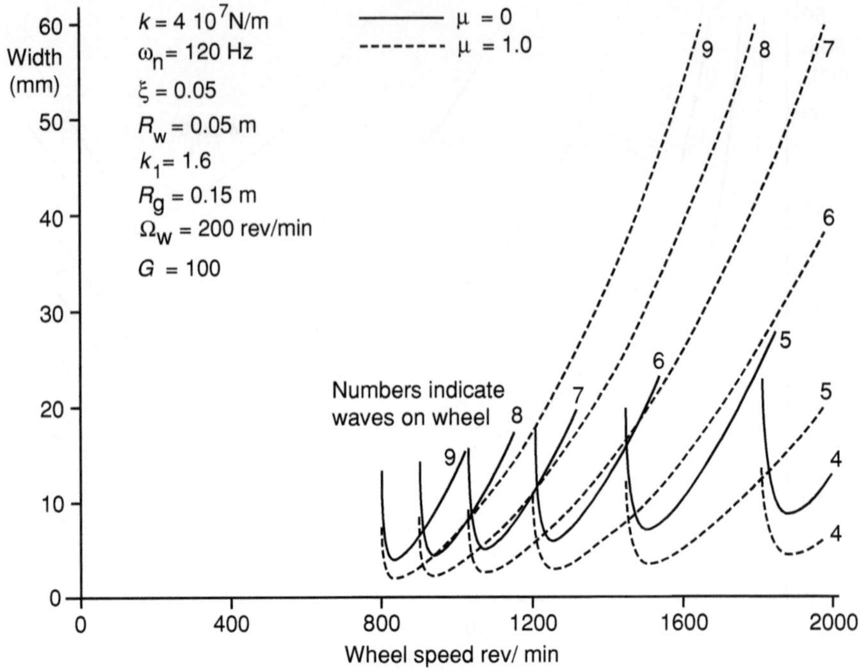

Fig. 5.20 Stability chart against wheel speed double regeneration—wheel chatter (Program 5.9)

was not that expected from regeneration. Also he found that the predicted growth rate increased continuously with width and did not reach a maximum as for the regenerative chatter case considered in Sect. 5.3.2. Most importantly, he noted that the mechanism of Arnold chatter in metal cutting, described in Sect. 1.2, depends on a force variation with speed that has a negative slope. The force model he used and that has been described in Sect. 5.2 has such a negative slope when force is plotted against wheel speed. Entwistle was not able to pursue any practical investigation of this form of chatter. However, over the years at conferences, questions have been raised about a so-called high-frequency chatter.

An invitation to write a paper [33] on chatter in grinding raising fundamental issues yet to be resolved led to a reconsideration of the possibility of non-regenerative chatter in grinding. Three possible types of non-regenerative chatter were considered [33], involving, respectively, torsion of the wheel, torsion of the work and transverse vibration with an inclined mode. Torsion of the wheel is the more likely in practice and needs to be considered.

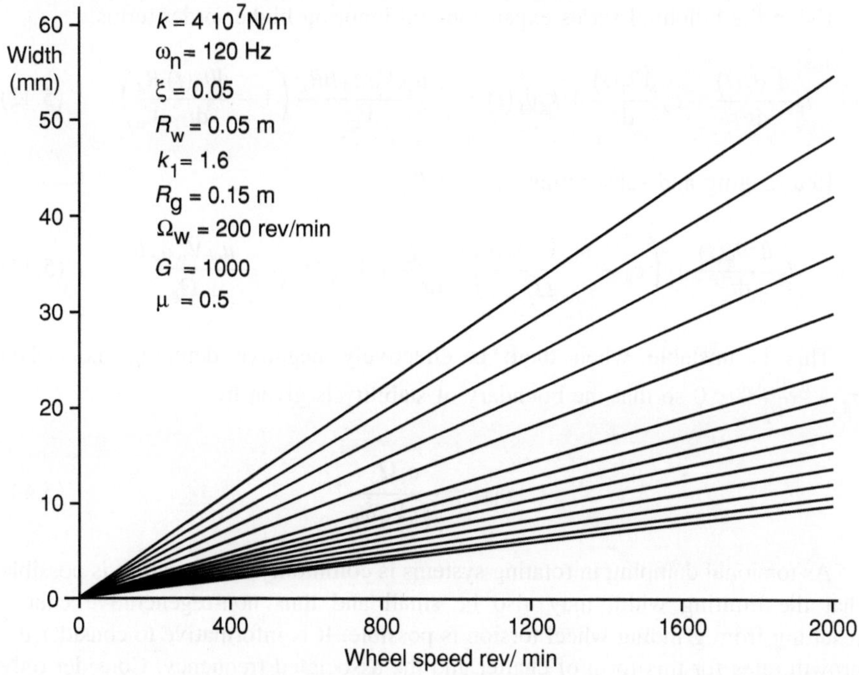

Fig. 5.21 Stability chart against wheel speed double regeneration—work chatter (Program 5.10)

5.3.8.1 Wheel Torsion

Following the practice of previous sections just one effect will be considered. Torsional vibration of the wheel alone will be assumed to occur, and it will also be assumed that the work speed remains constant, there is no transverse vibration, and there are no waves left on either the work or wheel surfaces. The equation of motion for the wheel torsional mode is then, since only $V_g(t)$ and $\theta_g(t)$ are varying, given by,

$$I_g \frac{d^2\theta_g(t)}{dt^2} + c_g \frac{d\theta_g(t)}{dt} + k_g\theta_g(t) = -P_t R_g = -\frac{u_{ch}\bar{V}_w\bar{\delta}_w b R_g}{V_g(t)} \tag{5.41}$$

where $V_g(t) = \bar{V}_g + \frac{d\theta_g(t)}{dt}R_g = \bar{V}_g\left(1 + \frac{d\theta_g(t)}{dt}\frac{R_g}{\bar{V}_g}\right)$ and hence,

$$I_g \frac{d^2\theta_g(t)}{dt^2} + c_g \frac{d\theta_g(t)}{dt} + k_g\theta_g(t) = -\frac{u_{ch}\bar{V}_w\bar{\delta}_w b R_g}{\bar{V}_g}\left(1 + \frac{d\theta_g(t)}{dt}\frac{R_g}{\bar{V}_g}\right)^{-1}$$

Using the binomial series expansion and ignoring higher-order terms,

$$I_g \frac{d^2\theta_g(t)}{dt^2} + c_g \frac{d\theta_g(t)}{dt} + k_g\theta_g(t) = -\frac{u_{ch}\bar{V}_w\bar{\delta}_w b R_g}{\bar{V}_g}\left(1 - \frac{d\theta_g(t)}{dt}\frac{R_g}{\bar{V}_g}\right) \tag{5.42}$$

Rearranging and substituting $\bar{V}_g = \bar{\Omega}_g R_g$

$$I_g \frac{d^2\theta_g(t)}{dt^2} + \left(c_g - \frac{u_{ch}\bar{V}_w\bar{\delta}_w b}{\bar{\Omega}_g^2}\right)\frac{d\theta_g(t)}{dt} + k_g\theta_g(t) = -\frac{u_{ch}\bar{V}_w\bar{\delta}_w b}{\bar{\Omega}_g} \tag{5.43}$$

This is unstable when there is effectively negative damping, i.e. when $c_g - \frac{u_{ch}\bar{V}_w\bar{\delta}_w b}{\bar{\Omega}_g^2} < 0$ so that the boundary of stability is given by

$$b_{lim} = \frac{c_g\bar{\Omega}_g^2}{u_{ch}\bar{V}_w\bar{\delta}_w} \tag{5.44}$$

As torsional damping in rotating systems is commonly very small, it is possible that the limiting width may also be small and thus non-regenerative chatter resulting from grinding wheel torsion is possible. It is informative to consider the growth rates for this form of chatter and the associated frequency. Consider only the varying components of Eq. 5.43 and substitute $\theta_g(t) = \Theta_g e^{(\sigma+i\omega)t}$ to obtain

$$I_g(\sigma + i\omega)^2 + \left(c_g - \frac{u_{ch}\bar{V}_w\bar{\delta}_w b}{\bar{\Omega}_g^2}\right)(\sigma + i\omega) + k_g = 0$$

Grouping real and imaginary terms yields two equations

$$I_g(\sigma^2 + \omega)^2 + \left(c_g - \frac{u_{ch}\bar{V}_w\bar{\delta}_w b}{\bar{\Omega}_g^2}\right)\sigma + k_g = 0 \tag{5.45}$$

$$I_g 2\sigma\omega + \left(c_g - \frac{u_{ch}\bar{V}_w\bar{\delta}_w b}{\bar{\Omega}_g^2}\right)\omega = 0 \quad \text{so that } \sigma = -\left(c_g - \frac{u_{ch}\bar{V}_w\bar{\delta}_w b}{\bar{\Omega}_g^2}\right)/2I_g \tag{5.46}$$

Thus, the growth rate becomes positive when the limiting width is exceeded and increases linearly with increasing width. At the stability boundary, the frequency of the vibration is from Eq. 5.45 with $\sigma = 0$ given by, $\omega = \sqrt{k_g/I_g}$ which is the undamped natural frequency of torsional vibration of the wheel. With increasing width, the growth rate rises linearly as shown in Eq. 5.46 and the frequency of vibration reduces. Substituting Eq. 5.46 in Eq. 5.45 gives the frequency of vibration as,

$$\omega^2 = \frac{k_g}{I_g} - \frac{1}{4I_g^2}\left(c_g - \frac{u_{ch}\bar{V}_w\bar{\delta}_w b}{\bar{\Omega}_g^2}\right)^2 \tag{5.47}$$

It is interesting to note that for the assumptions made above the frequency of any torsional vibration of the wheel will be either at the undamped natural frequency (at the stability boundary) or at a lower frequency.

5.3.8.2 Work Torsion

Entwistle [33] also considered the possibility of non-regenerative chatter arising from torsion of the work. In Sect. 5.3.6, concerning torsional vibration of the work, it was shown that this may improve chatter performance by effectively adding damping. It was concluded that if the direction of the work speed was reversed, the effect of torsion would be different. In fact, Entwistle [33] has shown that reversing the direction of the work to be opposite to that of the wheel may lead to non-regenerative chatter as it effectively introduces negative damping. The simplest illustration of this occurs by assuming torsional vibration of the work alone with constant wheel speed. There is no transverse vibration and there are no waves left on either the work or wheel surfaces. Equation 5.25 then becomes,

$$I_w\frac{d^2\theta_w(t)}{dt^2} + c_w\frac{d\theta_w(t)}{dt} + k_w\theta_w(t) = P_t R_w \tag{5.48}$$

Note that the sign of the applied torque has been reversed to be positive. Substituting for P_t from Eq. 5.1

$$I_w\frac{d^2\theta_w(t)}{dt^2} + c_w\frac{d\theta_w(t)}{dt} + k_w\theta_w(t) = \frac{u_{ch}V_w R_w\bar{\delta}_w b}{\bar{V}_g}$$

and the work surface speed is given by

$$V_w = \bar{V}_w + \theta_w R_w = \bar{V}_w + \frac{d\theta_w(t)}{dt}$$

so that

$$I_w\frac{d^2\theta_w(t)}{dt^2} + c_w\frac{d\theta_w(t)}{dt} + k_w\theta_w(t) = \frac{u_{ch}R_w\bar{\delta}_w b}{\bar{V}_g}\left(\bar{V}_w + \frac{d\theta_w(t)}{dt}R_w\right)$$

and rearranging

$$I_w\frac{d^2\theta_w(t)}{dt^2} + \left(c_w - \frac{u_{ch}R_w^2\bar{\delta}_w b}{\bar{V}_g}\right)\frac{d\theta_w(t)}{dt} + k_w\theta_w(t) = \frac{u_{ch}\bar{V}_w R_w\bar{\delta}_w b}{\bar{V}_g}$$

The right-hand side controls the mean values. The left-hand side indicates that negative damping and instability will occur when

$$c_{\mathrm{w}} - \frac{u_{\mathrm{ch}} R_{\mathrm{w}}^2 \bar{\delta}_{\mathrm{w}} b}{\bar{V}_{\mathrm{g}}} = 0$$

and rearranging and substituting from the previous definitions

$$b_{\mathrm{lim}} = \frac{c_{\mathrm{w}} \bar{V}_{\mathrm{g}}}{u_{\mathrm{ch}} R_{\mathrm{w}}^2 \bar{\delta}_{\mathrm{w}}} \tag{5.49}$$

As with the wheel torsion case, the frequency at the stability boundary will be the torsional natural frequency. The level of damping may be very small as this is often the case for torsion and so non-regenerative chatter may occur for small widths. However, there is a simple solution: reverse the direction of rotation of the work.

5.3.8.3 Transverse Vibration

The final possibility considered by Entwistle [33] was of non-regenerative chatter arising from transverse vibration of an inclined mode. The effects of an inclined mode have been discussed in Sect. 5.3.5, where the example of plunge grinding was investigated. Entwistle found that when roll grinding with no overlap, it was theoretically possible for non-regenerative chatter to arise. Regeneration on the wheel was not included in the model. As with the non-regenerative examples above, the growth rate rose linearly with width of cut. However, for realistic parameter values, the limiting width was large and so in practice, such chatter would be unlikely. It was also found that when even a small overlap was included, regeneration on the wheel predominated. This is why there was no evidence of non-regenerative chatter when the inclined mode was considered previously.

5.3.9 Review of Assumptions

The assumptions made in developing the simplified physical model are the same as those of Entwistle [1] and his comments are reproduced here with some modifications.

In general, attention is focused on the conditions at the stability boundary (neutral stability) where it is assumed that an increase in the width of the cutting zone will lead to unstable amplitude growth and a decrease will lead to the decay in amplitude of any existing oscillations.

Given that the force that excites the degrees of freedom is itself a function of any resulting motions (Eq. 5.2), the differential equations forming the model will be nonlinear. Furthermore, due to such phenomena as wear and grinding wheel metallic loading, the system will be time variant. It was assumed that the time variation in parameters is slow and can therefore be approximated by steady-state values. In relation to nonlinearities, Liapounov has shown [4, 26, 27] that in a small enough neighbourhood to critical (or equilibrium) points, the phase portraits of the nonlinear and linearised differential equations will be qualitatively similar. Hence, in this chapter, the solutions to the equations have been assumed to be complex exponentials as though the equations were linear.

In all cases, the amplitudes of vibration were considered to be small and variations were linearised about their operating point.

Though the torsional vibration of the work will cause changes in the surface speed of the work, it was implicitly assumed that the amplitudes were sufficiently small to not significantly affect the rotational period of the work.

The geometry of the contact zone is complex as the work and the wheel both possess local compliance and some mutual conformity exists. It was assumed that the contact zone could be considered to be a line contact. The effects of the geometry in the cutting zone are implicit in any grinding force model, and hence, this simplification is unlikely to affect the essential dynamic behaviour of the modelled system.

It was assumed that the compliance of the contact zone could be represented by a contact stiffness proportional to the width of cut.

The response of the machine structure was assumed to be that of a single mode. In reality, such systems will have multiple degrees of freedom. However, it is not unusual for one mode to have a larger dynamic response than others. Similarly, the torsional response of the work and wheel was also assumed to be that of a single mode.

It was assumed that the waves left on both the work and the wheel were not attenuated by any form of interference or deformations. Such attenuation is a function of amplitude and may not be active at the very small amplitudes present at the onset of chatter. It should be noted that the overlap factor has essentially the same effect as wave attenuation on the work and so could be studied if thought necessary.

With so many assumptions involving approximations, it might be considered that the real situation would be very different to the simplified models used in this chapter. However, most of the major effects that have been modelled are found to exist in practice.

5.4 Experimental Results

There are numerous papers with results from experiments involving chatter in grinding, and some of these are reviewed by Inasaki [2]. However, there is a problem in making use of many of the results of previous research outside the specific operating conditions under which they were measured. Not all researchers seem to be aware of all of the significant parameters of principal significance considered in this chapter (and there may well be other additional parameters). Some parameters were therefore not measured and are not reported. As an example, it is hard to find a publication where the presence of torsional vibration was investigated. Also the condition of the wheel is continuously changing and its state is never exactly known. Added to this is the uncertainty about the machine response under working conditions which may change with time, as for example when traversing in roll grinding.

A search of the literature will find that structural improvements that reduce the maximum negative in-phase component of the chatter receptance are always effective. In general, the conclusions from the simple model of Sect. 5.3.1 are found to apply: reduce the chip thickness and/or the work speed and increase the wheel speed. Grinding wheels with reduced hardness are also found to be less likely to chatter because of the contact stiffness effect. However, the contact stiffness increases with the width of cut. The solution proposed in Sect. 5.3.4 was the use of flexible grinding wheels with a stiffness that does not depend on the width of cut. These wheels were predicted to improve the chatter receptance for the whole machine, work and wheel system. Thus, these prevent regenerative chatter even when the regeneration is caused by waves on the work.

5.4.1 Flexible Grinding Wheels

Great improvements to chatter performance have been reported when using flexible grinding wheels. In Sect. 5.3.4, the predicted improvements from using flexible grinding wheels were considered. The initial experimental work was done by Sexton [28, 29], and further work is reported by Bzymek [30]. It needs to be stressed that the theoretical background to using flexible tools (see Sect. 4.4) is important. It is not simply a question of introducing a flexibility, but also of not adding a new low-frequency mode. If a grinding wheel is mounted on a flexible hub, it will not be effective in stopping chatter, but is likely to make it worse. The mass between the cutting/grinding contact and the flexibility must be as small as possible. Thus, so that chatter is less likely to occur, any additional mode of vibration that is the result of the introduced flexibility should have a high natural frequency.

The tests described by Sexton [28, 29] relate to CBN wheels. For such a wheel, there is usually only a thin layer containing the abrasive grits and this is of the order of 3 mm thickness. Initially, [28] the design used an abrasive rim mounted on a rigid hub via a flexible coupling. The rim was then able to vibrate relative to the hub and since the mass of the rim was low, a high natural frequency was maintained. The flexibility and damping could have been provided in several ways. The one chosen was to mount the CBN rim on neoprene rubber pads used in shear

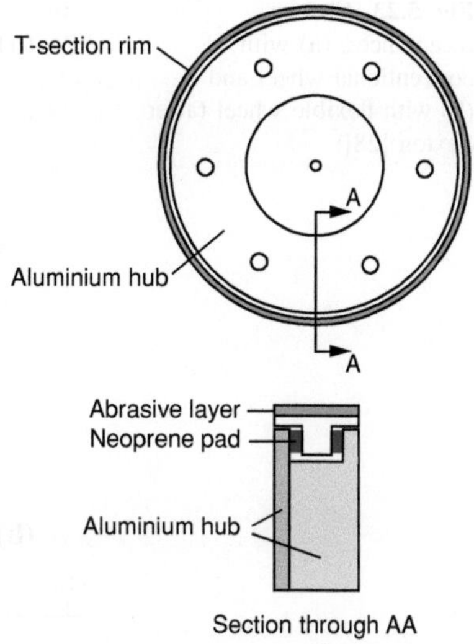

Fig. 5.22 Flexible wheel design, after Sexton [28]

since it was found that their damping was much higher in shear than in compression.

The final flexible wheel design is shown in Fig. 5.22. The neoprene pads were circular in cross section and the flexibility could be adjusted simply by changing the number of the pads. The rim was manufactured from a conventional wheel and as such, the "T" section was of Bakelite. From the measured mass of the rim and the stiffness and damping characteristics of the neoprene pads, it was predicted that a natural frequency, involving the rim on the rubber, of 550 Hz could be achieved with forty pads and that the damping ratio would be approximately 0.2. The flexibility provided by the pads would then be 2×10^{-7} m/N and this, for most grinding machines, would give a significant shift of the chatter receptance in the positive in-phase direction. On mounting the wheel hub "rigidly" and exciting the rim with conventional excitation equipment, it was found that the wheel frequency was 526 Hz, the damping ratio 0.2, but the static flexibility was 6×10^{-7} m/N. The additional flexibility arose from the rim not behaving as a rigid body and there was also some local deflection of the wheel in the zone where the force was applied. The net effect was that the wheel, when mounted on a machine, exhibited a flexibility greater than that supplied by the neoprene pads but having a natural frequency commensurate with the lower flexibility of the pads, rather than the higher flexibility of the total system.

Since, ultimately, the performance of the flexible CBN wheel was to be compared with a conventional CBN wheel, it was necessary to ensure, as far as possible, that the wheels were identical in all respects other than the modified flexibility. Thus, two nominally identical wheels were used, one of which was later modified to increase its radial flexibility. The wheels were resin bond with Bakelite hubs and contained De Beers ABN 360, a cubic boron nitride grit, nickel clad to 60 % by weight for use with resin bonds. The grit size was 120/140 US mesh, and

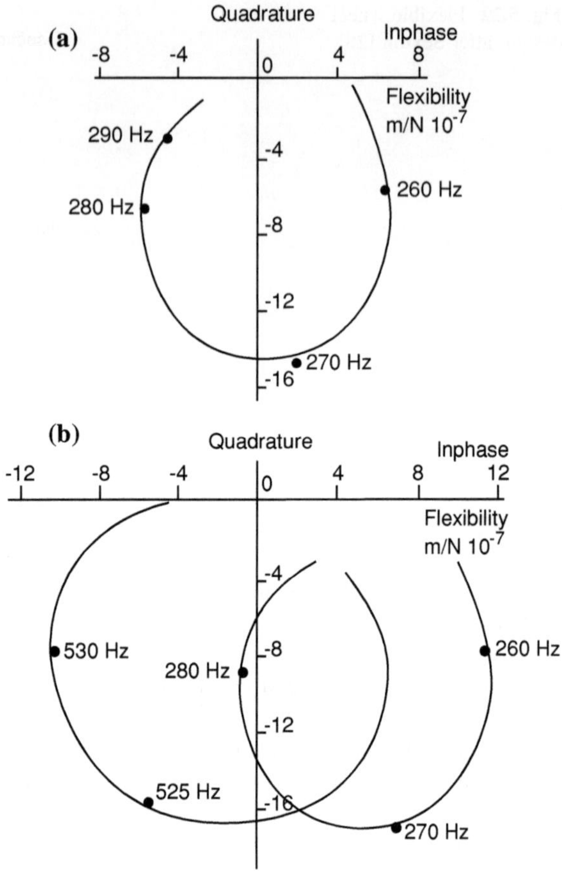

Fig. 5.23 Chatter receptances, (**a**) with conventional wheel and (**b**) with flexible wheel (after Sexton [28])

the concentration was 75 (75 concentration = 18.75 % by volume). The overall dimensions of the wheels were 300 mm diameter by 30 mm wide with the resin bond matrix, containing the abrasive, 3 mm deep.

Before modifying one of the wheels, both were used to grind a hardened EN31 work-piece, and it was found that there was no significant difference with respect to the surface finish, vibration levels or development of lobes on the wheel periphery. It was concluded that both wheels were nominally identical and, should any improvement in performance be found with the modified wheel, the improvement would result from the modification and would not be due to any inherent difference between the original wheels. The flexible wheel characteristics (with hub held "rigidly") were measured as $\omega_n = 526$ Hz, $k = 1.7 \times 10^6$ N/m and $\xi = 0.2$. The unmodified wheel characteristics were measured as $\omega_n = 1{,}850$ Hz, $k = 15 \times 10^6$ N/m and $\xi = 0.1$.

The grinding trials were carried out on a Churchill OX cylindrical grinder. The work was a 300-mm-long tube of hardened EN31 of external diameter 135 mm and wall thickness 15 mm. The vibration characteristics of the grinding machine, work and wheel were measured, and the chatter receptances (including the machine response) obtained for both the conventional and the flexible wheel are shown in Fig. 5.23. It is evident that the likely chatter frequency with the conventional wheel is around 280 Hz and the maximum negative in-phase component is -6.0×10^{-7} m/N. For the modified wheel, the desired effect on the total machine response has been achieved and the machine mode has been moved to give a maximum negative in-phase component of -1.0×10^{-7} m/N again at around 280 Hz. The wheel mode shows a significant loop in the response but is unlikely to cause chatter because of the high frequency, i.e. 530 Hz. For the roll grinding tests, the wheel speed was 1,510 rev/min, the work speed 67 rev/min, the traverse rate 670 mm/min, and the in-feed was 2.5 μm at each end of the work.

After approximately 30 min, it was evident that the conventional wheel was chattering. This was initially indicated by the very distinctive sound of chatter and confirmed by the vibration and force signals. The test was stopped and a circularity of the wheel measured with the wheel in situ. It was found that definite lobes, eleven in number, had formed around the wheel periphery, the crest to valley height of the lobes being approximately 5 μm. The grinding test was then continued under the same grinding conditions. After a further hour, the chatter vibration had reached such a level as to render further grinding impossible and was stopped. When the flexible wheel was used in this test, it showed no tendency to chatter at all after grinding under the same conditions and for the same length of time.

It was concluded that flexible grinding wheels could be used to prevent chatter from starting, although the experimental design was not suitable for mass production. There was also a concern about the long-term stability of the neoprene pad material. The next design described by Sexton [29] used a hub material that is not as dense as used for conventional wheels. The material that was chosen was "Retimet" manufactured by Dunlop Aviation who had been promoting its use in the aerospace industry for some years. Retimet, a flexible, porous, nickel material, had the appearance of aerated plastic (foam). The increased flexibility, compared to the non-aerated version, arose from the cellular structure. The material was available in a number of grades, based on its porosity, and it had both high damping and a high thermal conductivity. While there was no information on its stiffness properties, it was considered a likely candidate material, allowing a certain amount of rim deflection and thus promoting stable grinding. A resin bond CBN wheel was manufactured using Retimet as the hub material. It was 175 mm in diameter, 9 mm wide and contained 100/120 US mesh CBN at 100 concentration. The radial static stiffness of the wheel was measured to be approximately 1.4×10^4 N/m per mm width. Thus, the stiffness for the full 9 mm width was 1.26×10^7 N/m. Apart from the choice of hub material, the wheel was of conventional construction, with both its natural frequency and damping being high.

The performance of this wheel was compared with a conventional CBN wheel whose parameters were identical to the Retimet wheel excepting that its hub was of Bakelite. The radial stiffness of this wheel was approximately 4.2×10^6 N/m per mm width. Thus, the stiffness for the full 9 mm width was 3.78×10^7 N/m.

While Sexton did not report the effective flexibility of the wheel (that in the analyses above is in series with the contact stiffness), following the notation of Sect. 5.3.4, we have for this conventional wheel,

$$\frac{1}{k_c b} = \frac{1}{3.78 \times 10^7}\ \mathrm{N/m} \tag{5.50}$$

and assuming the same contact stiffness for the flexible wheel, we have

$$\frac{1}{k_f} + \frac{1}{k_c b} = \frac{1}{1.26 \times 10^7}\ \mathrm{N/m} \tag{5.51}$$

Thus subtracting Eq. 5.50 from Eq. 5.51,

$$\frac{1}{k_f} = \frac{1}{1.26 \times 10^7} - \frac{1}{3.78 \times 10^7}\ \mathrm{m/N} \quad \text{so that } k_f = 1.89 \times 10^7\ \mathrm{N/m}$$

Grinding tests were performed using the conventional CBN wheel. After grinding M2 high-speed steel for six hours on a Jones and Shipman 540 surface grinder with a wheel speed of 35 m/s and a cross-feed of 1.7 mm, chatter was clearly present. This performance was then compared to the Retimet wheel. It ground under identical conditions for 12 hours with no evidence of chatter.

It might be thought that, by employing a wheel with high radial flexibility, it would not be possible to produce geometrically accurate work-pieces and that the wheel would require substantially longer to spark-out. In fact, there was no problem with the geometric accuracy of the work and the spark-out of the flexible wheel was, in fact, quicker than that of a conventional wheel.

To further prove the effectiveness of reducing the radial flexibility of a wheel and the suitability of using Retimet for the hub material, a second larger wheel was manufactured using Retimet. This wheel was 250 mm in diameter and 12 mm wide and trials with it confirmed that it was equally as successful as the smaller Retimet wheel.

In order to illustrate further how low the flexibilities of the two Retimet wheels were in comparison with conventional ultra-hard wheels, the radial stiffnesses of a large number of wheels were measured. The radial stiffness of wheels with conventional hub materials, such as phenolic aluminium and Bakelite, fell in the range of 4–10 $\times$ 10^6 N/m per mm wheel width.

It should not be thought that Retimet is the only hub material that, by reason of its low flexibility, promotes stable grinding. In fact, Sexton [29] also manufactured a CBN wheel with a chipboard (glue-bonded timber-particle board) hub and this, although not a conventional engineering material, proved very successful because of its high flexibility and damping.

The question must be asked as to why such flexible wheels are not more widely used. The suspicion is that the need for high natural frequencies has not been rightly understood and wheels mounted on flexible hubs, but with low natural

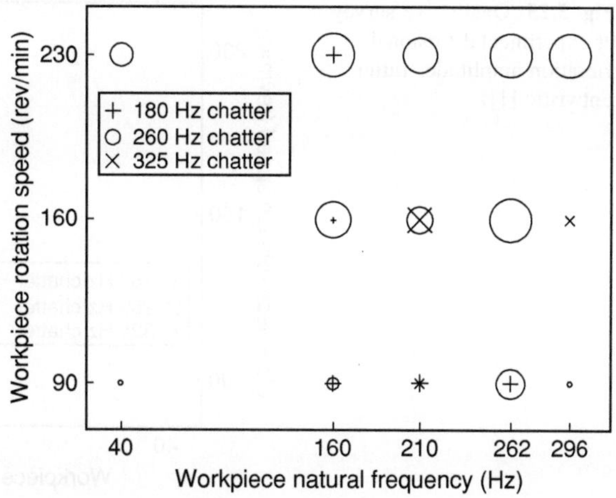

Fig. 5.24 Qualitative survey of experimental transverse vibration amplitudes (after Entwistle [1])

frequencies, make matters worse. Also Retimet may not have been the best material for the mass production of such wheels.

5.4.2 Torsional Flexibility

This is perhaps the area of most contention. Torsional effects in metal cutting were the matter of much debate in the 60s and 70s but were generally agreed to be not significant. The suspicion is that it has been assumed that the same would apply to grinding. However, as described in Sect. 5.3.6, there is in theory the possibility of a significant effect arising from torsion of the work and its drive. It appears that Entwistle [1, 31, 32] alone has conducted research on this matter. For his doctoral thesis, he went to great lengths to have several grinding set-ups, where the only difference was the torsional characteristic of the work/drive system. He was also able to change the torsional natural frequency while not changing the transverse response. His tests involved plunge grinding and measuring the transverse and torsional spectra after grinding for 1,000 s with a feed rate of approximately 7.5 μm/rev. The grinding wheel speed was 3,870 rev/min.

Figures 5.24 and 5.25 show some of his results. For each combination of work torsional natural frequency and work speed, a symbol is plotted, the size of which is proportional to the approximate dB magnitude of the vibration. Figure 5.24 shows the amplitudes of transverse vibration and Fig. 5.25 the amplitudes of torsional vibration. A different symbol is used to represent each of the three chatter frequencies that were found to be present. Now if the torsional characteristics of the work have no effect, it would be expected that for each of the three work speeds (*y* axis), the results would not change with the torsional natural frequency (*x* axis). However, it is clear that they do and the greatest change is at the lowest torsional frequency (40 Hz) and the lower work speeds. The simplified model used

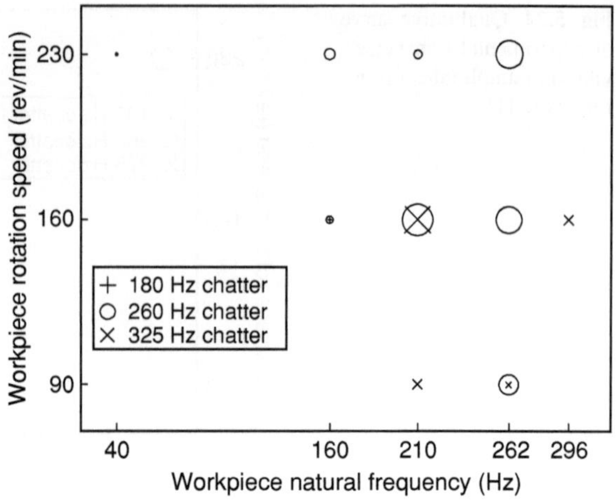

Fig. 5.25 Qualitative survey of experimental torsional vibration amplitudes (after Entwistle [1])

to investigate work torsion (Sect. 5.3.6) indicated that those would be the significant parameters.

The theoretical model also predicted that any torsional effect would depend on the in-feed. The value used was high (7.5 μm/rev), and thus, the effects may have been magnified compared to normal practice. In his conclusions, Entwistle [1, 32] states, *Recapitulating, the aim of the experimental programme was to determine whether the presence of workpiece torsional compliance influenced the chatter characteristics of a plunge grinding system operating beyond the limit of stability. The results summarised in this paper/thesis show that the torsional compliance does alter the chatter performance of the system and, in a small sub-set of cases tested, suppressed chatter growth completely. This is an observation which has not been previously reported in the literature.*

It seems that torsional vibration of the work could suppress chatter but more work needs to be done to confirm this and the particular conditions that might achieve this.

5.5 Centreless Grinding

In contrast to the grinding operations considered above, centreless grinding has received relatively little attention, particularly with respect to chatter [34–37]. Figure 5.26 shows a representation of a typical centreless plunge grinding set-up. One of the main advantages of the process is that round work-pieces may be achieved from initially non-round stock. The major difference, compared to conventional grinding, is that the work is effectively supported at three contacts and it is possible to have regeneration effects at each of these contacts. Most of the early work on instability in centreless grinding focused on what is termed "geometric instability" where lobes (i.e. surface waves) were produced on the work. These

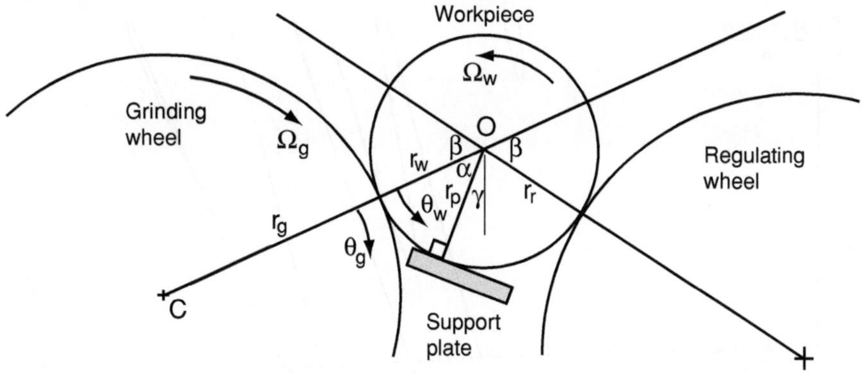

Fig. 5.26 Model of centreless grinding

Fig. 5.27 Dynamic model of structure and grinding wheel (after Pearce [39])

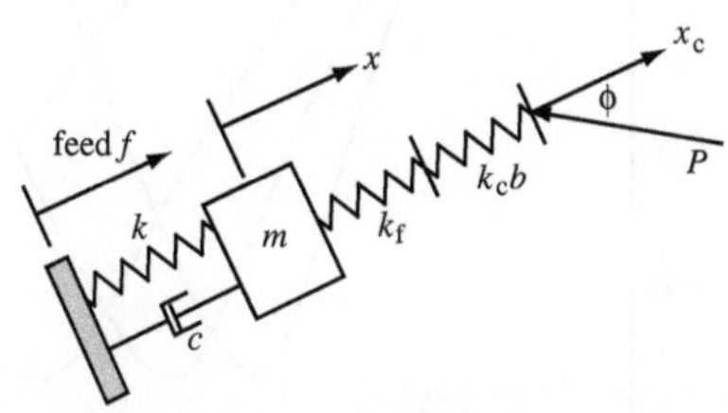

lobes were often at a frequency much less than any natural frequencies of the structure. However, there were also examples of higher-frequency instability and this was often called chatter and regarded as distinct from geometric instability. An excellent literature review is given in the paper by Li [38].

In some recent papers, Pearce [39, 40] has shown that chatter and geometric instability are governed by the same equations and seamlessly merge. Also it was shown that flexible grinding wheels could make the process more unstable. As this is the first example of such wheels making matters worse, it is necessary to consider why. The structural model used by Pearce follows the philosophy in this book wherein the simplest model, that allows significant parameters to be investigated, should be used. Thus, the dynamic characteristics were represented by a single mode with both a flexible wheel and contact stiffness as shown in Fig. 5.27.

The method of modelling was the same as for growth rates, Sect. 5.3.2. The rate of decay of any initial out-of-roundness of the work [41] is as important in centreless grinding as any instability. As there are three contacts with the work, the model is very complex particularly as the normal to each contact is in a different direction. After a great deal of mathematics, Pearce [39] showed that it was possible to produce a stability chart of the type found for conventional grinding. An example is shown in Fig. 5.28. The axes are in non-dimensional terms but are effectively the width at the boundary of stability against work speed. Each of the curves relates to a number of waves on the work. It was found, for the conditions

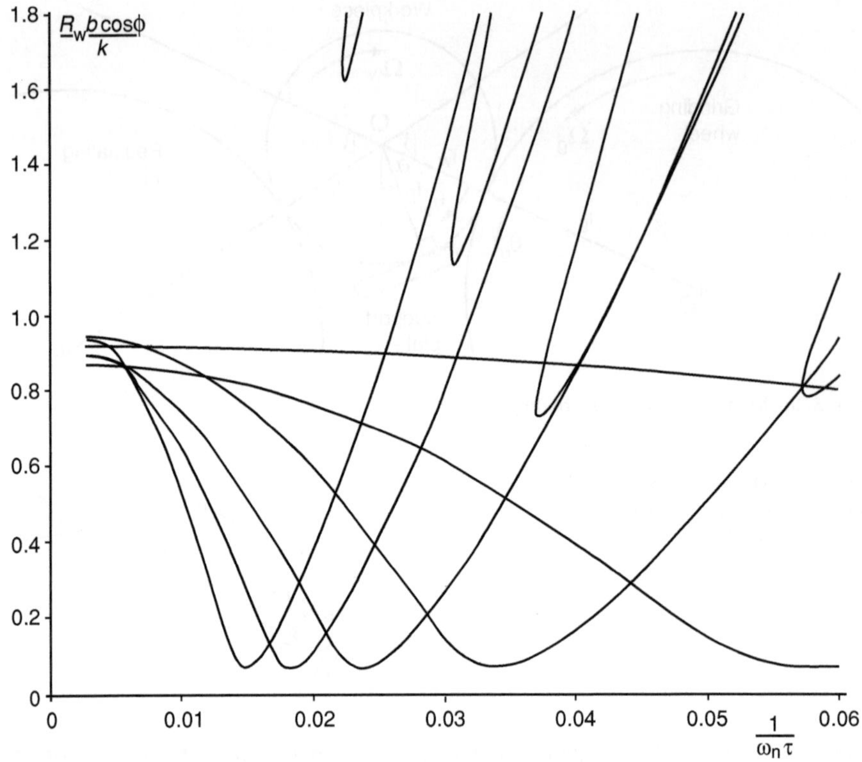

Fig. 5.28 Typical stability chart for centreless grinding (after Pearce [39])

used to generate Fig. 5.28, that only odd number stability lobes extended to low speeds (geometric instability). However, under different conditions, it was only the even number that did so. This is because of the phasing introduced by the distribution of the contact points around the periphery of the work. It should also be noted that the stability lobes are in some respects similar to those for conventional grinding. The smallest unstable widths are controlled by the dynamic response of the system. However, the maximum negative in-phase component is not necessarily the determining factor for the smallest unstable width.

Pearce [39, 40] investigated the significance of the many parameters that influence the process. The graphical approach of Gurney [14] gave many insights into why each parameter was significant. This was particularly true of the effect on chatter when using a flexible wheel. At the boundary of stability, there was assumed to be a constant-amplitude wave on the surface of the work. As for conventional grinding, this wave causes a regeneration effect at the contact with the wheel. However, the surface wave also causes the centre of the work to move when it contacts the support plate and also the regulating wheel. As the work centre is moved, the chip thickness varies and there are regeneration effects caused

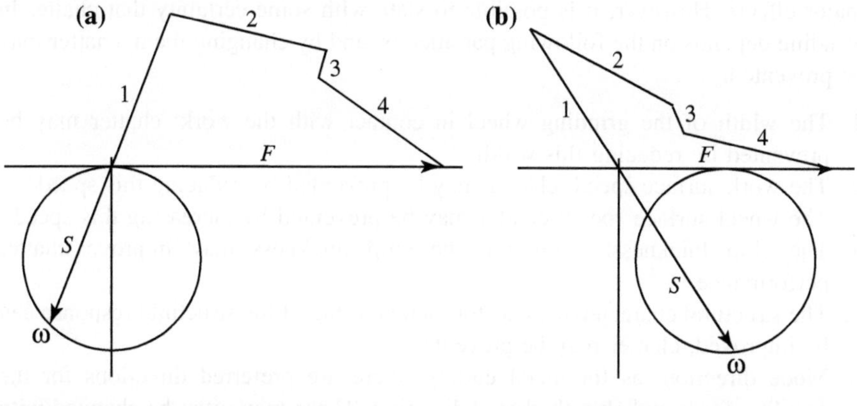

Fig. 5.29 Response locus and force vectors for (**a**) without a flexible wheel and (**b**) with a flexible wheel (after Pearce [40])

by the two extra contacts. These regeneration effects have different phases and magnitudes depending on the location and angles of the support plate and also of the regulating wheel.

If a response locus is drawn with the associated force vectors, the effect of wheel flexibility can be understood. Figure 5.29a shows the response locus without wheel flexibility. Figure 5.29b shows the response locus shifted to the right as a result of the wheel flexibility. There are four force vectors on each diagram: the non-regenerative force (number 1), and the three regenerative forces caused by each of the three contacts; the contact with the wheel (number 2); the contact with the support plate (number 3); the contact with the regulating wheel (number 4). It is apparent that chatter is not restricted to negative in-phase values of the response. This is a major departure from all the regenerative chatter theories developed previously. In the case of centreless grinding, moving the response locus in the positive real direction (by using a flexible wheel) does not necessarily improve matters but may make the process more unstable.

5.6 Conclusions

Simplified models have been used to illustrate many of the major effects in grinding chatter. If all the effects investigated are included in a single model and along with torsion of the wheel and multiple modes in different directions, then the number of parameters becomes overwhelming. Also it is possible to have more complex models of the grinding force [20, 42] and these add further to the number of parameters.

The possibility of torsional vibration of the wheel was considered by Entwistle [1] and Hesterman [43] included the torsional flexibility of a flexible wheel itself. As the models get more complex, it becomes increasingly difficult to determine the major effects. However, it is possible to state with some certainty that chatter in grinding depends on the following parameters, and by changing them, chatter may be prevented.

1. The width of the grinding wheel in contact with the work: chatter may be prevented by reducing this width.
2. The work surface speed: chatter may be prevented by reducing this speed.
3. The wheel surface speed: chatter may be prevented by increasing this speed.
4. The chip thickness: increasing the chip thickness may improve chatter performance
5. The structural characteristics: as for metal cutting, if the structural response can be improved, chatter may be prevented.
6. Mode direction: as for metal cutting, there are preferred directions for the grinding force and chip thickness direction. These may often be changed with advantage.
7. The hardness of the wheel: softer wheels are less likely to chatter because of reduced contact stiffness.
8. Flexible wheels: except for centreless grinding, these can be used to prevent chatter.

References

1. Entwistle RD (1997) Torsional compliance and chatter in grinding, PhD Dissertation, The University of Western Australia
2. Hahn RS (1954) On the theory of regenerative chatter in precision-grinding operations. Trans Am Soc Mech Eng 76(1):593–597
3. Hahn RS (1953) Metal cutting chatter and its elimination. Trans Am Soc Mech Eng 75:1073–1080
4. Minorsky N (1942) Self-excited oscillations in dynamical systems possessing retarded action. J Appl Mech 9:65–71
5. Ansoff HI (1949) Stability of linear oscillating systems with constant time lag. Trans ASME J Appl Mech 71:158–164
6. Landberg P (1957) Experiments on grinding. Microtechnic 11(1):18–25
7. Doi S (1958) An experimental study of chatter vibrations in grinding operations. Trans Am Soc Mech Eng 133–140
8. Lurie GB (1959) Vibrations in ginding. Mach Tool 6:17–19
9. Hahn RS (1959) Vibrations of flexible precision grinding spindles. Trans Am Soc Mech Eng J Eng Ind 81:201–206
10. Puzanov VV (1960) The nature of vibrations occurring during plunge grinding. Mach Tool 31(2):23–26
11. Sweeney G, Tobias SA (1963) An algebraic method for the determination of the dynamic stability of machine tools. In: Proceedings of international research in production engineering, American Society of Mechanical Engineers, Pittsburgh, pp 475–485
12. Sweeney G (1964) Dynamics of grinding. PhD Thesis, University of Birmingham
13. Gurney JP (1965) An analysis of surface wave instability in grinding. Trans Am Soc Mech Eng J Eng Ind 7(2):198–209

14. Gurney JP, Tobias SA (1962) A graphical analysis of regenerative machine tool instability. Trans Am Soc Mech Eng J Eng Ind 84:103–112
15. Snoeys R, Vanherck P (1967) Forced and self-excited vibrations in plunge grinding. In: Proceedings of the international conference on manufacturing technology, University of Michigan
16. Merritt HE (1965) Theory of self-excited machine tool chatter. Trans Am Soc Mech Eng J Eng Ind 87(4):447–454
17. Snoeys R, Brown D (1969) Dominating parameters in grinding wheel and workpiece regenerative chatter. In: Proceedings of the 10th MTDR conference, pp 325–348
18. Tlusty J, Spacek L (1954) Self-excited vibrations in machine tools. Czech Naklad CSAV, Prague
19. Brown D (1976) Grinding dynamics. PhD Thesis, University of Cincinnati
20. Inasaki I, Karpuschewski B, Lee HS (2001) Grinding chatter—origin and suppression. Ann CIRP 50(2):515–534
21. Chiu N, Malkin S (1993) Computer simulation for cylindrical plunge grinding. Ann CIRP 42(1):383–387
22. Drew SJ, Stone BJ (1997) Torsional (rotational) vibration—excitation of small rotating machines. J Sound Vib 201(4):437–463
23. Drew SJ, Mannan MA, Ong KL, Stone BJ (1999) An Investigation of In-process measurement of ground surfaces in the presence of vibration. Int J Mach Tools Manuf 39(12):1841–1861
24. Drew SJ, Mannan MA, Ong KL, Stone BJ (2001) The measurement of forces in grinding in the presence of vibration. Int J Mach Tools Manuf 41(4):509–520
25. Qureshi RA, Mannan MA, Drew SJ, Stone BJ (2002) A comparison of grinding forces arising from oscillating workpiece speed and chip thickness. Trans North Am Manuf Res Inst SME 30:169–176
26. Bartalucci B, Lisini GG (1970) Grinding at variable speed. In: 11th MTDR conference, pp 633–652
27. Walas SM (1991) Modelling with differential equations in chemical engineering. Butterworth-Heinemann, Stoneham
28. Sexton JS, Howes TD, Stone BJ (1982) The use of increased wheel flexibility to improve chatter performance in grinding. Proc IMechE 196(25):291–300
29. Sexton JS, Stone BJ (1981) The development of an ultra-hard abrasive grinding wheel which suppresses chatter. Ann CIRP 30(1):215–218
30. Bzymek ZM, Song G, Howes TD, Garrett RE (1994) Design of flexible grinding wheel with variable thickness hub. Trans ASME J Eng Ind 116(2):260–262
31. Entwistle RD, Stone BJ (1996) Torsional compliance in grinding chatter. In: Proceedings of eleventh annual meeting of the American Society for Precision Engineering, 9–14 Nov , pp 20–423
32. Entwistle RD, Stone BJ (1997) The effect of workpiece torsional flexibility on chatter performance in cylindrical grinding. In: Proceedings of the fifth international congress on sound and vibration, University of Adelaide, South Australia
33. Entwistle RD, Stone BJ (2013) Fundamental issues in self-excited chatter in grinding. J Mach Eng 13(3):26–50
34. Rowe WB, Barash MM, Koenigsberger F (1965) Some roundness characteristics of centreless grinding. In J MTDR 5:203–215
35. Furukawa Y, Miyashita M, Shiozaki S (1971) Vibration analysis of work rounding mechanism in centerless grinding. Int J MTDR 11:145–175
36. Miyashita M, Hashimoto F, Kanai A (1982) Diagram for selecting chatter free conditions on centerless grinding. Ann CIRP 31(1):221–223
37. Hashimoto F, Zhou SS, Lahoti GD, Miyashita M (2000) Stability diagram for chatter free center-less grinding and its application in machine development. Ann CIRP 49(1):225–230
38. Li H, Shin YC (2007) A time domain dynamic simulation model for stability prediction of infeed centerless grinding processes. J Manuf Sci Eng Am Soc Mech Eng 129(3):539–550

39. Pearce TRA, Stone BJ (2011) Unstable vibration in centreless grinding: part A. Geometric instability or chatter? Proc IMechE Part B J Eng Manuf 225(8):1227–1243
40. Pearce TRA, Stone BJ (2011) Unstable vibration in centreless grinding: part B. Graphical method. Proc IMechE Part B J Eng Manuf 225(8):1245–1254
41. Harrison AJL, Pearce TRA (2002) Prediction of lobe growth and decay in centreless grinding based on geometric considerations. Proc IMechE Part B, Journal of Engineering Manufacture 216(9):1201–1216
42. Tonshoff HK, Peters J, Inasaki I, Paul T (1992) Modelling and simulation of grinding processes. Ann CIRP 41(2):677–688
43. Hesterman DC, Stone BJ (2002) Improved model of chatter in grinding, including torsional effects. Proc IMechE Part K 216(2):169–180

Chapter 6
Conclusions

6.1 Introduction

The objective in the previous chapters was to determine methods of preventing chatter by using simple models to determine the most significant parameters. The methods used were all analytical. Also all of the methods proposed to stop chatter can be termed passive as the solution, once determined, did not change with time and had no external energy source. There are methods of both investigating and preventing chatter that are not analytical and do involve non-passive solutions. Before final conclusions can be drawn, it is necessary to describe three such approaches,

- Time-domain modelling
- Varying speed
- Active control.

Each of these topics is extensive, but a brief description will have to suffice.

6.2 Time-Domain Modelling

Time-domain modelling does not assume a steady sinusoidal motion with an exponential growth rate. It therefore has the advantage that any transient effects are predicted. Some typical publications concerned with the simulation of chatter are listed in the references [1–3]. Earlier references were cited for examples of time-domain programs. In Sect. 1.3.4, a time-domain model of chatter in turning was described and Fig. 1.18 was an illustration from the simulation program. This program allowed an animation of chatter to be observed for a selected width and work speed that were selected from a stability chart. This simulation confirmed the accuracy of the stability boundary and allowed some interesting observations on the build-up of chatter to be made. The simulation included the effect of loss of contact when the tool vibration was excessive. Also in Sect. 2.5, a time-domain

B. Stone, *Chatter and Machine Tools*, DOI: 10.1007/978-3-319-05236-6_6,

program was referenced that simulated mode-coupled chatter. It needs to be stressed that such simulations are no substitute for experimental measurements. Some authors use their time-domain solutions as confirmation of the analytical analysis. This is valid, in so far as the equations governing both are the same. However, this "numerical validation" does not guarantee that the models are accurate representations of the real-machining process.

One of the advantages of the simulation of Chap. 1 is that it allows various ideas to be investigated. The simulation runs at a much lower speed than a real machine and is thus very visual. It is possible to see what happens if the work speed is suddenly changed. This is because the program parameters are not reset when the speed is changed but the existing work surface is maintained. Why change the speed? From the understanding gained in the previous chapters, a varying speed effect was noted in Sect. 3.5.2.2. It will be found that suddenly changing the speed will quite frequently stop chatter. This is because chatter occurs when regeneration of the surface wave excites the structure at the same frequency that was present when the surface wave was created. If the speed is changed, the surface wave excites the structure at a different frequency. This causes the existing vibration to decay and a new vibration to start to grow. If the speed is changed frequently and by a sufficient amount, chatter is suppressed.

It is therefore necessary to see whether varying the speed is effective on a real machine. Clearly the speed cannot be changed instantly.

6.3 Varying Speed

There have been several papers that report investigations of varying speed in both metal cutting and grinding [4–14]. When applied to metal cutting, the main problem has been the need to change the speed more rapidly than is possible practically. There are also several other drawbacks:

- There can be large amplitude transients before the process becomes stable.
- The surface finish varies with speed, and when turning a cylinder, the effect of the speed variation can be observed along the axis as a change in the surface finish in the feed direction.
- Cutting forces are not completely independent of speed and so the mean force varies with speed and causes the machine deflection to vary and this shows up on the work's dimensions.

The situation with grinding is more promising, especially if chatter is governed by waves on the wheel. In this case, the vibration growth rates are much slower and so the rate of change of speed is not a problem. The problem of varying surface finish remains an issue.

While investigating flexible grinding wheels, Sexton [15] found another potential means of suppressing chatter by causing changes with time. It was initially difficult to obtain chatter while roll grinding. In the event, this was found to be the result of the roll being supported very differently at each end. The chatter receptance at each end of the roll resulted in chatter at different frequencies when plunge grinding. When traversing, the surface waves on the wheel that commenced to grow while at one end were subsequently removed at the other end and vice versa.

Changing machine response in time is worth investigating further as a means of suppressing chatter.

6.4 Active Control

Researchers with a control background have investigated using active control to prevent chatter. Initially this was called "adaptive control" but subsequently changed to "active control". If the onset of chatter can be measured, it is possible in theory to use the measurement as a control feedback to apply a stabilising force or motion. This approach has been applied with some success [16–21]. However, it does seem that a small threshold amplitude is required to ensure the feedback operates. Also the examples appear to be machine/operation-specific and not applicable to a wide range of machines and machining operations. They also involve more cost than the passive methods and devices described in this book. Also as active devices have an energy source, there is always the danger that they may cause an instability, albeit at a different frequency.

6.5 Final Conclusions

From the author's experience of solving chatter problems, it has been interesting to observe that the blame for the problem is either laid at the door of the manufacturer or the user. Users will blame the machine and its manufacturer. Manufacturers will often say that the problem is the result of the incorrect use of the machine.

First, it is important to remember that machine tool manufacturers aim to produce machines that will produce geometrically accurate components with good surface finish. This requires that the machines have good static stiffness (for geometrical accuracy) and thermal expansion affects are minimised. The machines must also have good dynamic characteristics. If this is not the case, accuracy may be affected and also surface finish. The most serious problem arising from poor dynamic characteristics is that of chatter which causes the surface finish to be unacceptable and has other serious consequences—tool wear and noise amongst the most important. This book has addressed the problem of chatter. The major parameters that affect chatter have been discussed and many solutions to chatter have been investigated. What is therefore to be done to stop chatter?

For machine designers, the strategy is to make the machine as stiff as possible but to remember that the dynamic characteristics require energy dissipation to reduce resonance effects. As shown in Chap. 4 (for a spindle), it is possible to have a high damping source but with little amplitude at that source and hence minimal energy dissipation. The major damping sources in machine tools are in metal--to-metal joints with oil present in the interface. Also bearings were shown in Chap. 4 to have more damping than is often appreciated. It is a design decision to balance stiffness at the damping locations with energy dissipation. Some very stiff machines have been found to have very severe resonant problems.

It is always possible to consider adding the energy dissipation after the machine has been designed and built and its dynamic characteristics measured. Vibration absorbers were once included in mass-produced machines. Cincinnati Dynapoise was an absorber in the over-arm of a horizontal milling machine and was used as a selling point. Vibration absorbers and their tuning have been described at length in Chap. 4. The use of absorbers is more often limited to boring bars with large length to diameter ratios. There is scope for their wider use.

This book has also described many methods of stopping chatter, not all of which appear to be well known. Thus, the selection of an appropriate high-performance milling cutter, flexible tool or grinding wheel can produce significant improvements in chatter performance.

Finally, there is no substitute for experience and experimental confirmation of models.

References

1. Smith S, TLusty J (1993) Efficient Simulation programs for chatter in milling. Ann CIRP 42(1):463–466
2. Biera J, Vinolas J, Nieto FJ (1997) Time-domain dynamic modelling of the external plunge grinding process. Int J Mach Tools Manuf 37(11):1555–1572
3. Weck M, Verhag E, Gather M (1973) Adaptive control for face-milling operations with strategies for avoiding chatter-vibrations and for automatic cut distribution. Ann CIRP 24(1):405–409
4. Bartalucci B, Lisini GG (1970) Grinding at variable speed. In: 11th MTDR Conference: 633–652
5. Takemura T, Kitamura T (1974) Active suppression of chatter by programmed variation of spindle speed. Ann CIRP 23(1):121–122
6. Inamura T, Sata T (1975) Stability analysis of cutting under varying spindle speed. J Fac Eng Univ Tokyo 33:13–29
7. Sexton JS, Stone BJ (1977) An experimental and analytical investigation of variable-speed cutting. MTIRA research report No 73
8. Sexton JS, Milne RD, Stone BJ (1977) A stability analysis of single-point machining with varying spindle speed. Appl Math Model 1(6):310–318
9. Sexton JS, Stone BJ (1978) The stability of machining with continuously varying spindle speed. Ann CIRP 27(1):321–326
10. Sexton JS, Stone BJ (1980) An investigation of the transient effects during variable speed cutting. J Mech Eng Sci 22(3):107–118
11. Jemielniak K, Widota A (1984) Suppression of self-excited vibration by the spindle speed variation method. Int J MTDR 24(3):207–214

12. Hoshi T, Matsumoto S, Mitsui S, Horiuchi O, Horiuchi Y (1986) Suppression of wheel regenerative grinding vibration by alternating wheel speed. J Jpn Soc Precis Eng 52(10):1802–1807
13. Al-Regib E, Ni J (2003) Programming spindle speed variation for machine tool chatter suppression. Int J Mach Tools Manuf 43(12):1229–1240
14. Zatarain M, Bediaga I, Munoa J, Lizarraide R (2008) Stability of milling processes with continuous spindle speed variation: Analysis in the frequency and time domains, and experimental correlation. Ann CIRP 57(1):379–384
15. Sexton JS, Stone BJ (1980) A method of reducing chatter in roll grinding. In: International conference on manufacturing engineering, Melbourne, pp 155–159
16. Klein RG, Nachtigal CL (1975) The application of active control to improve boring bar performance. J Dyn Sys Meas Control 97(2):179–183
17. Mei Z, Yang S, Shi H, Chang S, Ehmann KF (1994) Active chatter suppression by on-line variation of the rake and clearance angles in turning-principles and experimental investigation. Int J Mach Tools Manuf 34(7):981–990
18. Tewani SG, Rouch KE, Walcott BL (1995) A study of cutting process stability of a boring bar with active dynamic absorber. Int J Mach Tools Manuf 35(1):91–108
19. Browning DR, Golioto I, Thompson NB (1997) Active chatter control system for long overhang boring bars. Proc SPIE Smart Struct Mater 3044:270–280
20. Pratt JR, Nayfeh AH (1997) Experimental system identification and active vibration control of a smart machine tool. In: Proceedings of 11th VPI&SU symposium on structural dynamics and controls, pp 389–398
21. Pratt JR, Nayfeh AH (2001) Chatter control and stability analysis of a cantilever boring bar under regenerative cutting conditions. Phil Trans R Soc Lond A 359(1781):759–792

Appendix A
Basic Vibration Theory

This appendix and those following are a summary of the vibration theory that is needed to understand the chatter theory described in this book. These appendices are not a substitute for a complete text book on vibration.

A system may vibrate when it is possible for energy to be converted from one form to another and back again. In mechanical systems, this is usually from the kinetic energy of motion to the stored energy in for example a spring. A simple vibrating system, that illustrates many of the fundamental concepts of vibration, consists of a rigid mass attached by a massless spring to a fixed abutment, see Fig. A.1. For one-degree-of-freedom vibration, the mass is constrained to move in one direction, so that one coordinate uniquely defines the position of the system. If there is an energy dissipation source, such as a viscous damper, then unless the system has a continuing excitation, the vibration will gradually decay as energy is converted to heat. Vibration is thus often considered as "transient" or "forced" vibration. A transient vibration is one that dies away with time due to energy dissipation. Usually, there is some initial disturbance and following this the system vibrates without any further input. This is called transient vibration. A forced vibration is usually defined as being one that is kept going by an external excitation.

A.1 Transient Vibration: Undamped

Consider the motion of the undamped spring/mass system, shown in Fig. A.1, when it is initially disturbed and then allowed to vibrate freely. The displacement of the mass with time, $x(t)$, is measured from the static equilibrium position, i.e. the rest position.

If at some time t, the mass is displaced an amount $x(t)$ in the positive direction shown, then for a linear spring, there will be a force on the mass from the spring of $-kx(t)$.

Thus from Newton's second law of motion using a free-body diagram as shown in Fig. A.1,

B. Stone, *Chatter and Machine Tools*, DOI: 10.1007/978-3-319-05236-6,

Fig. A.1 Spring/mass system and free-body diagram

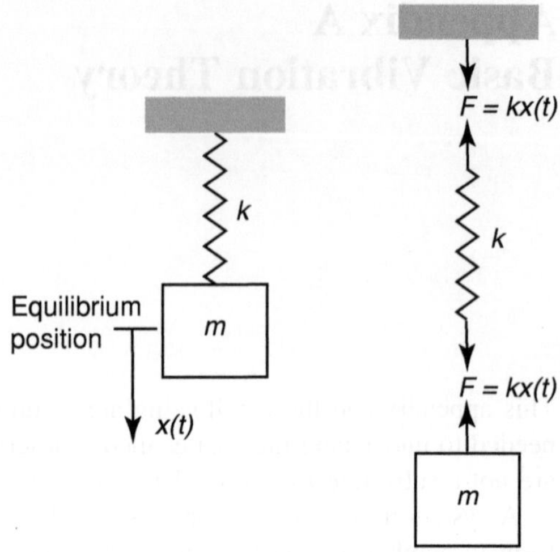

$$m\frac{d^2x(t)}{dt^2} = -kx(t) \tag{A.1}$$

Equation A.1 is called the equation of motion. The equation is unchanged if gravity effects are included and $x(t)$ is still defined relative to the static equilibrium position. Rearranging and dividing by m gives,

$$\frac{d^2x(t)}{dt^2} + \omega_n^2 x(t) = 0 \quad \text{where } \omega_n = \sqrt{\frac{k}{m}}$$

The solution will be of the form,

$$x(t) = Ae^{\alpha_1 t} + Be^{\alpha_2 t}$$

where the values of A, B, α_1 and α_2 may be found from the initial conditions $x(0)$ and $dx(0)/dt$. Thus, substituting $x(t) = Ae^{\alpha t}$ gives,

$$Ae^{\alpha t}(\alpha^2 + \omega_n^2) = 0 \quad \therefore \alpha^2 + \omega_n^2 = 0$$

And hence, $\alpha^2 = -\omega_n^2$ so that $\alpha_1 = +i\omega_n$ and $\alpha_2 = -i\omega_n$ where $i = \sqrt{-1}$
Thus,

$$x(t) = Ae^{+i\omega_n t} + Be^{-i\omega_n t}$$

Rearranging gives,

$$x(t) = \frac{(A+B)}{2}\left(e^{+i\omega_n t} + e^{-i\omega_n t}\right) + \frac{(A-B)}{2}\left(e^{+i\omega_n t} - e^{-i\omega_n t}\right)$$

So that $x(t) = C\cos\omega_n t + D\sin\omega_n t\ldots$ (a)
where $C = A + B$ and $D = (A - B)/i = -i(A - B)$.

When $t = 0$ $x(t) = x(0)$, so that substituting in (a) gives $x(0) = C$

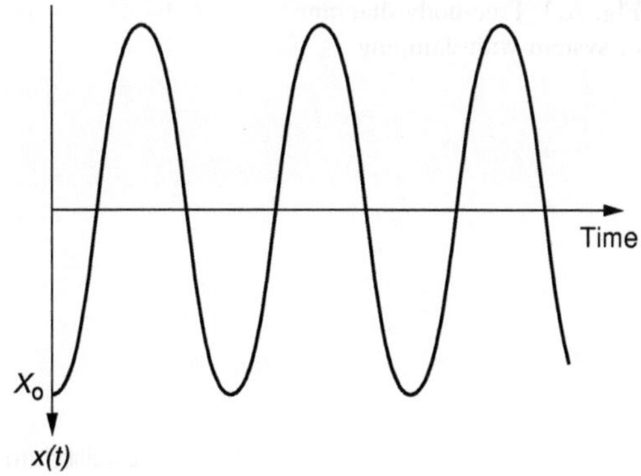

Fig. A.2 Transient vibration with no damping (Program A.1)

Also when $t = 0$ $\mathrm{d}x(t)/\mathrm{d}t = \mathrm{d}x(0)/\mathrm{d}t$, so that differentiating (a) and substituting gives $\mathrm{d}x(t)/\mathrm{d}t = \omega_n D$

So that finally we have,

$$x(t) = x(0)\cos\omega_n t + \frac{\frac{\mathrm{d}x(0)}{\mathrm{d}t}}{\omega_n}\sin\omega_n t \tag{A.2}$$

For example with an initial displacement $x(0)$ but with no initial velocity, so that $\mathrm{d}x(0)/\mathrm{d}t = 0$, the motion is a non-decaying sinusoidal vibration of amplitude $x(0)$ and a frequency ω_n (Fig. A.2).

$$x(t) = x(0)\cos\omega_n t$$

If the time between successive peaks of the waveform is T (s), then the frequency is $\omega_n = 1/T$ cycles/s (normally called Hertz—abbreviated to Hz). The frequency ω_n is conventionally termed the undamped natural frequency where as defined above,

$$\omega_n = \sqrt{\frac{k}{m}} \tag{A.3}$$

If k is in N/m and m is in kg, then the units of ω_n are in rad/s and since there are 2π radians in a cycle, to convert to Hz, we need to divide by 2π.

Note that the value of the undamped natural frequency does not depend on the initial conditions or the amplitude of the motion, but is a fixed property of the system.

A.2 Transient Vibration: Damped

Now let us add a viscous damper having a damping coefficient c to the spring/mass system previously considered, see Fig. A.3. This form of damping provides a force proportional to velocity and opposing the motion.

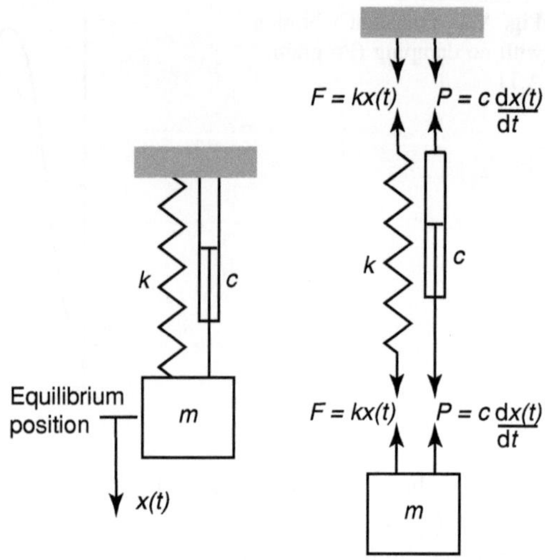

Fig. A.3 Free-body diagram of system with damping

Drawing a free-body diagram and applying Newton's second law of motion gives,

$$m\frac{\mathrm{d}^2x(t)}{\mathrm{d}t^2} = -kx(t) - c\frac{\mathrm{d}x(t)}{\mathrm{d}t} \tag{A.4}$$

and hence,

$$m\frac{\mathrm{d}^2x(t)}{\mathrm{d}t^2} + c\frac{\mathrm{d}x(t)}{\mathrm{d}t} + kx(t) = 0 \tag{A.5}$$

A.2.1 *Laplace Transform Solution*

Laplace transforms are used here, as this will be helpful when the transients arising at the start of forced vibration are considered. Taking Laplace transforms of Eq. A.5 yields,

$$m\left[-\frac{\mathrm{d}x(0)}{\mathrm{d}t} - sx(0) + s^2X(s)\right] + c[-x(0) + sX(s)] + kX(s) = 0$$

Rearranging this gives

$$X(s) = \frac{m\left(\frac{\mathrm{d}x(0)}{\mathrm{d}t} + sx(0)\right) + cx(0)}{ms^2 + cs + k} \tag{A.6}$$

For no damping ($c = 0$), we can check that the result is the same as that obtained previously—Eq. A.2. Thus, when $c = 0$ Eq. A.6 becomes,

$$X(s) = \frac{m\left(\frac{dx(0)}{dt} + sx(0)\right)}{ms^2 + k} = \frac{sx(0)}{s^2 + k/m} + \frac{\frac{dx(0)}{dt}}{s^2 + k/m}$$

Now taking Inverse Laplace Transforms and substituting $\omega_n = \sqrt{\frac{k}{m}}$ as before gives Eq. A.2 again,

$$x(t) = x(0)\cos\omega_n t + \frac{\frac{dx(0)}{dt}}{\omega_n}\sin\omega_n t \tag{A.2}$$

A.2.2 Equations in Non-dimensional Form

It is common to write the basic equation of motion in terms of ω_n and another parameter ξ, called the viscous damping ratio, which is defined as,

$\xi = \frac{c}{2\sqrt{km}}$ (The significance of ξ will become apparent later).

Thus, if Eq. A.5 is divided throughout by m, we obtain,

$$\frac{d^2x(t)}{dt^2} + 2\xi\omega_n\frac{dx(t)}{dt} + \omega_n^2 x(t) = 0 \tag{A.7}$$

If Laplace transforms are taken, we obtain

$$X(s) = \frac{\frac{dx(0)}{dt} + 2\xi\omega_n x(0)}{s^2 + 2\xi\omega_n s + \omega_n^2} + \frac{sx(0)}{s^2 + 2\xi\omega_n s + \omega_n^2} \tag{A.8}$$

Now the form of the solution depends on the value of ξ so that different solutions are obtained depending on whether $\xi < 1$, $\xi = 1$ or $\xi > 1$. We have already considered the case where c and hence ξ is zero. The solution was a non-decaying oscillation of frequency ω_n.

$$x(t) = x(0)\cos\omega_n t + \frac{\frac{dx(0)}{dt}}{\omega_n}\sin\omega_n t \tag{A.2}$$

A.2.3 Case When $\xi < 1$

When $\xi < 1$ the Inverse Laplace Transform of Eq. A.8 gives,

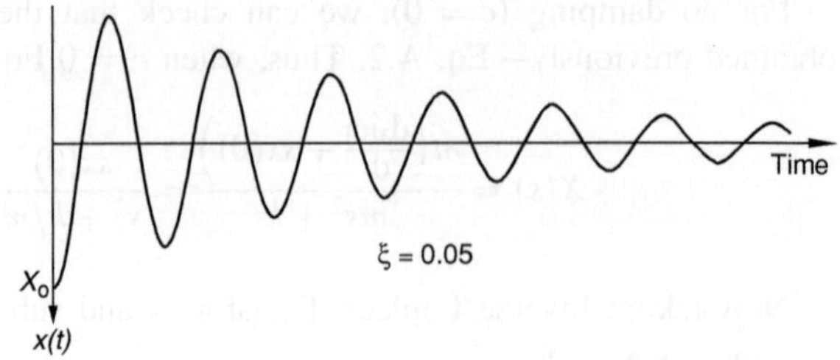

Fig. A.4 Transient vibration with $\xi < 1$ (Program A.2)

$$x(t) = e^{-\xi\omega_n t}\left[x(0)\cos\,\omega_n\sqrt{(1-\xi^2)}t + \frac{\left[\frac{dx(0)}{dt} + \xi\omega_n x(0)\right]\sin\,\omega_n\sqrt{(1-\xi^2)}t}{\omega_n\sqrt{(1-\xi^2)}}\right] \tag{A.9}$$

This is an exponentially decaying oscillation. If again, the specific example of an initial displacement and no initial velocity is taken, i.e. $x(0) = X_0$ and $dx(0)/dt = 0$, then

$$x(t) = X_0 e^{-\xi\omega_n t}\left[\cos\,\omega_n\sqrt{(1-\xi^2)}t + \frac{\xi\sin\,\omega_n\sqrt{(1-\xi^2)}t}{\sqrt{(1-\xi^2)}}\right] \tag{A.10}$$

And the damped natural frequency is

$$\omega_D = \omega_n\sqrt{(1-\xi^2)} \tag{A.11}$$

A typical variation of $x(t)$ with t is shown in Fig. A.4.

The greater the value of ξ, the more rapid will be the decay of the oscillation. As $\xi \to 1$, then the solution tends to have no oscillation. In fact, $\xi = 1$ is the smallest value of ξ that does not give any oscillation and this is thus termed the critical damping ratio. From the definition of ξ $(= c/2\sqrt{km})$, it is clear that when $\xi = 1$ the critical value of $c_c = 2\sqrt{km}$, so that $\xi = c/c_c$.

A.2.4 Case When $\xi = 1$

When $\xi = 1$, the Inverse Laplace Transform of Eq. A.8 gives,

$$x(t) = x(0)e^{-\omega_n t} + \left[\frac{dx(0)}{dt} + \omega_n x(0)\right]te^{-\omega_n t} \tag{A.12}$$

This is a non-oscillatory motion and for the case considered previously, i.e. $x(0) = X_0$ and $dx(0)/dt = 0$; then,

$$x(t) = x(0)(1 + \omega_n t)e^{-\omega_n t} \tag{A.13}$$

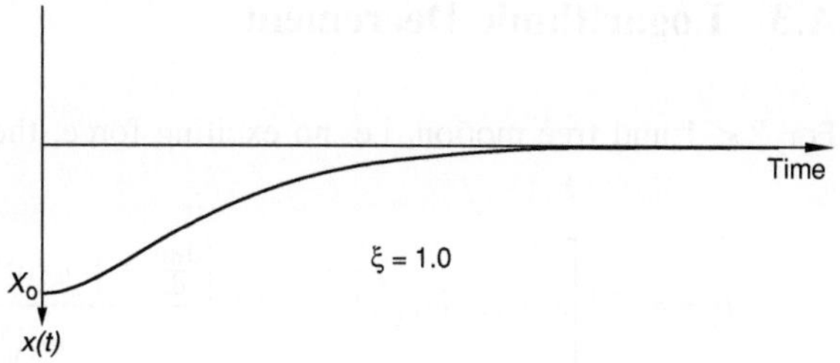

Fig. A.5 Transient vibration with $\xi = 1$ (Program A.3)

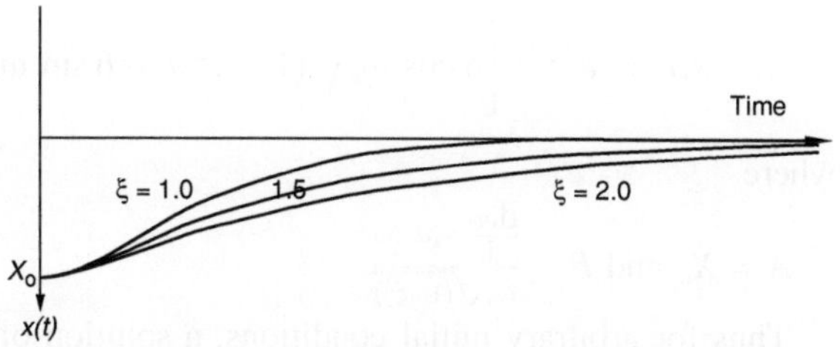

Fig. A.6 Transient vibration with $\xi > 1$ (Program A.4)

The variation of x(t) with t is shown in Fig. A.5. If the value of ξ exceeds 1.0, then another mathematical solution is obtained.

A.2.5 *Case When ξ > 1*

When $\xi > 1$, the Inverse Laplace Transform of Eq. A.8 gives,

$$x(t) = e^{-\xi\omega_n t}\left[x(0)\cosh\omega_n\sqrt{(\xi^2-1)}t + \frac{\left[\frac{dx(0)}{dt} + \xi\omega_n x(0)\right]\sinh\omega_n\sqrt{(\xi^2-1)}t}{\omega_n\sqrt{(\xi^2-1)}}\right] \tag{A.14}$$

This is a non-oscillatory decaying motion and for the case considered previously, i.e. $x(0) = X_0$ and $dx(0)/dt = 0$; then,

$$x(t) = X_0 e^{-\xi\omega_n t}\left[\cosh\omega_n\sqrt{(\xi^2-1)}t + \frac{\xi\sinh\omega_n\sqrt{(\xi^2-1)}t}{\sqrt{(\xi^2-1)}}\right] \tag{A.15}$$

Typical variations of x(t) with t for various values of ξ are shown in Fig. A.6.

For real engineering structures, it would be extremely unusual for ξ to exceed unity, and thus, any transient vibration is normally oscillatory and decaying. It is possible if $\xi < 1$, to calculate ξ from measured transients. This approach uses the logarithmic decrement.

A.3 Logarithmic Decrement

For $\xi < 1$ and free motion, i.e. no exciting force, the solution was shown to be

$$x(t) = e^{-\xi\omega_n t}\left[x(0)\cos\,\omega_n\sqrt{(1-\xi^2)}t + \frac{\left[\frac{\mathrm{d}x(0)}{\mathrm{d}t} + \xi\omega_n x(0)\right]\sin\,\omega_n\sqrt{(1-\xi^2)}t}{\omega_n\sqrt{(1-\xi^2)}}\right] \tag{A.9}$$

And this may be written as

$$x(t) = e^{-\xi\omega_n t}\left[A\cos\,\omega_n\sqrt{(1-\xi^2)}t + B\sin\,\omega_n\sqrt{(1-\xi^2)}t\right] \tag{A.16}$$

where

$A = X_0$ and $B = \frac{\frac{\mathrm{d}x(0)}{\mathrm{d}t} + \xi\omega_n X_0}{\omega_n\sqrt{(1-\xi^2)}}$

Thus for arbitrary initial conditions, a solution of the form given in Eq. A.16 will apply. Now Eq. A.16 may be manipulated as follows.

$$\begin{aligned} x(t) &= e^{-\xi\omega_n t}\sqrt{(\mathrm{A}^2-\mathrm{B}^2)}\left[\begin{array}{l}\frac{\mathrm{A}}{\sqrt{(\mathrm{A}^2-\mathrm{B}^2)}}\cos\,\omega_n\sqrt{(1-\xi^2)}t \\ +\frac{\mathrm{B}}{\sqrt{(\mathrm{A}^2-\mathrm{B}^2)}}\sin\,\omega_n\sqrt{(1-\xi^2)}t\end{array}\right] \\ &= e^{-\xi\omega_n t}\sqrt{(\mathrm{A}^2-\mathrm{B}^2)}\left[\sin\phi\cos\,\omega_n\sqrt{(1-\xi^2)}t + \cos\phi\sin\,\omega_n\sqrt{(1-\xi^2)}t\right] \\ &= \mathrm{C}e^{-\xi\omega_n t}\sin\left[\omega_n\sqrt{(1-\xi^2)}t + \phi\right] \end{aligned} \tag{A.17}$$

where $\tan\phi = \mathrm{A/B}$ and $\mathrm{C} = \sqrt{\mathrm{A}^2 + \mathrm{B}^2}$

Equation A.17 represents a decaying oscillation of frequency $\omega_n\sqrt{(1-\xi^2)}$, the damped natural frequency.

Now consider successive maximum amplitudes in the vibration as shown in Fig. A.7.

At a maximum, $\frac{\mathrm{d}x(t)}{\mathrm{d}t} = 0$. Differentiating Eq. A.17,

$$\frac{\mathrm{d}x(t)}{\mathrm{d}t} = \mathrm{C}\left[\begin{array}{c}-\xi\omega_n e^{-\xi\omega_n t}\sin\left[\omega_n\sqrt{(1-\xi^2)}t + \phi\right] \\ +\omega_n\sqrt{(1-\xi^2)}e^{-\xi\omega_n t}\cos\left[\omega_n\sqrt{(1-\xi^2)}t + \phi\right]\end{array}\right]$$

Fig. A.7 Transient vibration with $\xi < 1$

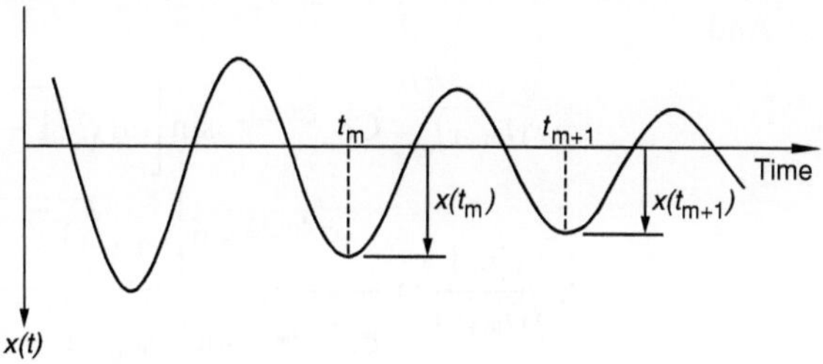

This is zero when

$$-\xi\omega_n \sin\left[\omega_n\sqrt{(1-\xi^2)}t+\phi\right]+\omega_n\sqrt{(1-\xi^2)}\cos\left[\omega_n\sqrt{(1-\xi^2)}t+\phi\right]=0$$

that is when $\tan\left[\omega_n\sqrt{(1-\xi^2)}t+\phi\right]=\frac{\sqrt{(1-\xi^2)}}{\xi}$

Which is when

$$\omega_n\sqrt{(1-\xi^2)}t+\phi=\tan^{-1}\left[\frac{\sqrt{(1-\xi^2)}}{\xi}\right]+n\pi$$

This represents alternate maxima and minima. Consider any two successive maxima or minima. Then,

$$\omega_n\sqrt{(1-\xi^2)}t_m+\phi=\tan^{-1}\left[\frac{\sqrt{(1-\xi^2)}}{\xi}\right]+2m\pi \tag{a}$$

And

$$\omega_n\sqrt{(1-\xi^2)}t_{m+1}+\phi=\tan^{-1}\left[\frac{\sqrt{(1-\xi^2)}}{\xi}\right]+2(m+1)\pi \tag{b}$$

(b)–(a) gives the 'period' of the damped oscillation (the time between successive zero crossings)

$$t_{m+1}-t_m=\frac{2\pi}{\omega_n\sqrt{(1-\xi^2)}} \tag{c}$$

Now

$$x(t_m)=Ce^{-\xi\omega_n t_m}\sin\left[\omega_n\sqrt{(1-\xi^2)}t_m+\phi\right]$$

And

$$x(t_{m+1}) = Ce^{-\xi\omega_n t_{m+1}} \sin\left[\omega_n\sqrt{(1-\xi^2)}t_{m+1} + \phi\right]$$

$$\therefore \frac{x(t_m)}{x(t_{m+1})} = \frac{e^{-\xi\omega_n t_m} \sin\left[\omega_n\sqrt{(1-\xi^2)}t_m + \phi\right]}{e^{-\xi\omega_n t_{m+1}} \sin\left[\omega_n\sqrt{(1-\xi^2)}t_{m+1} + \phi\right]}$$

But since $\tan\left[\omega_n\sqrt{(1-\xi^2)}t + \phi\right] = \frac{\sqrt{(1-\xi^2)}}{\xi}$ for $t = t_m$ and t_{m+1}

$$\sin\left[\omega_n\sqrt{(1-\xi^2)}t_{m+1} + \phi\right] = \sin\left[\omega_n\sqrt{(1-\xi^2)}t_m + \phi\right]$$

$$\therefore \frac{x(t_m)}{x(t_{m+1})} = \frac{e^{-\xi\omega_n t_m}}{e^{-\xi\omega_n t_{m+1}}} = e^{\xi\omega_n(t_{m+1}-t_m)}$$

And substituting for $t_{m+1} - t_m$ from (c),

$$\frac{x(t_m)}{x(t_{m+1})} = e^{\left(\frac{2\pi\xi}{\sqrt{(1-\xi^2)}}\right)}$$

The log decrement δ is defined as

$$\delta = \log_e\left(\frac{x(t_m)}{x(t_{m+1})}\right) = \left(\frac{2\pi\xi}{\sqrt{(1-\xi^2)}}\right) \tag{A.18}$$

If δ is measured experimentally, we may determine ξ in terms of δ as

$$\delta = \left(\frac{2\pi\xi}{\sqrt{(1-\xi^2)}}\right)$$

$$\therefore \delta^2 = \frac{4\pi^2\xi^2}{1-\xi^2}$$

$$\therefore \delta^2 - \xi^2\delta^2 = 4\pi^2\xi^2$$

$$\therefore \xi^2(4\pi^2 + \delta^2) = \delta^2$$

And hence,

$$\xi = \frac{\delta}{\sqrt{(4\pi^2 + \delta^2)}} \tag{A.19}$$

Thus, it is possible from a measured decay trace to determine δ and calculate the damping ratio ξ.

A.4 Forced Vibration: External Force

If the mass is subjected to a force *F*(*t*) acting in the positive *x*(*t*) direction as shown in Fig. A.8, then the equation of motion, Eq. A.5, becomes,

$$m\frac{\mathrm{d}^2x(t)}{\mathrm{d}t^2}+c\frac{\mathrm{d}x(t)}{\mathrm{d}t}+kx(t)=F(t) \tag{A.20}$$

Taking Laplace Transforms

$$m\left[-\frac{\mathrm{d}x(0)}{\mathrm{d}t}-sx(0)+s^2X(s)\right]+c[-x(0)+sX(s)]+kX(s)=LF(t)$$

$$X(s)=\frac{LF(t)}{ms^2+cs+k}+\frac{m\left(\frac{\mathrm{d}x(0)}{\mathrm{d}t}+sx(0)\right)+cx(0)}{ms^2+cs+k} \tag{A.21}$$

The latter term is identical to that already examined—Eq. A.6, i.e. there is a transient component of the response which is dependent on the initial conditions.

The solution of *x*(*t*) is thus the superposition of a decaying vibration ($\xi < 1$) arising from the initial conditions plus a solution arising from the applied force. We shall now examine this component of the solution.

$$X(s)=\frac{LF(t)}{ms^2+cs+k}$$

Frequently *F*(*t*) is cyclic in nature. Thus, the solution when $F(t) = F \sin \omega t$ is commonly examined. This is also relevant since many excitation forces may be separated into sinusoidal components via Fourier series.

Now $LF(t) = LF \sin \omega t = \frac{F}{s^2+\omega^2}$

Therefore,

$$X(s)=\frac{F}{(s^2+\omega^2)(ms^2+cs+k)} \tag{A.22}$$

Expanding via partial fractions

$$X(s)=\frac{\mathrm{A}s+\mathrm{B}}{(s^2+\omega^2)}+\frac{\mathrm{C}s+\mathrm{D}}{(ms^2+cs+k)} \tag{A.23}$$

where the second term will yield an exponential decaying motion of the kind already encountered in Eq. A.6. This is a transient motion arising from the fact that taking the one-sided Laplace Transform implies that the forcing function is zero when $t < 0$. Thus, the sinusoidal excitation commences at $t = 0$, and there is a transient, associated with the 'start-up' of the excitation, that decays with time. A typical start-up transient is shown in Fig. A.9, which has the initial conditions zero, i.e. the system is at rest in the equilibrium position when $t = 0$. In some real applications, these transients can be very significant, e.g. with pneumatic drills and rotating machinery.

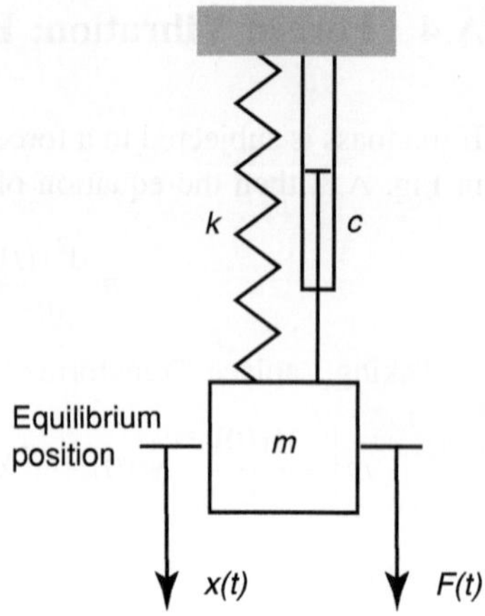

Fig. A.8 Forced excitation

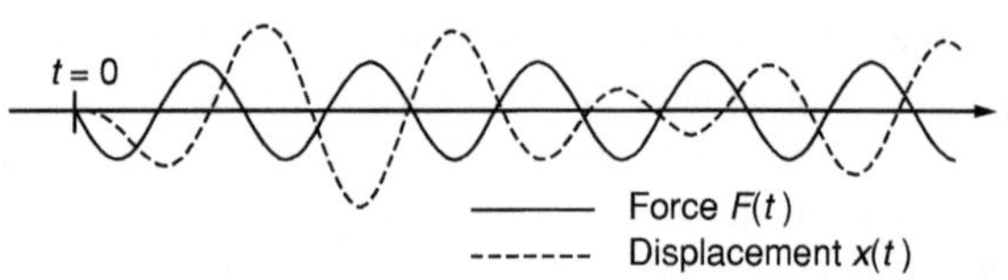

Fig. A.9 Forced vibration showing initial transient (Program A.5)

When considering sinusoidal excitation, we are normally interested in the steady-state motion that results after the transients have died away and this is given by the first term in Eq. A.23. Thus, we are interested in A and B. From Eq. A.22 and Eq. A.23, if the coefficients of s^3, s^2, s and the constant term are compared, it is possible to show that

$$\mathrm{A} = \frac{-\omega c F}{(k - m\omega^2)^2 + \omega^2 c^2}$$

$$\mathrm{B} = \frac{\omega(k - m\omega^2)F}{(k - m\omega^2)^2 + \omega^2 c^2}$$

The steady-state component of $x(t)$ is given by taking the Inverse Laplace Transform of,

$$X(s) = \frac{\mathrm{A}s + \mathrm{B}}{(s^2 + \omega^2)}$$

Thus, $X(t) = \mathrm{A}\cos\omega t + \frac{\mathrm{B}}{\omega}\sin\omega t$
And substituting for A and B gives

$$x(t) = \frac{F(-\omega c \cos\omega t + (k - m\omega^2)\sin\omega t)}{(k - m\omega^2)^2 + \omega^2 c^2}$$

So that after some manipulation,

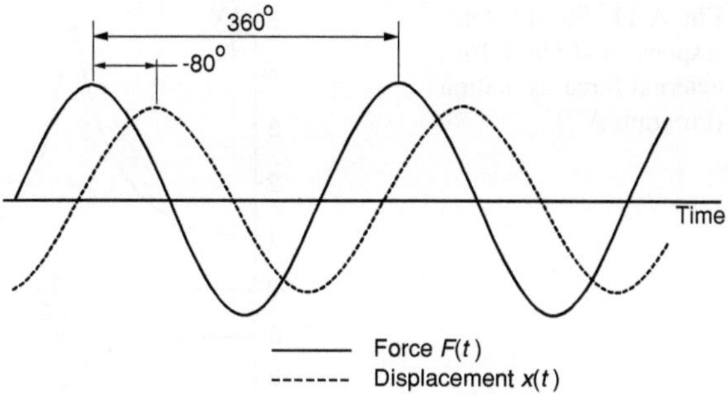

Fig. A.10 Definition of phase (Program A.6)

$$x(t) = \frac{F \sin(\omega t + \phi)}{\left((k - m\omega^2)^2 + \omega^2 c^2\right)^{1/2}} \tag{A.24}$$

where

$$\tan \phi = \frac{-\omega c}{(k - m\omega^2)} \tag{A.25}$$

It is conventional to represent $x(t)$ as $X \sin(\omega t + \phi)$, where X is the amplitude of the response and ϕ the phase angle. Figure A.10 shows a typical trace of force and displacement against time. The excitation force is $F \sin \omega t$ and the resulting displacement has a phase ϕ of -80°, i.e. a phase lag of 80°. The phase between two successive peaks is defined as 360° and is thus not dependent on frequency.

Thus from Eq. A.24,

$$X = \frac{F}{\left((k - m\omega^2)^2 + \omega^2 c^2\right)^{1/2}}$$

and this may be non-dimensionalised by multiplying throughout by k and dividing by F. Thus,

$$\frac{kX}{F} = \frac{1}{\left(\left(1 - \frac{\omega^2}{\omega_n^2}\right)^2 + 4\xi^2 \frac{\omega^2}{\omega_n^2}\right)^{1/2}} \quad \text{and} \quad \tan \phi = \frac{-2\xi\omega/\omega_n}{\left(1 - \frac{\omega^2}{\omega_n^2}\right)} \tag{A.26}$$

The variation of kX/F and ϕ as functions of ω/ω_n is shown in Fig. A.11 for $\xi = 0.1$.

The main points to note are as follows:

1. $\frac{kX}{F} \to 1$ as $\frac{\omega}{\omega_n} \to 0$ this is generally known as the quasi-static condition as $\frac{X}{F} \to \frac{1}{k}$.
2. $\frac{kX}{F} \to 0$ as $\frac{\omega}{\omega_n} \to \infty$.
3. Resonance occurs, i.e. kX/F is a maximum, when $\mathrm{d}(kX/F)/\mathrm{d}\omega$ is zero. This can be shown to be when,

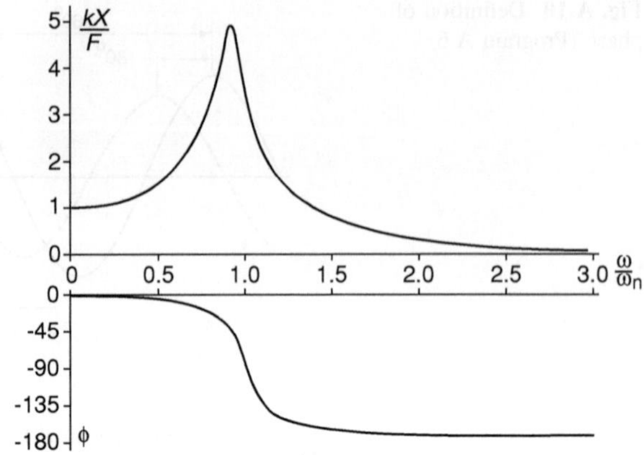

Fig. A.11 Steady-state response and phase for external force excitation (Program A.7)

$$\omega_{\mathrm{r}} = \omega_{\mathrm{n}}\sqrt{(1 - 2\xi^2)}$$

However, note that there is no real solution for ω_{r} when $\xi > 1/\sqrt{2}$, i.e. the amplitude of the response continuously falls with an increase of excitation frequency.

4. The final point of interest is the response amplitude at resonance, i.e. when $\omega = \omega_{\mathrm{r}} = \omega_{\mathrm{n}}\sqrt{(1 - 2\xi^2)}$

$$\frac{kX}{F} = \frac{1}{\left(\left(1 - (1 - 2\xi^2)\right)^2 + 4\xi^2(1 - 2\xi^2)\right)^{1/2}}$$

$$\frac{kX}{F} = \frac{1}{(4\xi^4 + 4\xi^2 - 8\xi^4)^{1/2}} = \frac{1}{(4\xi^2 - 4\xi^4)^{1/2}} = \frac{1}{2\xi(1 - \xi^2)^{1/2}}$$

which for small values of ξ is equal to $1/2\xi$.

The method of approach thus far has been to use Laplace Transforms. However, when only the steady-state response is required, it is possible to obtain this more directly using complex numbers and using $F e^{i\omega t}$ to represent the exciting force and a consequential steady-state response of $X e^{i\omega t}$. This method does not provide the transient solutions due to the initial conditions at the commencement of the excitation. If the reader is familiar with the complex number representation, then the next section is not relevant.

Fig. A.12 Space representation of complex variable i

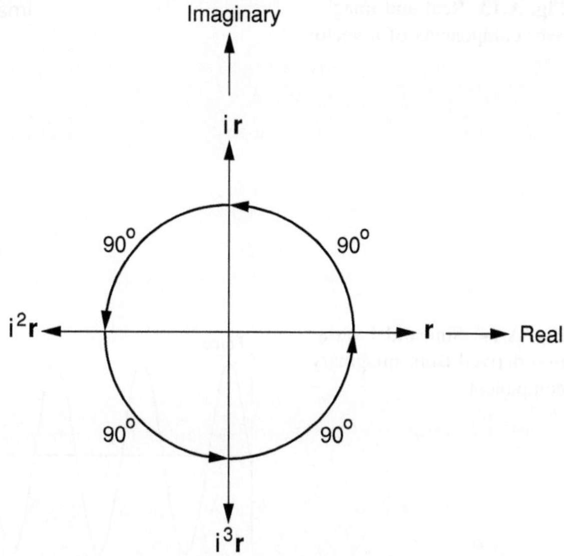

A.5 Complex Number Representation of Vibration

There are advantages in using complex numbers for the analysis of vibration, particularly when steady-state sinusoidal vibration is considered. The complex variable i (some texts use j rather than i) is defined as $\sqrt{-1}$. A space interpretation of i may be useful, consider Fig. A.12.

A phasor **r** is shown in Fig. A.12. Now let multiplication by i cause i**r** to rotate 90° anticlockwise relative to **r**. Multiply by i again and the result is a further rotation anticlockwise of 90°. Thus, $\mathrm{i}^2\mathbf{r}$ is opposite to **r**. It therefore follows that

$$\mathrm{i}^2\mathbf{r} = -\mathbf{r}$$

$\therefore \mathrm{i}^2 = -1$ and hence, $\mathrm{i} = \sqrt{-1}$

Thus, multiplication by the complex variable i can be used to represent an anticlockwise rotation of the phasor by 90°.

Referring to Fig. A.12 again, the phasor **r** is defined as being in the real direction (in-phase) and i**r** is in the complex direction (sometimes called the imaginary direction or the quadrature direction).

Now consider a general phasor and its components in the real (in-phase) and imaginary (90° leading) directions as shown in Fig. A.13.

The phasor **r** may then be represented by a component in the real direction and a component in the imaginary direction indicated by i, thus

$$\mathbf{r} = r\cos\,\theta + \mathrm{i}r\sin\,\theta$$

$$\therefore \mathbf{r} = r(\cos\,\theta + \mathrm{i}\sin\,\theta)\ \text{ and hence }\ \mathbf{r} = r\mathrm{e}^{\mathrm{i}\theta}$$

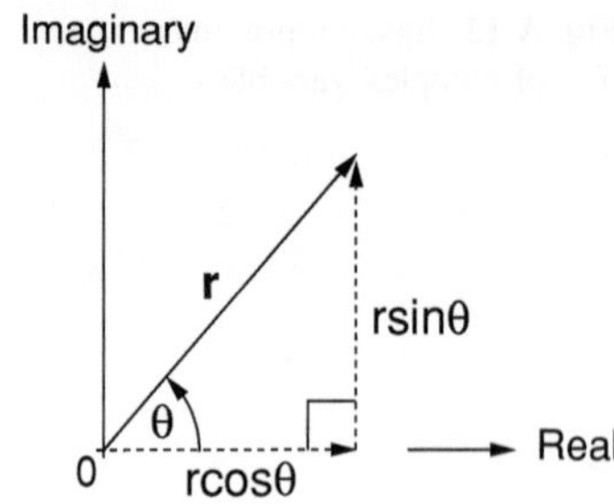

Fig. A.13 Real and imaginary components of a vector

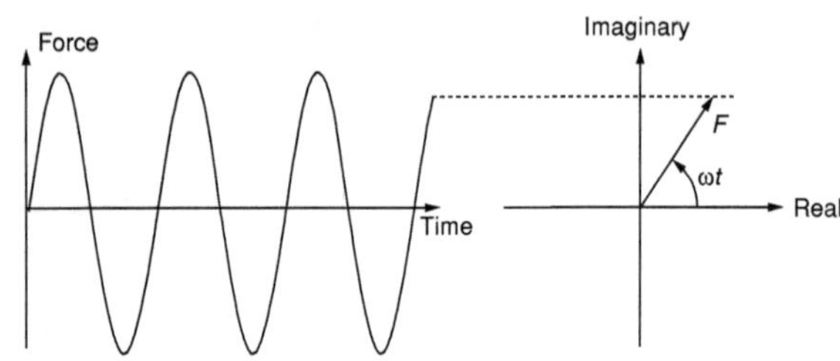

Fig. A.14 Sinusoidal variation derived from imaginary component

It is possible to represent a sinusoidally varying force $F \sin \omega t$, using complex numbers. Thus, for example, if we define the force variation with time as $F\mathrm{e}^{\mathrm{i}\omega t}$, then using the equations already derived,

$$F\mathrm{e}^{\mathrm{i}\omega t} = F \cos \omega t + \mathrm{i}F \sin \omega t$$

The imaginary component of $F\mathrm{e}^{\mathrm{i}\omega t}$ is thus a sinusoidal variation with time. This is shown graphically in Fig. A.14.

At this stage, it may not be clear where the advantage of using the complex variable arises. This will become apparent as we consider a sinusoidal excitation force and the sinusoidal response. In general, the response will lag behind the excitation. If the force is $F \sin \omega t$, then for a linear system, the response will be of the form $X \sin(\omega t + \phi)$ where ϕ is given by Eq. A.25. However, both the excitation and the response may be represented by rotating phasors. This is illustrated in Fig. A.15, with the force being represented by $\mathrm{Fe}^{\mathrm{i}\omega t}$ and the response by $\mathrm{Xe}^{\mathrm{i}(\omega t+\phi)}$. Note that positive angles are anticlockwise, but as ϕ from Eq. A.25 will be negative, it is shown negative in Fig. A.15.

We are normally interested in the ratio of the response to the excitation. Thus, if we take F as the phase reference, we have,

$$\frac{\mathrm{Xe}^{\mathrm{i}(\omega t+\phi)}}{\mathrm{Fe}^{\mathrm{i}\omega t}} = \frac{\mathrm{Xe}^{\mathrm{i}\phi} e^{\mathrm{i}\omega t}}{\mathrm{Fe}^{\mathrm{i}\omega t}} = \frac{\mathrm{Xe}^{\mathrm{i}\phi}}{F}$$

In this form, the force is the phase reference and we may draw F along the real axis and the response as a phasor lagging by some phase angle. At any instant, this is the same as freezing the rotating vectors of Fig. A.15 and then rotating them so that the force vector lies along the real axis as shown in Fig. A.16.

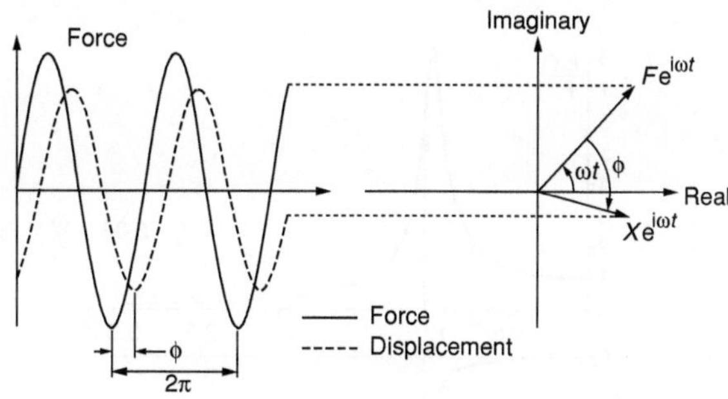

Fig. A.15 Representation of force and displacement by rotating vectors (Program A.8)

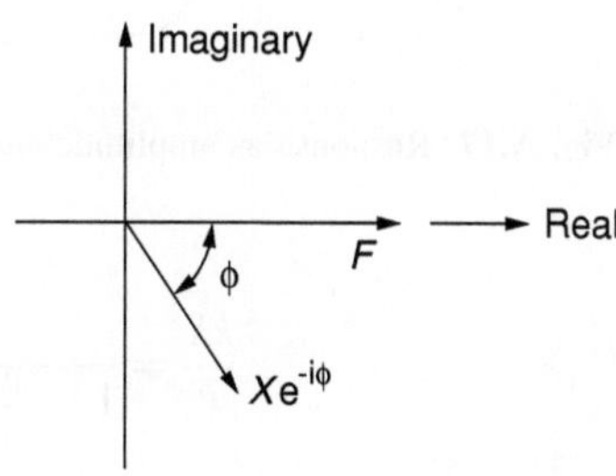

Fig. A.16 Response with excitation force *F* in the reference (real) direction

To illustrate the benefits of using complex numbers for vibration analysis, we shall return to the forced excitation of the spring/mass system with damping.

A.6 Forced Vibration: Complex Representation

Consider again the equation of motion, Eq. A.19, this becomes

$$m\frac{d^2x(t)}{dt^2} + c\frac{dx(t)}{dt} + kx(t) = \mathrm{F}e^{i\omega t} \tag{A.27}$$

Substituting $x(t) = \mathrm{X}e^{i\omega t}$ gives

$$-m\omega^2\mathrm{X}e^{i\omega t} + i\omega c\mathrm{X}e^{i\omega t} + k\mathrm{X}e^{i\omega t} = \mathrm{F}e^{i\omega t}$$

And thus,

$$\frac{X}{F} = \frac{1}{k - m\omega^2 + i\omega c}$$

This is a complex expression, the amplitude of which is the response amplitude and the phase of which indicates the phase between displacement and force. If the equation is non-dimensionalised by multiplying throughout by *k,* we obtain,

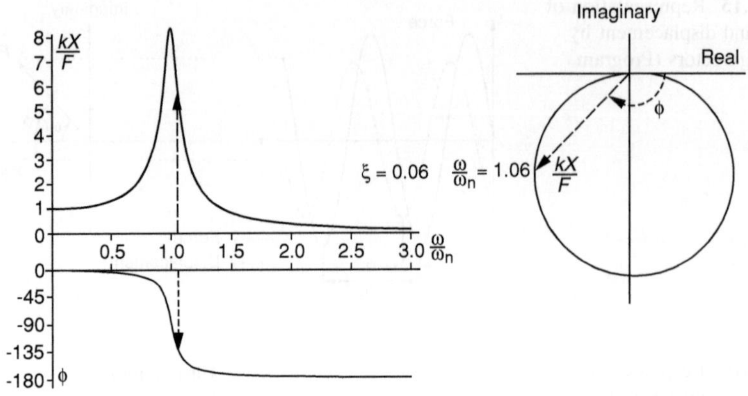

Fig. A.17 Response as amplitude and phase and also response locus (Program A.9)

$$\frac{kX}{F} = \frac{1}{1 - \frac{\omega^2}{\omega_n^2} + i2\xi\frac{\omega}{\omega_n}} = \frac{1 - \frac{\omega^2}{\omega_n^2} - i2\xi\frac{\omega}{\omega_n}}{\left(1 - \frac{\omega^2}{\omega_n^2}\right)^2 + 4\xi^2\frac{\omega^2}{\omega_n^2}}$$

This has the same response amplitude and phase lag as before,

$$\frac{kX}{F} = \frac{1}{\left(\left(1 - \frac{\omega^2}{\omega_n^2}\right)^2 + 4\xi^2\frac{\omega^2}{\omega_n^2}\right)^{1/2}} \quad \text{and} \quad \tan\phi = \frac{-2\xi\omega/\omega_n}{\left(1 - \frac{\omega^2}{\omega_n^2}\right)} \tag{A.26}$$

It is clear that this method allows the steady-state response to be obtained far more simply than using the method of Sect. A.4, and thus, it is the preferred method for steady-state analysis. Of great interest for chatter theory is the concept of a response locus.

A.6.1 Response Locus

For some applications, the response is presented as a polar frequency response plot that combines the response amplitude and phase in one diagram. This is illustrated in Fig. A.17. At each frequency of excitation, the response amplitude and the phase are plotted with phase angles positive anticlockwise from the *kX/F* axis. Each point on the locus represents, for the associated frequency, the amplitude (radius) and phase (angle) of the response.

For the particular case of chatter in machine tools, the in-phase (i.e. the real component of the response) is of great interest. This is shown in Fig. A.18. It will be found that for chatter, the maximum negative in-phase component determines whether machining will be unconditionally stable or not.

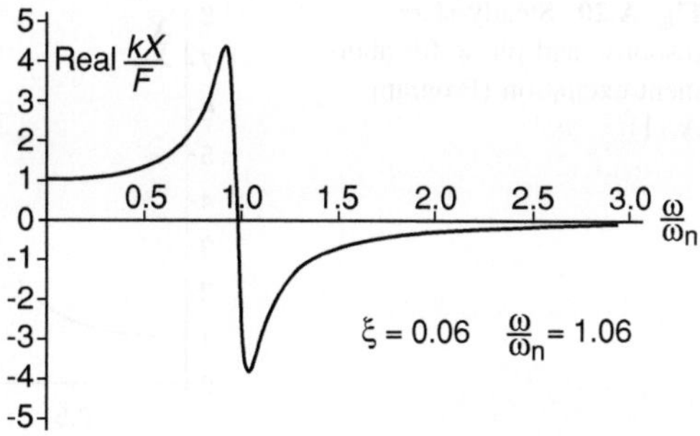

Fig. 18 In-phase component of the response (Program A.10)

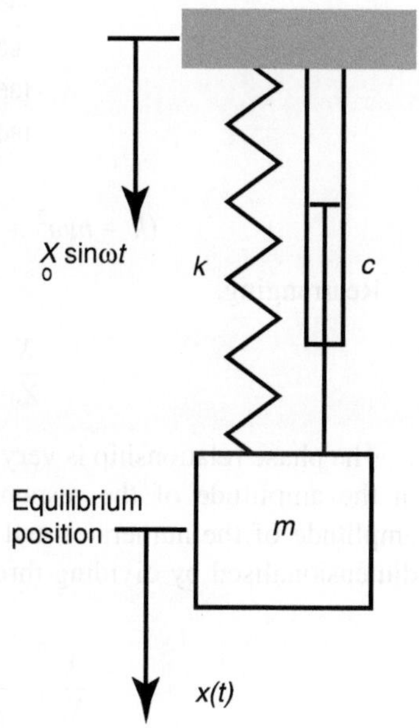

Fig. A.19 Abutment excitation

A.7 Forced Vibration: Abutment Excitation

Sinusoidal excitation by the abutment is shown in the Fig. A.19. Applying Newton's second law of motion gives,

$$m\frac{d^2x(t)}{dt^2} = k(x_0(t) - x(t)) + c\left(\frac{dx_0(t)}{dt} - \frac{dx(t)}{dt}\right) \tag{A.28}$$

If a Laplace Transform solution was completed, it would be found that there is an initial 'start-up' transient as for the excitation by a sinusoidal force.

For this example, it is much simpler to find the steady-state response by taking the input excitation to be $x_o = X_o e^{i\omega t}$ and the response to be $x(t) = Xe^{i\omega t}$. On substituting in Eq. A.25, this gives,

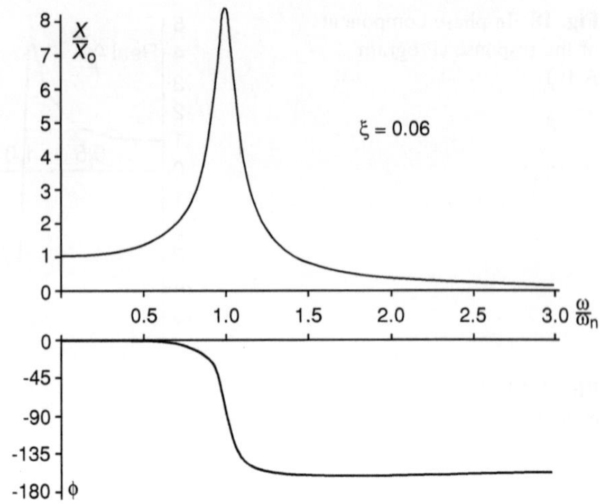

Fig. A.20 Steady-state response and phase for abutment excitation (Program A.11)

$$(k - m\omega^2 + \mathrm{i}\omega c)X\mathrm{e}^{\mathrm{i}\omega t} = (k + \mathrm{i}\omega c)X_0\mathrm{e}^{\mathrm{i}\omega t}$$

Rearranging,

$$\frac{X}{X_0} = \frac{(k + \mathrm{i}\omega c)}{(k - m\omega^2 + \mathrm{i}\omega c)}$$

The phase relationship is very complicated but normally we would be interested in the amplitude of the response and this may be determined by finding the amplitude of the numerator and denominator functions. If the result is also non-dimensionalised by dividing throughout by k, the following result is obtained,

$$\frac{X}{X_0} = \frac{\left(1 + 4\xi^2 \frac{\omega^2}{\omega_n^2}\right)^{1/2}}{\left(\left(1 - \frac{\omega^2}{\omega_n^2}\right)^2 + 4\xi^2 \frac{\omega^2}{\omega_n^2}\right)^{1/2}} \tag{A.29}$$

and the phase is given by,

$$\tan\phi = \frac{-2\xi \frac{\omega^3}{\omega_n^3}}{\left(1 - \frac{\omega^2}{\omega_n^2}\right) + 4\xi^2 \frac{\omega^2}{\omega_n^2}}$$

Typical response and phase curves are shown in Fig. A.20.

The main points to note are as follows:

1. $\frac{X}{X_0} \to 1$ as $\frac{\omega}{\omega_n} \to 0$, i.e. almost no deflection across the spring.
2. $\frac{X}{X_0} = 1$ when $\frac{\omega}{\omega_n} = \sqrt{2}$ and is independent of the value of ξ.
3. $\frac{X}{X_0} \to 0$ as $\frac{\omega}{\omega_n} \to \infty$, when vibration isolation occurs.

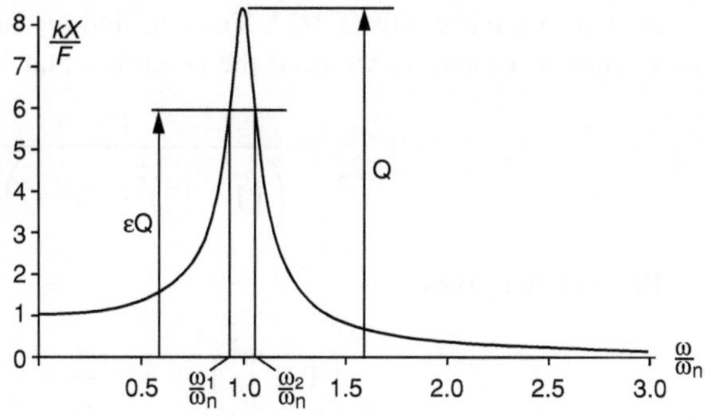

Fig. A.21 Typical response curve

4. Resonance occurs, i.e. X/X_0 is a maximum, when $d(X/X_0)/d\omega$ is zero. This can be shown to be when

$$\omega_r = \omega_n \sqrt{\frac{-1 + \sqrt{1 + 8\xi^2}}{4\xi^2}}$$

Note that this resonance occurs for all values of ξ.

A.8 Dynamic Magnification Factor (Q)

One of the areas of confusion in the vibration area is the multiplicity of methods of expressing quantities of damping. So far we have encountered the following:

1. c the viscous damping coefficient.
2. ξ the viscous damping ratio defined as $\xi = c/2\sqrt{km}$
3. δ the logarithmic decrement, defined as $\delta = 2\pi/\sqrt{(1-\xi^2)}$

Another common measure of damping is the dynamic magnification factor Q. The ratio of the amplitude at resonance to the quasi-static (low-frequency) amplitude when excited by an oscillating force is termed the dynamic magnification factor Q. For lightly damped structures, Q may be shown to equal to $1/2\xi$. This is a particularly useful measure as it may be determined from the steady-state response curve using the bandwidth of the resonance peak as shown below.

Figure A.21 shows a typical resonance curve and at the frequencies ω_1 and ω_2 the response is ε times the response at resonance. $\omega_2 - \omega_1$ is called the bandwidth ($\Delta\omega$) and may be used to determine Q. As has been shown previously,

$$\frac{kX}{F} = \frac{1}{\left(\left(1 - \frac{\omega^2}{\omega_n^2}\right)^2 + 4\xi^2 \frac{\omega^2}{\omega_n^2}\right)^{1/2}}$$

And at resonance, this is $1/2\xi$. Thus, to find ω_1 and ω_2 when the response is ε times that at resonance, we need the condition that

$$\frac{\varepsilon}{2\xi} = \frac{1}{\left(\left(1 - \frac{\omega^2}{\omega_n^2}\right)^2 + 4\xi^2 \frac{\omega^2}{\omega_n^2}\right)^{1/2}}$$

Rearranging gives

$$\left(1 - \frac{\omega^2}{\omega_n^2}\right)^2 + 4\xi^2 \frac{\omega^2}{\omega_n^2} = \frac{4\xi^2}{\varepsilon^2}$$

If we put $q = \omega/\omega_n$, then

$$q^4 + q^2(4\xi^2 - 2) + 1 - \frac{4\xi^2}{\varepsilon^2} = 0$$

$$\therefore q^2 = \frac{-(4\xi^2 - 2) \pm \sqrt{(4\xi^2 - 2)^2 - 4\left(1 - \frac{4\xi^2}{\varepsilon^2}\right)}}{2}$$

and

$$q^2 = 1 - 2\xi^2 \pm 2\xi\sqrt{\xi^2 - 1 + \frac{1}{\varepsilon^2}}$$

Now if ξ is small and ε is not too near unity,

$$\therefore q^2 = 1 \pm 2\xi\sqrt{\frac{1}{\varepsilon^2} - 1}$$

Thus,

$$\frac{\omega_1^2}{\omega_n^2} = 1 - 2\xi\sqrt{\frac{1}{\varepsilon^2} - 1} \tag{A.30}$$

and

$$\frac{\omega_2^2}{\omega_n^2} = 1 + 2\xi\sqrt{\frac{1}{\varepsilon^2} - 1} \tag{A.31}$$

Now

$$\frac{\omega_2 - \omega_1}{\omega_n} = \frac{\omega_2^2 - \omega_1^2}{\omega_n(\omega_2 + \omega_1)}$$

And $\omega_2 + \omega_1$ is approximately equal to $2\omega_n$, therefore taking Eqs. A.31–Eq. A.30 gives,

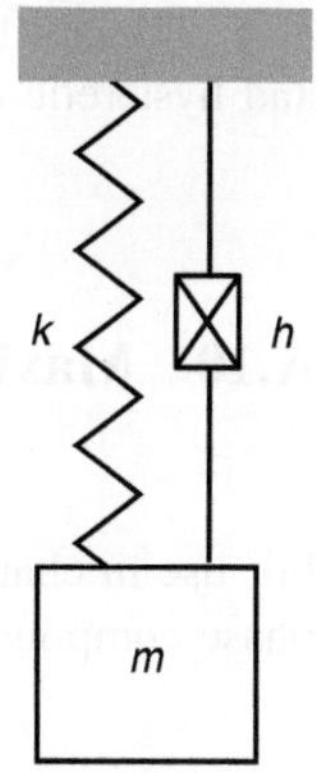

Fig. A.22 Representation of hysteretic damping

$$\frac{\Delta\omega}{\omega_n} = \frac{\omega_2 - \omega_1}{\omega_n} = \frac{\omega_2^2 - \omega_1^2}{2\omega_n^2} = \frac{\left(1 + 2\xi\sqrt{\frac{1}{\varepsilon^2} - 1}\right) - \left(1 - 2\xi\sqrt{\frac{1}{\varepsilon^2} - 1}\right)}{2}$$
$$= 2\xi\sqrt{\frac{1}{\varepsilon^2} - 1}$$

It is common to measure $\Delta\omega$ when $\varepsilon = 1/\sqrt{2}$ (this is called the 3-dB point as $20\log(1/\sqrt{2}) = -3$) and also to approximate $\omega_r = \omega_n$ so that

$$\frac{\Delta\omega}{\omega_r} = 2\xi \tag{A.32}$$

This is a very useful measure of damping as it may be applied to a response with several resonances and will give the equivalent damping ratio for each resonance.

A.9 Hysteretic Damping

Thus far damping has been assumed to be viscous in nature; that is, there is a retarding force that depends on velocity. Under steady-state vibration conditions with a displacement $Xe^{i\omega t}$, the viscous damping force is therefore $i\omega cXe^{i\omega t}$. This has an increasing magnitude (ωc) as the vibration frequency increases. However, in many structures and materials, the damping is not found to increase with frequency. It is therefore common to replace ωc with h so that the damping force does not depend on frequency but only on the amplitude of the vibration. This form of damping is referred to as hysteretic and the symbol used to represent it is shown in Fig. A.22.

The hysteretic damping model is only used for steady-state vibration and is not used for transients. Where hysteretic damping is used, it is a simple matter to replace ωc with h. There is also the equivalent of the viscous damping ratio

$\xi = c/2\sqrt{km}$; the hysteretic damping ratio is defined as $\alpha = h/k$. Thus, a spring and hysteretic damper are often represented for steady-state vibration as,

$$k + \mathrm{i}h = k(1 + \mathrm{i}h/k) = k(1 + \mathrm{i}\alpha)$$

A.10 Maximum Negative In-Phase Response

For use in chatter predictions, it is important to know the maximum negative in-phase component of the response. Now

$$\frac{X}{F} = \frac{1}{k - m\omega^2 + \mathrm{i}\omega c} = \frac{k - m\omega^2 - \mathrm{i}\omega c}{(k - m\omega^2)^2 + \omega^2 c^2}$$

So that the in-phase (real) part is

$$\left(\frac{X}{F}\right)_{\text{Real}} = \frac{k - m\omega^2}{(k - m\omega^2)^2 + \omega^2 c^2} = \left(k - m\omega^2\right)\left(\left(k - m\omega^2\right)^2 + \omega^2 c^2\right)^{-1}$$

To find the maximum negative (or positive) or value, differentiate with respect to time and equate to zero, so that

$$\begin{aligned}\frac{\mathrm{d}}{\mathrm{d}\omega}\left(\frac{X}{F}\right)_{\text{Real}} &= -2m\omega\left(\left(k - m\omega^2\right)^2 + \omega^2 c^2\right)^{-1} \\ &\quad - \left(k - m\omega^2\right)\left(\left(k - m\omega^2\right)^2 + \omega^2 c^2\right)^{-2}\left(2\left(k - m\omega^2\right)(-2m\omega) + 2\omega c^2\right) \\ &= \frac{-2m\omega\left(\left(k - m\omega^2\right)^2 + \omega^2 c^2\right) - (k - m\omega^2)(2(k - m\omega^2)(-2m\omega) + 2\omega c^2)}{\left(\left(k - m\omega^2\right)^2 + \omega^2 c^2\right)^2}\end{aligned}$$

This is zero when the numerator is zero, i.e. when

$$-2m\omega\left(\left(k - m\omega^2\right)^2 + \omega^2 c^2\right) - \left(k - m\omega^2\right)\left(2\left(k - m\omega^2\right)(-2m\omega) + 2\omega c^2\right) = 0$$

Expanding

$$\begin{aligned}&- 2m\left(k^2 - 2mk\omega^2 + m^2\omega^4 + \omega^2 c^2\right) + 4m\left(k^2 - 2mk\omega^2 + m^2\omega^4\right) \\ &- 2c^2\left(k - m\omega^2\right) = 0\end{aligned}$$

Grouping terms

$$m^3\omega^4 - 2m^2 k\omega^2 + mk^2 - c^2 k = 0$$

Therefore,

$$\omega^2 = \frac{m^2 k \pm \sqrt{m^4 k^2 - m^3(mk^2 - c^2 k)}}{m^3} = \frac{k \pm \sqrt{c^2 k/m}}{m}$$

Using non-dimensional terms,

$$\omega^2 = \frac{k}{m} \pm \sqrt{\frac{k}{m}\frac{c^2}{m^2}} = \omega_n^2 \pm \sqrt{4\xi^2\omega_n^4} = \omega_n^2(1 \pm 2\xi)$$

Since the maximum negative in-phase response is above ω_n,

$$\omega = \omega_n\sqrt{1 + 2\xi}$$

The value of the maximum negative in-phase response is

$$\left(\frac{X}{F}\right)_{\text{Real}} = \frac{k - m\omega_n^2(1+2\xi)}{\left(k - m\omega_n^2(1+2\xi)\right)^2 + c^2\omega_n^2(1+2\xi)}$$
$$= \frac{-2\xi}{k\left(4\xi^2 + 4\xi^2(1+2\xi)\right)} = \frac{-1}{4k\xi(1+\xi)} = \frac{-m}{c\left(2\sqrt{mk} + c\right)}$$

For machine tool chatter, it is possible to have a negative mode (see Sect. 2.3) and it will then be necessary to find the maximum positive in-phase response.

$\omega = \omega_n\sqrt{1 - 2\xi}$ gives the maximum positive in-phase response.

And the value of the maximum positive in-phase response is

$$\left(\frac{X}{F}\right)_{\text{Real}} = \frac{k - m\omega_n^2(1-2\xi)}{\left(k - m\omega_n^2(1-2\xi)\right)^2 + c^2\omega_n^2(1-2\xi)}$$
$$= \frac{2\xi}{k\left(4\xi^2 + 4\xi^2(1-2\xi)\right)} = \frac{1}{4k\xi(1-\xi)} = \frac{m}{c\left(2\sqrt{mk} - c\right)}$$

A.11 Conclusions

It has been shown that vibration may occur because of some initial conditions/disturbances and also because of excitation by cyclic forces. In general, cures for unwanted vibration may be achieved by,

(a) Removing the initial conditions/disturbance that cause the transient vibration.
(b) Removing the cyclic excitation.
(c) Increasing the damping to reduce the response at resonance.
(d) Avoiding resonant conditions by either changing the resonant frequency or the frequency of excitation.
(e) Employing vibration isolation, i.e. making the resonant frequency much less than the excitation frequency. Also note that for abutment excitation, increasing damping is not good for vibration isolation.
(f) For chatter performance, the situation is more complex (See Chaps. 1 and 2).

Appendix B
Two-Degrees-of-Freedom Vibration

The spring/mass system considered in Appendix A was constrained to move in a single axial direction. It followed that the system had one degree of freedom, as it is possible to uniquely define the position of the system by one coordinate, i.e. x, most vibratory systems have many degrees of freedom and therefore more complex models are required to model their vibration behaviour. It is useful to consider a two-degrees-of-freedom system, as this may give an understanding that may helpfully be extrapolated to systems with many degrees of freedom. Initially, we will consider systems without damping because,

1. The mathematics is easier.
2. In practice, the damping is often small.
3. The main results are not too dependent on damping.
4. Damping may be considered later, either quantitatively or qualitatively.

The approach to be adopted initially will not be mathematically rigorous, but gives natural frequencies very simply, and a more rigorous method will be adopted subsequently.

B.1 Natural Frequency Calculation

The simplified approach will be first applied to the single-degree-of-freedom system already considered and shown in Fig. B.1.

B.1.1 One Degree of Freedom

Since there is no damping, the system will vibrate, if disturbed, at its natural frequency and this vibration will be non-decaying. The motion may therefore be assumed to be of the form $x(t) = X \sin \omega t$. The equation of motion with no external excitation is,

B. Stone, *Chatter and Machine Tools*, DOI: 10.1007/978-3-319-05236-6,

Fig. B.1 Undamped one-degree-of-freedom system

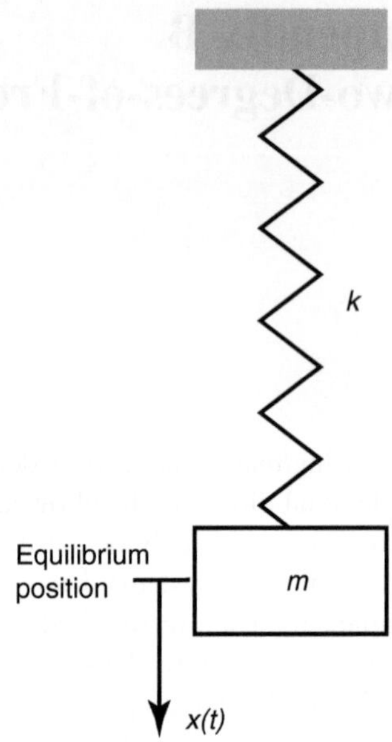

$$m\frac{d^2x(t)}{dt^2} + kx(t) = 0$$

which is known to be the differential equation describing simple harmonic motion.

Substituting the assumed form of $x(t)$ gives

$$-m\omega^2 X \sin \omega t + kX \sin \omega t = 0$$

As $\sin \omega t$ is not zero for all t

$$X(k - m\omega^2) = 0$$

Now either $X = 0$ and then there is no motion, i.e. the system is at rest, or X exists and there is a non-decaying vibration $X \sin \omega t$ if

$$(k - m\omega^2) = 0$$

i.e. when $\omega = \omega_n = \sqrt{\frac{k}{m}}$, the undamped natural frequency as found in Appendix A.

Thus, we have determined the natural frequency by assuming sinusoidal motion under free (i.e. not forced) vibration conditions and then finding, in this case, the single frequency at which this is possible. This definition is the undamped natural frequency.

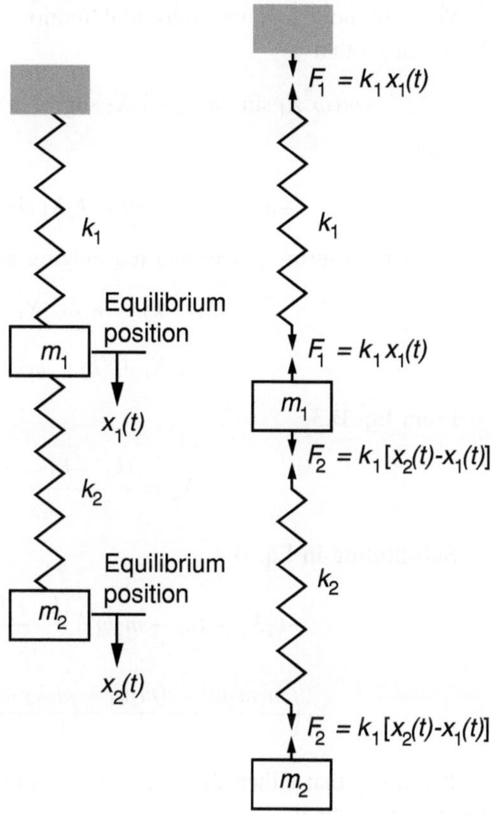

Fig. B.2 Axial two-degrees-of-freedom system and free-body diagrams

B.1.2 Two Degrees of Freedom

The same approach may be applied to a two-degree-of-freedom system, once the equations of motion are known. Consider the two-degree-of-freedom system shown in Fig. B.2. The two coordinates x_1 and x_2 uniquely define the position of the system, if it is constrained to move axially.

Consider the deflected position at some time t. From the free-body diagram of the mass m_1 and applying Newton's second law, we obtain the equation of motion of the mass m_1 as,

$$m_1 \frac{\mathrm{d}^2 x_1}{\mathrm{d}t^2} = -k_1 x_1 + k_2(x_2 - x_1) \tag{B.1}$$

and similarly, the equation of motion of the mass m_2 is

$$m_2 \frac{\mathrm{d}^2 x_2}{\mathrm{d}t^2} = -k_2(x_2 - x_1) \tag{B.2}$$

We will now assume sinusoidal motion such that $x_1 = X_1 \sin \omega t$ and $x_2 = X_2 \sin \omega t$ so that

$$-m_1\omega^2 X_1 \sin \omega t = -k_1 X_1 \sin \omega t - k_2 X_1 \sin \omega t + k_2 X_2 \sin \omega t$$

And

$$-m_2\omega^2 X_2 \sin \omega t = k_2 X_1 \sin \omega t - k_2 X_2 \sin \omega t$$

The sin ωt terms cancel and rearranging gives

$$(k_1 + k_2 - m_1\omega^2)X_1 - k_2 X_2 = 0 \tag{B.3}$$

$$-k_2 X_1 + (k_2 - m_2\omega^2)X_2 = 0 \tag{B.4}$$

From Eq. B.3,

$$X_2 = \frac{(k_1 + k_2 - m_1\omega^2)}{k_2} X_1 \tag{B.5}$$

Substituting in Eq. B.4,

$$-k_2 X_1 + (k_2 - m_2\omega^2)\frac{(k_1 + k_2 - m_1\omega^2)}{k_2} X_1 = 0$$

$$\therefore X_1 \left(\frac{m_1 m_2 \omega^4 - (m_2 k_1 + m_2 k_2 + m_1 k_2)\omega^2 + k_1 k_2}{k_2} \right) = 0$$

It follows that either $X_1 = 0$, and no motion exists, or motion of the assumed form may occur if,

$$m_1 m_2 \omega^4 - (m_2 k_1 + m_2 k_2 + m_1 k_2)\omega^2 + k_1 k_2 = 0 \tag{B.6}$$

This is a quadratic equation in ω^2 and thus gives two frequencies, at which sinusoidal and non-decaying motion at a single frequency may occur without being forced. That is, there are two natural frequencies ω_{n1} and ω_{n2}. As a particular example take the situation where $m_1 = m_2 = m$ and $k_1 = k_2 = k$. Then Eq. B.6 becomes,

$$m^2\omega^4 - 3mk\omega^2 + k^2 = 0 \tag{B.7}$$

Solving this equation gives the two natural frequencies as,

$$\omega_{n1} = 0.618\sqrt{\frac{k}{m}} \text{ and } \omega_{n2} = 1.618\sqrt{\frac{k}{m}}$$

(It must be noted that the recurrence of the particular sequence of numbers "618" in this example is not to be taken as indicating that this will normally be the case, rather it is the result of the particular example being investigated).

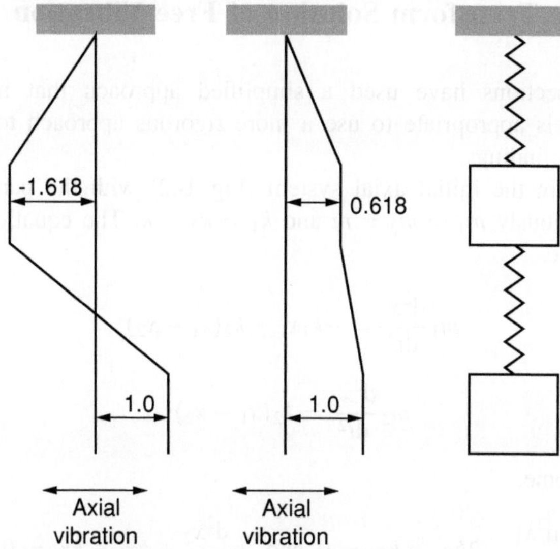

Fig. B.3 Mode shapes—deflection shown perpendicular to the real direction (Program B.1)

For each frequency, there is an associated ratio of X_1/X_2. Rearranging Eq. (B.5) with $m_1 = m_2 = m$ and $k_1 = k_2 = k$,

$$\frac{X_1}{X_2} = \frac{k}{(2k - m\omega^2)}$$

$$\text{For } \omega_{n1} = 0.618\sqrt{\frac{k}{m}} \quad \frac{X_1}{X_2} = \frac{k}{(2k - 0.382k)} = 0.618$$

$$\text{And for } \omega_{n2} = 1.618\sqrt{\frac{k}{m}} \quad \frac{X_1}{X_2} = \frac{k}{(2k - 2.618k)} = -1.618$$

These ratios are called mode shapes, or Eigen vectors. (Again it must be stressed that the recurrence of the particular sequence of numbers "618" in this example is not to be taken as indicating that this will normally be the case).

Mode plots are shown in Fig. B.3. In this figure, the axial motion amplitude at each position is shown perpendicular to the axis of vibration for clarity. It is informative to observe the motion and the animation program will allow this.

Thus, there are two modes of vibration with natural frequencies ω_{n1} and ω_{n2} and having associated mode shapes. It should also be noted that the natural frequencies are not dependent on the amplitude of vibration though the ratio of amplitudes (the mode shape) is fixed for a particular natural frequency.

B.2 Laplace Transform Solution of Free Vibration

The previous sections have used a simplified approach that minimises the mathematics. It is appropriate to use a more rigorous approach to confirm and extend the main findings.

Consider again the initial axial system (Fig. B.2) with the particular values considered previously $m_1 = m_2 = m$ and $k_1 = k_2 = k$. The equations of motion were found to be,

$$m_1 \frac{\mathrm{d}^2 x_1}{\mathrm{d}t^2} = -k_1 x_1 - k_2(x_1 - x_2) \tag{B.1}$$

$$m_2 \frac{\mathrm{d}^2 x_2}{\mathrm{d}t^2} = k_2(x_1 - x_2) \tag{B.2}$$

which thus become,

$$m\frac{\mathrm{d}^2 x_1}{\mathrm{d}t^2} + 2kx_1 - kx_2 = 0 \text{ and } m\frac{\mathrm{d}^2 x_2}{\mathrm{d}t^2} + kx_2 - kx_1 = 0$$

Taking Laplace Transforms

$$m\left[-\frac{\mathrm{d}x_1(0)}{\mathrm{d}t} - sx_1(0) + s^2 X_1(s)\right] + 2kX_1(s) - kX_2(s) = 0$$

$$m\left[-\frac{\mathrm{d}x_2(0)}{\mathrm{d}t} - sx_2(0) + s^2 X_2(s)\right] + kX_2(s) - kX_1(s) = 0$$

If as an example, the initial conditions are taken to be

$$\frac{\mathrm{d}x_2(0)}{\mathrm{d}t} = x_2(0) = \frac{\mathrm{d}x_1(0)}{\mathrm{d}t} = 0 \quad \text{and} \quad x_1(0) = X_0$$

Then

$$X_1(s)(ms^2 + 2k) + X_2(s)(-k) = msX_0 \tag{B.8}$$

And

$$X_1(s)(-k) + X_2(s)(ms^2 + k) = 0 \tag{B.9}$$

From Eq. B.9,

$$X_2(s) = \frac{k}{ms^2 + k} X_1(s)$$

Substituting in Eq. B.8,

$$X_1(s)(ms^2 + 2k) - \frac{k^2}{ms^2 + k} X_1(s) = msX_0$$

$$\therefore X_1(s) = \frac{(ms^2 + k)ms}{(ms^2 + 2k)(ms^2 + k) - k^2} X_0$$

$$\therefore X_1(s) = \frac{m^2s^3 + kms}{m^2s^4 + 3kms^2 + k^2} X_0$$

After a great deal of mathematics, this becomes,

$$X_1(s) = \frac{0.276X_0s}{s^2 + \left(0.618\sqrt{\frac{k}{m}}\right)^2} + \frac{0.724X_0s}{s^2 + \left(1.618\sqrt{\frac{k}{m}}\right)^2}$$

Taking Inverse Laplace Transforms,

$$x_1 = 0.276X_0 \cos\left(0.618\sqrt{\frac{k}{m}}t\right) + 0.724X_0 \cos\left(1.618\sqrt{\frac{k}{m}}t\right)$$

It was shown in Sect. B.1.2 that $\omega_{n1} = 0.618\sqrt{\frac{k}{m}}$ and $\omega_{n2} = 1.618\sqrt{\frac{k}{m}}$ are the two natural frequencies of the system, so that

$$x_1 = 0.276X_0 \cos(\omega_{n1}t) + 0.724X_0 \cos(\omega_{n2}t)$$

The motion is thus the superposition of vibration at the two natural frequencies. If a similar analysis is done to derive x_2, then the following equation is obtained,

$$x_2 = 0.447X_0 \cos(\omega_{n1}t) - 0.447X_0 \cos(\omega_{n2}t)$$

Again the motion is the superposition of vibration at the two natural frequencies. In addition, the ratio x_1/x_2 of the components at the frequency ω_{n1} is $0.276X_o/0.447X_o = 0.618$, the first mode shape, and at the frequency ω_{n2}, the ratio x_1/x_2 is $0.724X_o/(-0.447X_o) = -1.618$, the second mode shape. Thus, both natural frequencies are superposed and each is vibrating in its mode shape. The variation of both x_1 and x_2 with $\omega_n t$ ($\omega_n = \sqrt{k/m}$) is shown in Fig. B.4.

Program B.3 allows the mode contributions and their superposition to be observed. This superposition of modes may be generalised and we may conclude that the free vibration of undamped systems will normally involve the superposition of all the natural frequencies vibrating in their mode shapes. The particular motion will depend on the initial conditions and the resulting amplitude weighting taken on by each mode.

B.3 Laplace Transform Solution of Forced Vibration

Consider again the axial system with the particular values considered previously ($m_1 = m_2 = m$ and $k_1 = k_2 = k$) and with a sinusoidal input on the abutment given by $x_o = X_o \sin \omega t$, as shown in Fig. B.5.

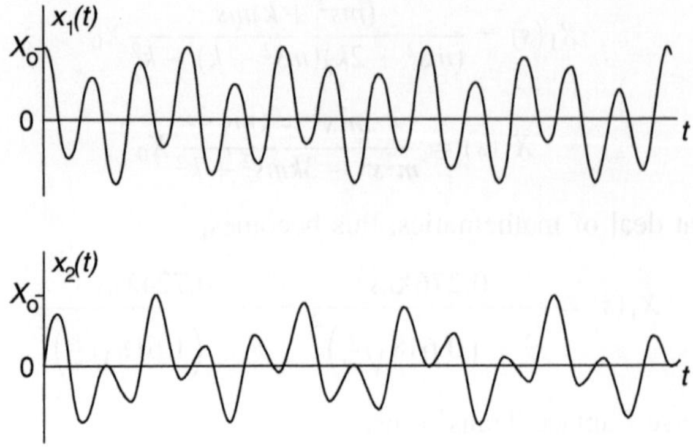

Fig. B.4 Transient motion of two-degrees-of-freedom system (Program B.2)

Fig. B.5 Abutment excitation

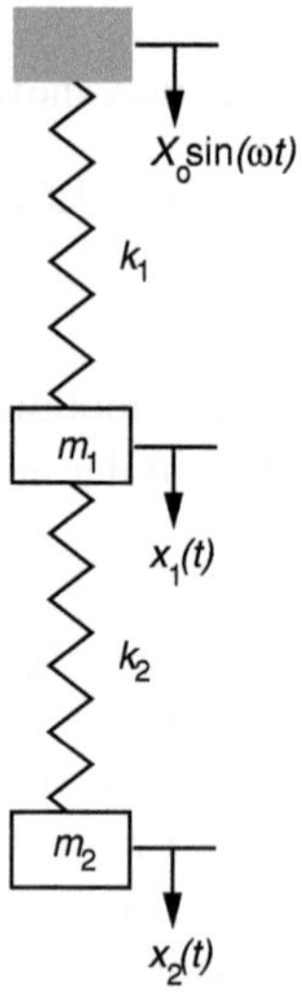

The equations of motion are

$$m\frac{d^2x_1}{dt^2} = k(x_o - x_1) - k(x_2 - x_1)$$

$$m\frac{d^2x_2}{dt^2} = -k(x_2 - x_1)$$

Rearranging and substituting for x_o

$$m\frac{d^2x_1}{dt^2} + 2kx_1 - kx_2 = kX_o \sin \omega t$$

$$m\frac{d^2x_2}{dt^2} + kx_2 - kx_1 = 0$$

If we take Laplace transforms with all initial conditions zero (otherwise, a solution of the form derived in the previous section will be superposed, resulting from the initial conditions exciting the natural frequencies) we obtain,

$$m[s^2X_1(s)] + 2kX_1(s) - kX_2(s) = kX_0\frac{\omega}{s^2+\omega^2} \tag{B.10}$$

$$m[s^2X_2(s)] + kX_2(s) - kX_1(s) = 0 \tag{B.11}$$

From Eq. B.11, $X_2(s) = \frac{kX_1(s)}{ms^2+k}$
And substituting in Eq. B.10,

$$m[s^2X_1(s)] + 2kX_1(s) - \frac{k^2}{ms^2+k}X_1(s) = kX_0\frac{\omega}{s^2+\omega^2}$$

$$\therefore X_1(s) = \frac{(ms^2+k)kX_0\omega}{(s^2+\omega^2)(m^2s^4+3kms^2+k^2)}$$

Taking partial fractions

$$X_1(s) = \frac{\mathrm{A}s+\mathrm{B}}{(s^2+\omega^2)} + \frac{\mathrm{C}s^3+\mathrm{D}s^2+\mathrm{E}s+\mathrm{F}}{(m^2s^4+3kms^2+k^2)}$$

The second term produces vibration at the two natural frequencies $\omega_{n1} = 0.618\sqrt{\frac{k}{m}}$ and $\omega_{n2} = 1.618\sqrt{\frac{k}{m}}$. This is caused by the abrupt start-up of the abutment motion. The steady-state solution comes from the first term and if the values of A and B are determined, then

$$\mathrm{A} = 0 \text{ and } \mathrm{B} = \frac{\omega(k-m\omega^2)kX_0}{(m^2\omega^4+3km\omega^2+k^2)}$$

And the steady-state solution is obtained from

$$X_1(s) = \frac{\omega(k-m\omega^2)kX_0}{(s^2+\omega^2)(m^2\omega^4+3km\omega^2+k^2)}$$

Taking Inverse Laplace Transforms

$$x_1 = \frac{(k-m\omega^2)kX_0\sin\omega t}{(m^2\omega^4+3km\omega^2+k^2)}$$

If we make $x_1 = X_1 \sin\omega t$ after a great deal of mathematics it may be shown that

$$\frac{X_1}{X_0} = \frac{0.724}{\left(1-\frac{\omega^2}{\omega_{n1}^2}\right)} + \frac{0.276}{\left(1-\frac{\omega^2}{\omega_{n2}^2}\right)} \tag{B.12}$$

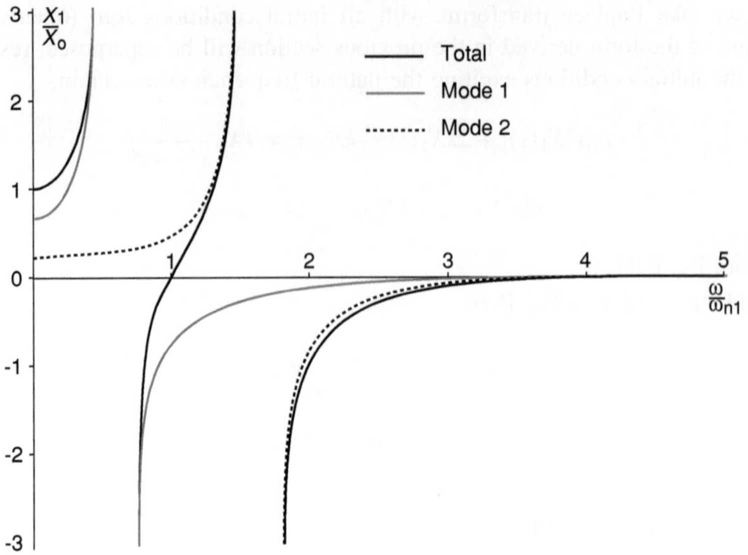

Fig. B.6 Steady-state response with modal components (Program B.4)

Now it has already been shown that the response of a single-degree-of-freedom system to abutment excitation is (see Appendix A.7),

$$\frac{X}{X_0} = \frac{\left(1 + 4\xi^2 \frac{\omega^2}{\omega_n^2}\right)^{1/2}}{\left(\left(1 - \frac{\omega^2}{\omega_n^2}\right)^2 + 4\xi^2 \frac{\omega^2}{\omega_n^2}\right)^{1/2}} \tag{A.29}$$

If the system is undamped ($\xi = 0$) this becomes,

$$\frac{X}{X_0} = \frac{1}{\left(1 - \frac{\omega^2}{\omega_n^2}\right)}$$

Thus, Eq. B.12 represents the superposition of the response of two single-degree-of-freedom systems having natural frequencies ω_{n1} and ω_{n2}, respectively. The response of the components and the superposition is shown in Fig. B.6 plotted against ω/ω_n ($\omega_n = \sqrt{k/m}$).

Now if the response of the mass m_2 had been found the following result would have been obtained as the steady-state response,

$$\frac{X_2}{X_0} = \frac{1.17}{\left(1 - \frac{\omega^2}{\omega_{n1}^2}\right)} - \frac{0.17}{\left(1 - \frac{\omega^2}{\omega_{n2}^2}\right)} \tag{B.13}$$

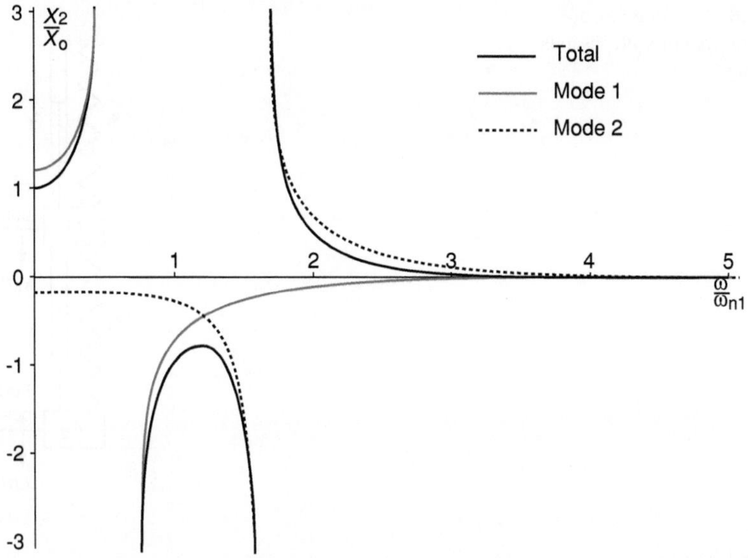

Fig. B.7 Steady-state response with modal components (Program B.4)

Again the motion is the superposition of the response of two single-degree-of-freedom systems. The individual responses and the superposition are shown in Fig. B.7.

In addition, the ratio of the first mode terms from Eqs. B.12 and B.13 gives X_1/X_2 as 0.724/1.17, i.e. 0.618, the first mode shape, and the ratio of the second mode terms gives X_1/X_2 as 0.276/−0.17, i.e. −1.618, the second mode shape.

B.4 Direct Derivation of Steady-State Response

Consider the previous example again, but let the excitation be represented by $x_0 = X_0 e^{i\omega t}$. The equations of motion are

$$m\frac{d^2x_1}{dt^2} + 2kx_1 - kx_2 = kX_0 e^{i\omega t}$$

$$m\frac{d^2x_2}{dt^2} + kx_2 - kx_1 = 0$$

In the steady-state, $x_1 = X_1 e^{i\omega t}$ and $x_2 = X_2 e^{i\omega t}$ so that

$$-m\omega^2 X_1 e^{i\omega t} + 2kX_1 e^{i\omega t} - kX_2 e^{i\omega t} = kX_0 e^{i\omega t}$$

$$-m\omega^2 X_2 e^{i\omega t} + kX_2 e^{i\omega t} - kX_1 e^{i\omega t} = 0$$

Fig. B.8 Two-degrees-of-freedom system with viscous damping

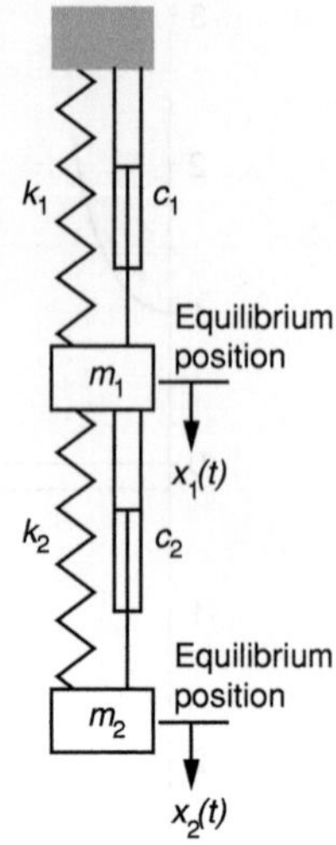

And hence,

$$(2k - m\omega^2)X_1 - kX_2 = kX_0 \tag{B.14}$$

$$-kX_1 + (k - m\omega^2)X_2 = 0 \tag{B.15}$$

From Eq. B.15,

$$X_2 = \frac{kX_1}{k - m\omega^2}$$

Substituting in Eq. B.14,

$$(2k - m\omega^2)X_1 - \frac{k^2X_1}{k - m\omega^2} = kX_0$$

And therefore,

$$\frac{X_1}{X_0} = \frac{(k - m\omega^2)k}{m^2\omega^4 + 3km\omega^2 + k^2} \tag{B.16}$$

which is the same result as obtained previously. However, the mathematics has been greatly reduced compared to using Laplace Transforms.

B.5 The Effects of Damping

Consider the system of Fig. B.2 but with viscous damping added as shown in Fig. B.8.

The equations of motion without any excitation are,

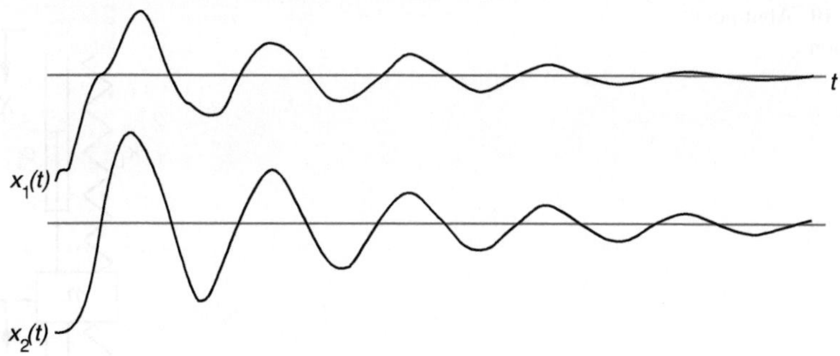

Fig. B.9 Damped transient response (Program B.5)

$$m_1 \frac{d^2x_1}{dt^2} = -k_1x_1 + k_2(x_2 - x_1) - c_1\frac{dx_1}{dt} + c_2\left(\frac{dx_2}{dt} - \frac{dx_1}{dt}\right)$$

$$m_2 \frac{d^2x_2}{dt^2} = -k_2(x_2 - x_1) - c_2\left(\frac{dx_2}{dt} - \frac{dx_1}{dt}\right)$$

These equations are difficult to solve without a computer. An animation program allows the motion to be observed for a range of initial conditions and system parameters. Except for particular values of the damping coefficients, it is not possible to separate the motion into modes that may be superimposed. A typical transient is shown in Fig. B.9.

B.5.1 Abutment Excitation

When a sinusoidal abutment excitation is applied (see Fig. B.10), the equations of motion become,

$$m_1 \frac{d^2x_1}{dt^2} = -k_1(x_1 - x_o) + k_2(x_2 - x_1) - c_1\left(\frac{dx_1}{dt} - \frac{dx_o}{dt}\right) + c_2\left(\frac{dx_2}{dt} - \frac{dx_1}{dt}\right)$$

and

$$m_2 \frac{d^2x_2}{dt^2} = -k_2(x_2 - x_1) - c_2\left(\frac{dx_2}{dt} - \frac{dx_1}{dt}\right)$$

and $x_o = X_o \sin \omega t$.

The mathematics is complex, but it is found that there is an initial transient before steady-state motion is achieved. An example is shown in Fig. B.11.

The steady-state solution to forced vibration is somewhat easier to obtain, as using an excitation and solution involving $e^{i\omega t}$ will give terms of the form $ic\omega e^{i\omega t}$ for each of the viscous damping terms. As there is normally a spring and viscous damper in parallel, these will produce terms of the form $(k + ic\omega)e^{i\omega t}$. Thus,

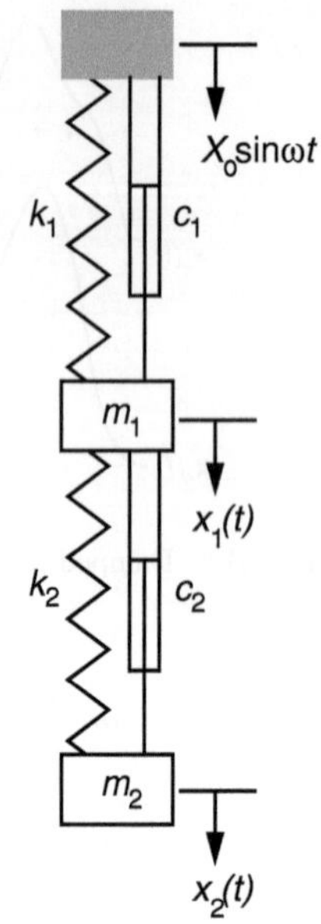

Fig. B.10 Abutment excitation

compared to the undamped case, it is only necessary to replace k with $(k + \mathrm{i}c\omega)$ in the final solution. Consider the system of Fig. B.8 with the particular values considered previously $m_1 = m_2 = m$ and $k_1 = k_2 = k$ and also with viscous dampers c in parallel with each of the springs. If an abutment excitation $X_o e^{\mathrm{i}\omega t}$ is again considered, the steady-state solution is obtained from Eq. B.16 by substituting $k = (k + \mathrm{i}c\omega)$. Thus,

$$\frac{X_1}{X_0} = \frac{(k + \mathrm{i}\omega c - m\omega^2)k}{m^2\omega^4 + 3(k + \mathrm{i}\omega c)m\omega^2 + (k + \mathrm{i}\omega c)^2} \tag{B.17}$$

This represents a complex response. If the magnitude alone is examined, it will be found to have two resonant peaks and for small values of c, these would be found to occur very close to the two undamped natural frequencies $\omega_{n1} = 0.618\sqrt{k/m}$ and $\omega_{n2} = 1.618\sqrt{k/m}$. Figure B.12 shows such a response for some particular numerical examples.

B.5.2 External Force

If the excitation is a sinusoidally varying force, as shown in Fig. B.13, similar results are obtained. Consider such a force applied to the mass m_1 as shown. The equations of motion are

$$m_1\frac{\mathrm{d}^2x_1}{\mathrm{d}t^2} = -k_1x_1 + k_2(x_2 - x_1) - c_1\frac{\mathrm{d}x_1}{\mathrm{d}t} + c_2\left(\frac{\mathrm{d}x_2}{\mathrm{d}t} - \frac{\mathrm{d}x_1}{\mathrm{d}t}\right) + \mathrm{F}\sin\omega t$$

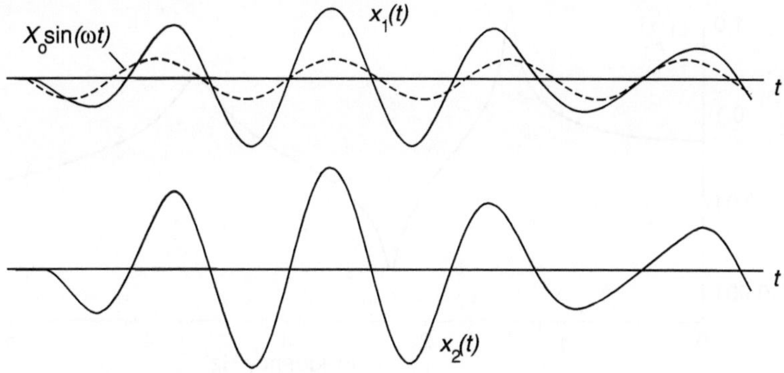

Fig. B.11 Start-up transients with sinusoidal excitation (Program B.6)

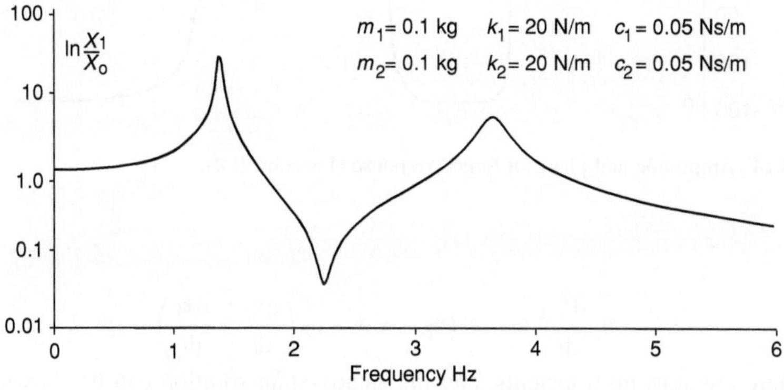

Fig. B.12 Response of damped two-degrees-of-freedom system (Program B.7)

Fig. B.13 Forced excitation of a two-degrees-of-freedom system

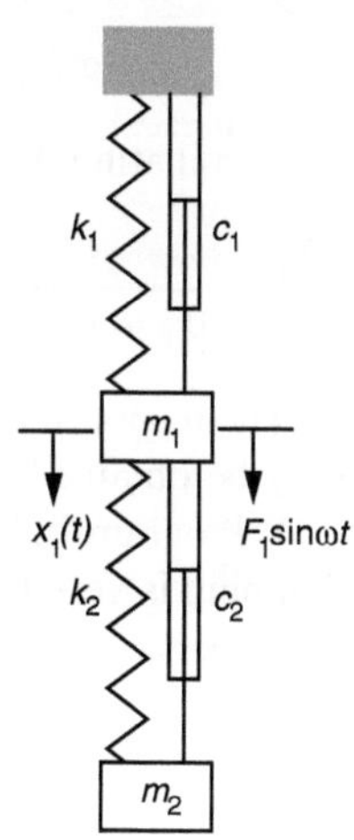

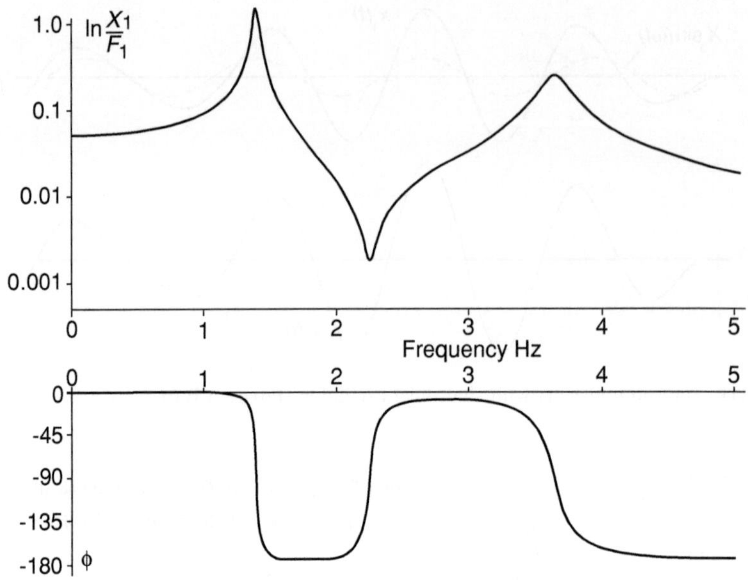

Fig. B.14 Amplitude and phase of forced response (Program B.8)

And

$$m_2 \frac{\mathrm{d}^2 x_2}{\mathrm{d}t^2} = -k_2(x_2 - x_1) - c_2\left(\frac{\mathrm{d}x_2}{\mathrm{d}t} - \frac{\mathrm{d}x_1}{\mathrm{d}t}\right)$$

There are start-up transients, and the steady-state solution can be shown to be found by using $F\mathrm{e}^{\mathrm{i}\omega t}$ and a response for the mass m_1 as $X_1\mathrm{e}^{\mathrm{i}\omega t}$ and mass m_2 as $X_2\mathrm{e}^{\mathrm{i}\omega t}$. Thus, the equations of motion yield,

$$-m_1\omega^2 X_1 = -k_1 X_1 + k_2(X_2 - X_1) - \mathrm{i}\omega c_1 X_1 + \mathrm{i}\omega c_2(X_2 - X_1) + F$$
$$-m_2\omega^2 X_2 = -k_2(X_2 - X_1) - \mathrm{i}\omega c_2(X_2 - X_1)$$

Eliminating X_2 from these equations gives,

$$\frac{X_1}{F} = \frac{k_2 - m_2\omega^2 + \mathrm{i}\omega c_2}{[k_1 + k_2 - m_1\omega^2 + \mathrm{i}\omega c_1 + \mathrm{i}\omega c_2][k_2 - m_2\omega^2 + \mathrm{i}\omega c_2] - [k_2 + \mathrm{i}\omega c_2]^2} \quad \text{(B.18)}$$

The response has an amplitude and phase, and using a computer program, these may be obtained. An example is shown in Fig. B.14.

If Program B.8 is used, it will be found that two resonant peaks occur unless the damping is very high. There is usually approximately an 180° phase change across a resonance.

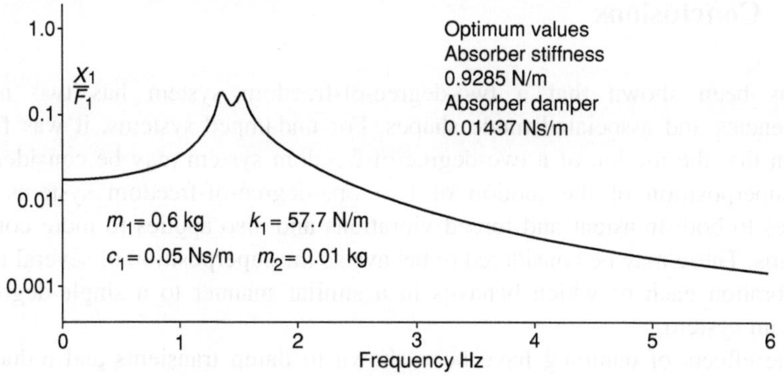

Fig. B.15 Response with optimised absorber (Program B.9)

B.6 Vibration Absorbers

The steady-state responses shown in Figs. B.12 and B.14 have two resonant peaks of different magnitudes. It is possible, by varying the values of the parameters, to make the two resonant peaks equal in magnitude by reducing one while increasing the other. In practice, one mode of an existing structure can be considered to be represented by the top spring/mass/damper (k_1 m_1 c_1) in Fig. B.13 and these parameter values are fixed. The lower spring/mass/damper (k_2 m_2 c_2) is called a vibration absorber, and the values may be changed.

When optimising the response of an existing system by adding a vibration absorber, it is conventional to minimise the maximum response. This involves choosing the optimum values for the absorber mass m_2, stiffness k_2 and damping c_2. It is found that the larger the absorber mass, the smaller the maximum response that may be achieved. Before computers, optimum values were found for the case with $c_1 = 0$, i.e. the system to which the absorber was added had no damping. For this case, the optimum values are given by

$$\frac{k_2}{k_1} = \frac{m_2/m_1}{[1 + m_2/m_1]^2} \quad \text{and} \quad \xi_2 = \frac{2c_2}{\sqrt{m_2 k_2}} = \sqrt{\frac{3m_2/m_1}{8[1 + m_2/m_1]^3}}$$

With the advent of computers, it is possible to find the optimum values for the general case including damping in the main system. For the system shown in Fig. B.13, the optimum response, for particular values of m_1, k_1, c_1 and m_2, is shown in Fig. B.15.

B.7 Conclusions

It has been shown that a two-degree-of-freedom system has two natural frequencies and associated mode shapes. For undamped systems, it was further shown that the motion of a two-degree-of-freedom system may be considered as the superposition of the motion of two one-degree-of-freedom systems. This applies to both transient and forced vibrations and also applies to more complex systems. These may be considered to behave as the superposition of several modes of vibration each of which behaves in a similar manner to a single-degree-of-freedom system.

The effects of damping have been shown to damp transients and reduce the amplitudes of vibration at resonance. Damping also can couple modes of vibration so that superposition is inaccurate.

Also it has been shown how the response to forced vibration may be improved by the use of vibration absorbers as the maximum response may be reduced. It is important to note that vibration absorbers used to stop chatter have to be optimised for that purpose (See Sect. 4.3).

Appendix C
A Systems Approach Using Receptances

As engineering systems have become more and more complex, so have the methods that have been developed for their analysis. This has resulted in the availability of mathematical models that may very accurately represent the real system. These allow engineers to examine possible problems at the drawing board stage or even existing problems with a working system. However, because the modelling is so complex, it is as difficult to comprehend the model as the original problem. Various parameters may be varied and the effect on the whole system determined but with so many parameters, which often interact, the task becomes daunting. When models were simpler, there was at least the advantage that the major parameters were isolated and it was easier to find possible ways of solving problems. Nevertheless, the disadvantage with the simple model was always the concern about the accuracy with which the real system was modelled. It therefore appears that an insuperable problem exists. Either the model is too simple but may be comprehensible, or the model accurately represents the system but is incomprehensible as far as finding a solution to a problem.

In many branches of engineering, the way forward has been to adopt a systems approach. Thus, a complex system is broken down into a number of subsystems that may be modelled separately. The problem then becomes one of the interactions between the subsystems. The effects of modifications to a single subsystem on the total system are more easily comprehended. Such a method may be developed for the general area of vibration. It is intended to introduce the relevant concepts that allow even complex problems to be amenable to solution using the systems approach. The concepts will be illustrated by simple examples.

C.1 Definition of Receptance

The systems approach uses receptances—the receptance α_{12} is defined as,

$$\alpha_{12} = \frac{X_1 \mathrm{e}^{\mathrm{i}\omega t}}{F_2 \mathrm{e}^{\mathrm{i}\omega t}} = \frac{X_1}{F_2} \tag{C.1}$$

B. Stone, *Chatter and Machine Tools*, DOI: 10.1007/978-3-319-05236-6,

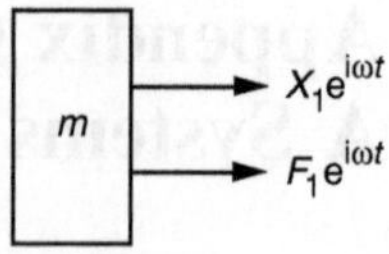

Fig. C.1 Excitation and response of a rigid mass

where X_1 is the steady-state response of a system at a position and in a direction defined by the subscript 1 and is often a complex number that indicates a phase with respect to the steady exciting force $F_2e^{i\omega t}$ that is applied to the system at a position and direction defined by the subscript 2. α_{12} is termed the cross-receptance, whereas α_{11} is called a direct receptance as then the force and displacement are at the same position and in the same direction. It is useful to consider some simple systems and derive their receptances. These receptances can then be used to build more complex systems.

C.1.1 Rigid Mass

Fig. C.1 shows the excitation of a rigid mass by an oscillating force. Let the force be applied through the centre of mass. Then, from Newton's second law,

$$F_1e^{i\omega t} = m\frac{d^2x_1}{dt^2}$$

Since $x_1 = X_1e^{i\omega t}$, we obtain $F_1e^{i\omega t} = -m\omega^2X_1e^{i\omega t}$
Therefore,

$$\alpha_{11} = \frac{X_1}{F_1} = -\frac{1}{m\omega^2} \tag{C.2}$$

C.1.2 Spring and Viscous Damper in Parallel

Figure C.2 shows the excitation of a spring and viscous damper. From Newton's second law,

$$F_1e^{i\omega t} = kx_1 + c\frac{dx_1}{dt}$$

Substituting $x_1 = X_1e^{i\omega t}$,

$$\alpha_{11} = \frac{X_1}{F_1} = \frac{1}{k + i\omega c} \tag{C.3}$$

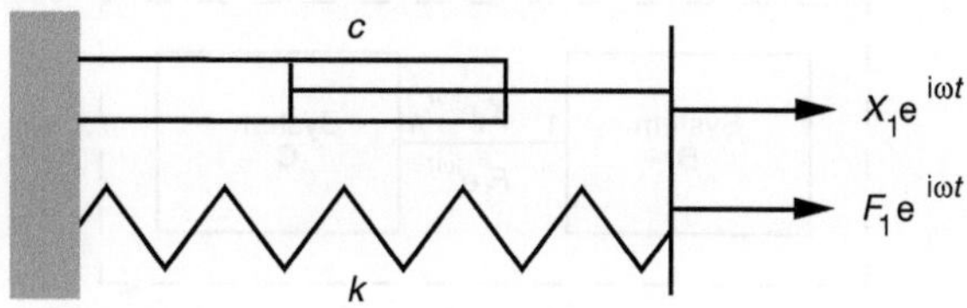

Fig. C.2 Excitation of a spring and viscous damper in parallel

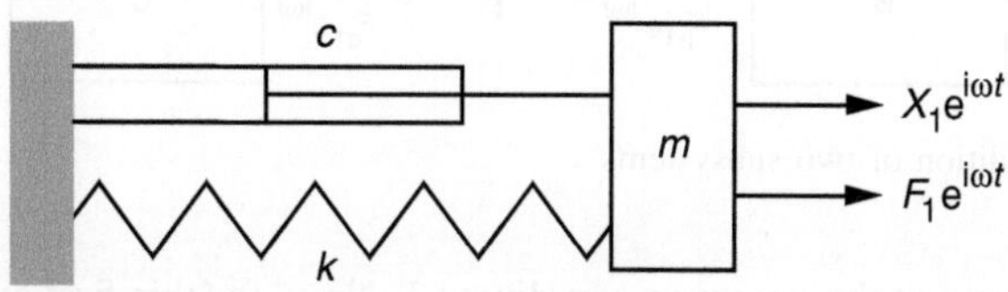

Fig. C.3 Excitation of a spring/mass with viscous damping

C.1.3 Spring/Mass System with Viscous Damping

Figure C.3 shows the excitation of a spring/mass system with viscous damping. From Newton's second law,

$$F_1 e^{i\omega t} = kx_1 + c\frac{dx_1}{dt} + m\frac{d^2x_1}{dt^2}$$

Substituting $x_1 = X_1 e^{i\omega t}$ gives

$$F_1 e^{i\omega t} = kX_1 e^{i\omega t} + i\omega c X_1 e^{i\omega t} - m\omega^2 X_1 e^{i\omega t}$$

$$\therefore \alpha_{11} = \frac{X_1}{F_1} = \frac{1}{k - m\omega^2 + i\omega c} \qquad \text{(C.4)}$$

The main point to note is that the receptances vary with frequency and also are complex numbers, which indicates a phase angle difference between X_1 and F_1.

Thus in general, $\alpha_{11} = a + ib = \sqrt{(a^2 + b^2)}e^{i\theta}$ where $\tan\theta = b/a$.

The main use of receptances is in predicting the response of complex systems by adding together simpler subsystems.

C.2 Addition of Two Systems

Consider the addition of two subsystems as shown in Fig. C.4. Let the original system be **B** in Fig. C.4 and let the additional system **C** be added at co-ordinate 1. Note the line joining the coordinates 1 on systems **B** and **C** is drawn for clarity here and in subsequent analyses. In practice, this line has zero length as the two

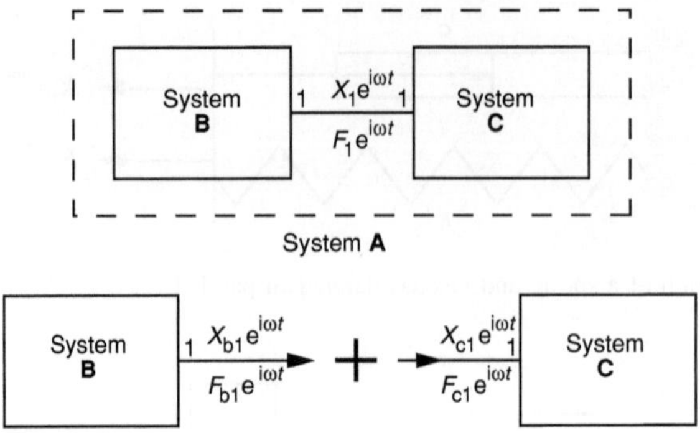

Fig. C.4 The addition of two subsystems

systems are joined at the common coordinate 1. Now, for this first example, let the excitation and response be measured at coordinate 1 so that it will be necessary to find the direct receptance α_{11} of the combined system **A** in terms of the direct receptances of the two subsystems, β_{11} and γ_{11}.

The procedure is to separate the combined system **A** into its component systems, i.e. subsystems **B** and **C** and introduce forces at the coordinate 1 on each subsystem so that the separate systems are behaving in an identical manner to that when they are joined. This separation is shown in Fig. C.4.

It is helpful to introduce an additional subscript (i.e. b or c) to indicate which subsystem is being considered.

Consider system **B** by definition

$$\beta_{11} = \frac{X_{b1}}{F_{b1}} \tag{C.5}$$

and for system **C** by definition

$$\gamma_{11} = \frac{X_{c1}}{F_{c1}} \tag{C.6}$$

For the separate systems to behave as when joined, all of the displacements are identical,

$$X_{b1} = X_{c1} = X_1 \tag{C.7}$$

and the sum of the forces on the subsystems at the join is equal to the force applied to the combined system at the join,

$$F_{b1} + F_{c1} = F_1 \tag{C.8}$$

i.e. the force F_1 is shared between the two subsystems.

It remains to determine X_1/F_1, i.e. α_{11}, from Eqs. C.5–C.8

From Eq. C.5, $F_{b1} = \frac{X_{b1}}{\beta_{11}}$ and substituting from Eq. C.7, $F_{b1} = \frac{X_1}{\beta_{11}}$

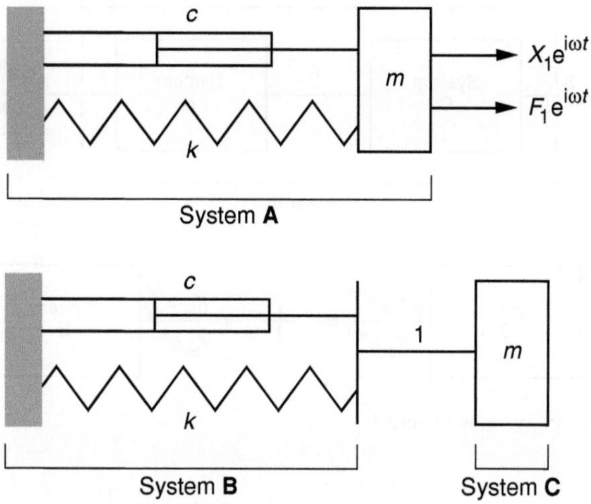

Fig. C.5 Separation of system into known components

From Eq. C.6, $F_{c1} = \frac{X_{c1}}{\gamma_{11}}$ and substituting from Eq. C.7, $F_{c1} = \frac{X_1}{\gamma_{11}}$
Substituting in the force equation, Eq. C.8, gives

$$\frac{X_1}{\beta_{11}} + \frac{X_1}{\gamma_{11}} = F_1$$

$$\therefore \frac{F_1}{X_1} = \frac{1}{\alpha_{11}} = \frac{1}{\beta_{11}} + \frac{1}{\gamma_{11}} \tag{C.9}$$

This result appears very simple, but it is important to remember that β_{11}, γ_{11} and α_{11} may be complex quantities because of phase and that for conventional systems, they will all be functions of frequency.

As a simple example of the addition of two systems, consider the simple spring/mass system with damping which was examined previously. This may be considered as the addition of a mass to a spring and dashpot as shown in Fig. C.3.

Since the mass is rigid, the force has the same effect whether it is applied at the join or on the free side of the mass. The receptances of the subsystems have been derived previously (Fig C.5).

Thus, for the mass

$$\gamma_{11} = -\frac{1}{m\omega^2}$$

and for the spring and damper

$$\beta_{11} = \frac{1}{k + i\omega c}$$

The receptance of the combined system is therefore given by

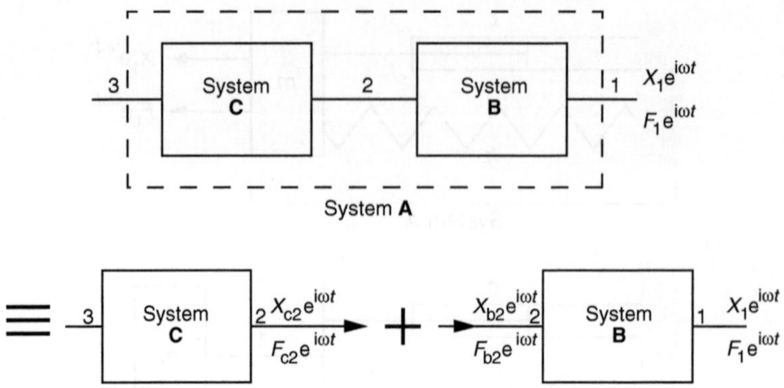

Fig. C.6 Addition of systems in series

$$\frac{1}{\alpha_{11}} = \frac{1}{\beta_{11}} + \frac{1}{\gamma_{11}} = k + i\omega c - m\omega^2$$

so that

$$\alpha_{11} = \frac{1}{k - m\omega^2 + i\omega c}$$

which is the response of a spring/mass system with viscous damping that was found previously.

C.3 Addition of Two Systems in Series

It is possible to derive further receptance addition formulae which allow more complex systems to be built. Consider a system **C** to which is added a system **B** at coordinate 2 as shown in Fig. C.6. To find the receptance of the combined system **A** at coordinate 1, the combined system is divided into its component subsystems and forces are introduced at the join so that the vibration remains unchanged compared to its assembled state.

Thus, for system **C** by definition

$$X_{c2} = \gamma_{22}F_{c2} \tag{C.10}$$

For system **B** since two forces are applied, the displacement at any co-ordinate will be the sum of the displacements caused by each force; that is, superposition is assumed to be possible. Thus,

$$X_{b2} = \beta_{22}F_{b2} + \beta_{21}F_1 \tag{C.11}$$

and

$$X_1 = \beta_{11}F_1 + \beta_{12}F_{b2} \tag{C.12}$$

Now for the systems to be identical when separated as when joined

$$X_{b2} = X_{c2} \tag{C.13}$$

and since there is no external force applied at 2,

$$F_{b2} + F_{c2} = 0 \tag{C.14}$$

Substitute $F_{c2} = -F_{b2}$ from Eq. C.14, for X_{b2} from Eq. C.11 and X_{c2} from Eq. C.10 in equation Eq. C.13. Thus,

$$\beta_{22}F_{b2} + \beta_{21}F_1 = -\gamma_{22}F_{b2}$$

$$\therefore F_{b2} = \frac{-\beta_{21}F_1}{\beta_{22} + \gamma_{22}} \tag{C.15}$$

and substituting for F_{b2} in Eq. C.12 gives $X_1 = \beta_{11}F_1 - \frac{\beta_{12}\beta_{21}F_1}{\beta_{22}+\gamma_{22}}$

For linear conservative systems, Maxwell's reciprocal theorem holds and $\beta_{12} = \beta_{21}$

$$\therefore \frac{X_1}{F_1} = \alpha_{11} = \beta_{11} - \frac{\beta_{12}^2}{\beta_{22} + \gamma_{22}} \tag{C.16}$$

It is also possible to find other receptances of the combined system. Thus, from Eq. C.11,

$$\begin{aligned} X_2 = X_{b2} = \beta_{22}F_{b2} + \beta_{21}F_1 &= -\frac{\beta_{21}\beta_{22}F_1}{\beta_{22} + \gamma_{22}} + \beta_{21}F_1 \\ \therefore \frac{X_2}{F_1} = \alpha_{21} = \beta_{21} - \frac{\beta_{21}\beta_{22}}{\beta_{22} + \gamma_{22}} \end{aligned} \tag{C.17}$$

Also for system C,

$$X_3 = X_{c3} = \gamma_{32}F_{c2} = -\gamma_{32}F_{b2} = \frac{\beta_{21}\gamma_{32}F_1}{\beta_{22} + \gamma_{22}}$$

$$\therefore \frac{X_3}{F_1} = \alpha_{31} = \frac{\beta_{21}\gamma_{32}}{\beta_{22} + \gamma_{22}} \tag{C.18}$$

C.4 Building Complex Systems

Consider the axial system shown in Fig. C.7 that has five degrees of freedom. We could write down the 5 equations of motion and after a considerable amount of maths determine the steady-state receptance *X*/*F*. Alternatively, we can apply the series addition of two systems, described in the previous section and shown in Fig. C.6. We can in fact do this four times in the following manner. The first

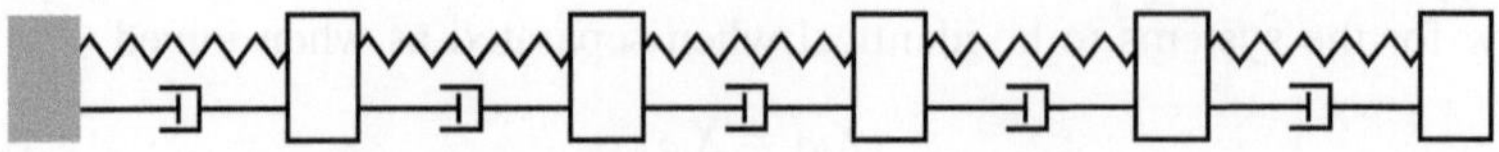

Fig. C.7 Axial system with five degrees of freedom

Fig. C.8 First subsystem

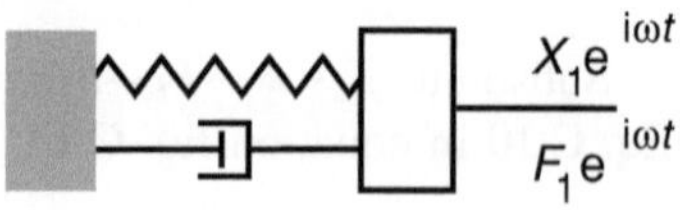

Fig. C.9 Second subsystem to be added

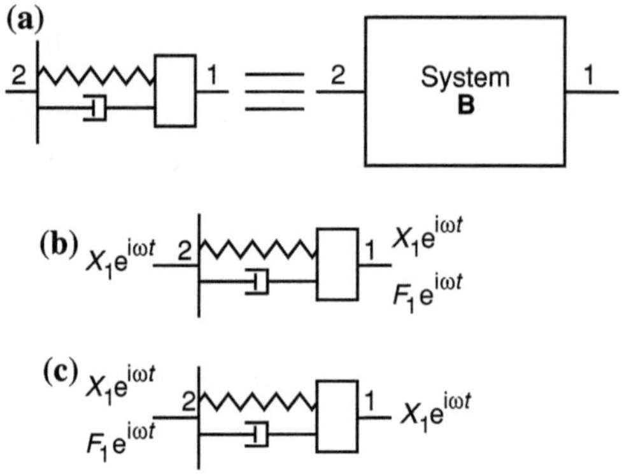

subsystem (system **C**) is as shown in Fig. C.8, and we have found the receptance for such a system to be

$$\gamma_{11} = \frac{1}{k - m\omega^2 + i\omega c} \tag{C.4}$$

Now we need to determine the receptances of the next subsystem that is shown in Fig. C.9a.

We need to find β_{11}, β_{12}, β_{22}. Consider first an oscillating force applied at coordinate 1 (Fig. C.9b). The spring and damper are considered to be massless so that this is the same as just the excitation of a mass. Thus, using Eq. B.2,

$$\beta_{11} = \frac{X_1}{F_1} = -\frac{1}{m\omega^2}$$

Also since there is no force on the massless spring/damper, there is no deflection across them and $X_2 = X_1$ so that

$$\beta_{21} = \frac{X_2}{F_1} = \frac{X_1}{F_1} = -\frac{1}{m\omega^2}$$

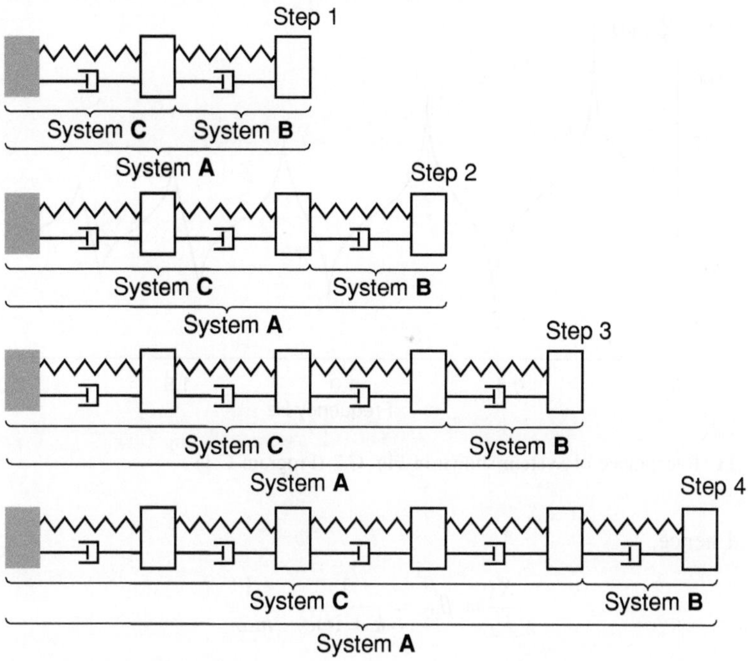

Fig. C.10 Successive series additions

Now consider an exciting force at coordinate 2 (Fig. C.9c). The spring/damper is massless so that there can be no resultant force on it. Thus, the force $F_2e^{i\omega t}$ also acts on the mass (so as to give an equal and opposite force on the spring/ damper). Thus,

$$\beta_{12} = \frac{X_1}{F_2} = -\frac{1}{m\omega^2} \quad \text{and} \quad \text{note } \beta_{12} = \beta_{21}$$

Now consider the deflection across the spring/damper

$$F_2 e^{i\omega t} = k(x_2 - x_1) + c\left(\frac{dx_2}{dt} - \frac{dx_1}{dt}\right)$$

And substituting $x_1 = X_1 e^{i\omega t}$ and $x_2 = X_2 e^{i\omega t}$ gives,

$$F_2 e^{i\omega t} = k(X_2 - X_1)e^{i\omega t} + i\omega c(X_2 - X_1)e^{i\omega t}$$

So that $\frac{X_2 - X_1}{F_2} = \frac{1}{k + i\omega c}$

$$\therefore \frac{X_2}{F_2} = \frac{1}{k + i\omega c} + \frac{X_1}{F_2}$$

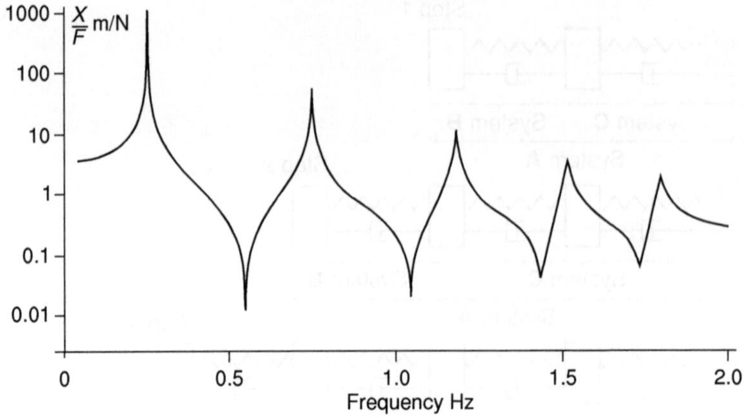

Fig. C.11 Receptance of system shown in Fig. C.7 (Program C.1)

And hence,

$$\frac{X_2}{F_2} = \beta_{22} = \frac{1}{k + \mathrm{i}\omega c} - \frac{1}{m\omega^2}$$

We are now in a position to consider the system of Fig. C.7 that may be constructed by a succession of series additions as shown in Fig. C.10. Step 1 involves a series addition of subsystem **C** and subsystem **B** to form a new system **A**. This system now becomes the new subsystem **C** for the next series addition in step 2. This is repeated until the complete system is constructed. Finally, we achieve the complete system and have the receptance at the right-hand end. The response of such a 5-degrees-of-freedom system is shown in Fig. C.11.

It is possible to determine the response of any mass resulting from excitation applied to any mass. This is achieved in the same way as finding the "mode" or deflected shapes.

C.5 Prediction of the System "Mode" Shapes

The mode shape should really be called the deflected shape when exciting at a particular frequency. Every point on an undamped vibrating structure will be either in-phase or out of phase with any other point. However, when the system is damped, the mode shapes will exhibit phase variations through the structure. In practice, it is not generally possible to excite a single mode. We may therefore be interested in the actual response and deflected shape at any frequency and that will include contributions from all of the modes.

The method we will use will be to give the last subsystem added an arbitrary unit displacement at the end. Consider a typical system made up of several

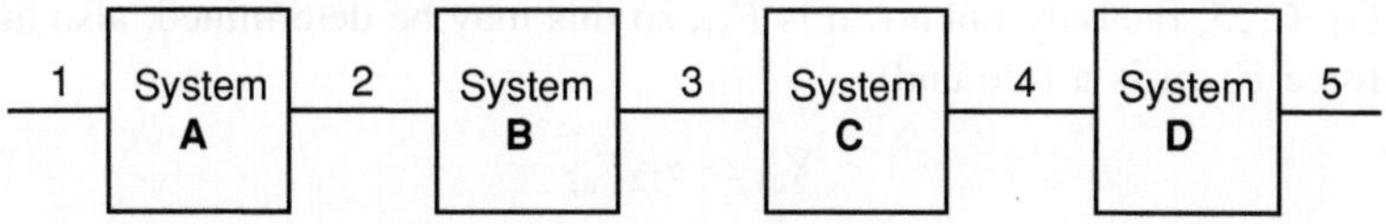

Fig. C.12 A complex system comprising four subsystems

subsystems (as shown in Fig. C.12). For the sake of this example, let us assume that there is an external force (F_2) applied at coordinate 2.

We set the amplitude at coordinate 5 on subsystem **D** to 1.0, i.e. $X_{d5} = 1.0$. Now since we know all the receptances for each subsystem, it follows that we would know the cross-receptance for the whole system that connects 2 and 5, i.e. β_{25} is known. We would thus be able to calculate F_2 necessary to give $X_{d5} = 1.0$, since

$$\beta_{25} = \frac{X_{d5}}{F_2}$$

For subsystem **D,** we note the equation,

$$X_{d5} = \delta_{45} F_{d4} \tag{C.19}$$

There is no term involving F_5 because this is a free end and so $F_5 = 0$. As we know X_{d5} and δ_{45}, we may easily determine F_{d4}. We may also calculate X_{d4} from

$$X_{d4} = \delta_{44} F_{d4} \tag{C.20}$$

Now consider subsystem **C**, $F_{c4} = -F_{d4}$ as there is no external force and from compatibility at the join $X_{c4} = X_{d4}$. Thus,

$$X_{c4} = (X_{d4}) = \gamma_{34} F_{c3} + \gamma_{44} F_{c4} \tag{C.21}$$

In Eq. C.21, the only unknown is F_{c3}, so this may be determined. Also

$$X_{c3} = \gamma_{33} F_{c3} + \gamma_{34} F_{c4} \tag{C.22}$$

and so $Xc3$ is now known.

Now consider subsystem **B**, $F_{b3} = -F_{c3}$ as there is no external force and $X_{b3} = X_{c3}$ from compatibility at the join. Thus,

$$X_{b3} = X_{c3} = \beta_{23} F_{b2} + \beta_{33} F_{b3} \tag{C.23}$$

In Eq. C.23, the only unknown is F_{b2}, so this may be determined. Also

$$X_{b2} = \beta_{23} F_{b2} + \beta_{33} F_{b3} \tag{C.24}$$

so that X_{b2} may be calculated.

Now consider subsystem **A**, $F_{a2} = F_2 - F_{b2}$ as there is an external force F_2 (that has been found above) and from compatibility at the join $X_{a2} = X_{b2}$. Thus,

$$X_{a2} = X_{b2} = \alpha_{12} F_{a1} + \alpha_{22} F_{a2} \tag{C.25}$$

In Eq. C.25, the only unknown is F_{a1}, so this may be determined, also as there is no force F_1 (it is a free end).

$$X_{a1} = \alpha_{12} F_{a2} \tag{C.26}$$

and so X_{a1} is now known.

This method lends itself to programming on a computer as the process is largely repetitive. At the end of the process, the forces at each end of every subsystem are known. Thus, if the internal receptances are known, it is possible to calculate any intermediate displacement that is required. This was the basis of the axial multidegrees-of-freedom Program C.1.

If at the start of the process, only the direct receptance at the point of excitation is known for the whole system, it is then appropriate to make the displacement unity at the excitation point and work out (one direction at a time) from that point. The procedure is identical to that described above.

C.6 Vibration Absorbers: Damping a Clamped/Free Bar

The basic concept behind vibration absorbers is well known and is described in Appendix B. A clamped/free bar is an approximate representation of many real engineering structures that are susceptible to transverse vibration. For example, a boring bar may be approximated by a clamped/free bar and often have little damping and hence a large resonant response. It is shown in Appendix B that a vibration absorber is a useful way of reducing resonant amplitudes.

Consider a uniform circular clamped/free bar as shown in Fig. C.13. The vibration absorber consists of a spring, mass and hysteretic damper.

For this example, the receptance at the end with the absorber added is given by,

$$\frac{1}{\alpha_{11}} = \frac{1}{\beta_{11}} + \frac{1}{\gamma_{11}}$$

and ignoring shear and rotary inertia effects, β_{11} the tip receptance of the bar, is given by [1],

$$\beta_{11} = -\frac{(\cos \lambda L \sinh \lambda L - \sin \lambda L \cosh \lambda \mathrm{L})}{EI\lambda^3(\cos \lambda L \cosh \lambda L + 1)} \quad \text{where } \lambda^4 = \frac{\rho A \omega^2}{EI}$$

And the absorber receptance is given by,

$$\gamma_{11} = \frac{1}{k(1 + \mathrm{i}\alpha)} - \frac{1}{m\omega^2}$$

A computer program is required to optimise the absorber for any given bar. In practice, the spring and damping are supplied by a rubber element and alpha is

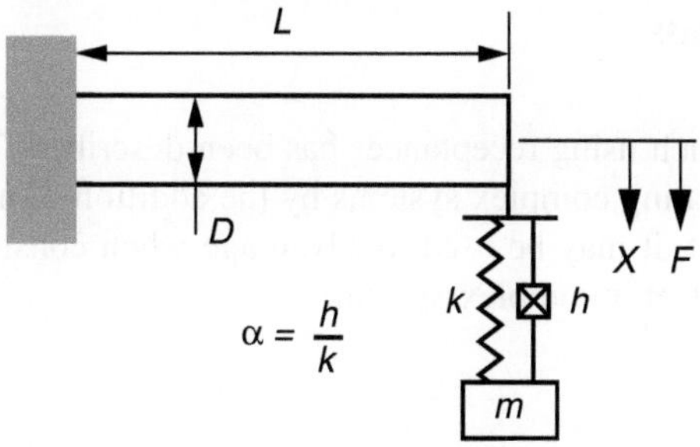

Fig. C.13 Clamped/free bar with a vibration absorber added

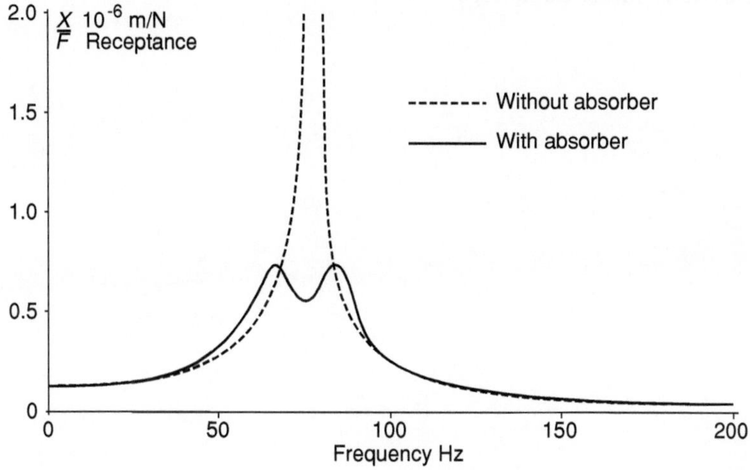

Fig. C.14 Bar response with and without absorber (Program C.2)

nearly constant over the frequency range of interest. Thus for a given absorber mass (the larger the better), there is only the parameter k to optimise. For a large boring bar of length $L = 1.2$ m, diameter $D = 0.15$ m using rubber with $\alpha = 0.25$ and an absorber mass of 2 kg, the optimum response (i.e. the peak response is minimised) is shown in Fig. C.14.

It should be noted that the bar without absorber has an infinite response at resonance as no damping has been included when determining the transverse receptances of bars. In practice, such bars have little damping so that the improvement that may be achieved by using an absorber is significant.

C.7 Conclusions

The systems approach using receptances has been described. This method has the advantage of generating complex systems by the addition of much simpler known systems. As a result, it may be used to advantage when considering the effects of changes to one part of a complex system.

Reference

[1] Bishop RED, Johnson DC (1960) The mechanics of vibration. Cambridge University Press, Cambridge

About the Author

Professor Stone completed his Ph.D at the University of Bristol in 1968 on the topic of 'Vibration absorbers for machine tools'. His Ph.D supervisor was Prof. Colin Andrew who in turn was the Ph.D student of Prof. S. A. Tobias. After completing his Ph.D, he worked for the British Machine Tool Industry Research Association (MTIRA) and spent over 3 years investigating vibration problems on the machine tools of the member companies of MTIRA. On his return, as a lecturer, to the University of Bristol in 1972, he continued his research on machine tool vibration. In particular, his research focus was on preventing or reducing vibration in machining processes. His research has led to two patents and over 150 publications. In 1981, he took up the chair of Mechanical Engineering at the University of Western Australia and this location resulted in an increase in contacts with Asia, most notably Singapore and Japan. In 1991, he was the first Australian to be invited as a visiting expert for the Japan Machine Tool Builders Association. In 2003, he was invited to give the first memorial lecture for the late Prof. S. Doi at the head offices of Mori Seiki. As well as his vibration research, he has many publications on the teaching of dynamics and was awarded the inaugural Australian National Teaching Award for Engineering in 1997. He uses animation programs to help students 'see' vibration and this text references/links to many such programs.

B. Stone, *Chatter and Machine Tools*, DOI: 10.1007/978-3-319-05236-6,

List of Computer Programs Included in extras.springer.com

Programs By Chapter

1. **Basic chatter theory**

   Program 1.1: Example of chatter on a lathe—noise as measured.
   Program 1.2: Simulation of chatter in turning.
   Program 1.3: Non-regenerative force.
   Program 1.4: Regenerative force.
   Program 1.5: Resultant force.
   Program 1.6: In phase response.
   Program 1.7: Force vectors and response.
   Program 1.8: Stability chart.
   Program 1.9: Animation with stability chart.
   Program 1.10: Stability chart with process damping.

2. **Extension of chatter theory**

   Program 2.1: Effect of having two modes.
   Program 2.2: Stability chart when two modes.
   Program 2.3: Stability chart when two modes and process damping.
   Program 2.4: Effect of the direction of the modes on the chatter receptance.
   Program 2.5: Effect of the direction of the modes on the real part of the chatter receptance.
   Program 2.6: Variation of $b_{\lim}$ with circumferential position of the tool.
   Program 2.7: Effect of a single mode direction on different milling configurations.
   Program 2.8: Effect of two mode directions on different milling configurations.
   Program 2.9: Stability chart for milling when one significant mode.
   Program 2.10: Boundary of stability for mode coupling chatter.
   Program 2.11: Comparison of mode coupling and regenerative chatter.
   Program 2.12: Animation of mode coupling chatter.

B. Stone, *Chatter and Machine Tools*, DOI: 10.1007/978-3-319-05236-6,

3. **High performance cutters**

   Program 3.1: Stability chart for an alternating pitch cutter.
   Program 3.2: Stability chart for a Strasman cutter.
   Program 3.3: Stability chart for trapezidal cutter, $e = 25\ \%$.
   Program 3.4: Stability chart for trapezidal cutter, $e = 65\ \%$.
   Program 3.5: Stability chart for broaching.
   Program 3.6: Stability chart for milling.

4. **Structural modifications**

   Program 4.1: Response of spindle.
   Program 4.2: In-phase response of spindle.
   Program 4.3: Animation of spindle vibration.
   Program 4.4: Optimised in-phase response with an absorber.
   Program 4.5: In-phase response when absorber optimised for peak response.
   Program 4.6: Optimised peak response of bar with absorber.
   Program 4.7: Optimised in-phase response of bar with absorber.
   Program 4.8: Optimised in-phase response of double bar.
   Program 4.9: In-phase response of a spindle with added inertia.
   Program 4.10: In-phase response with a flexible tool.

5. **Grinding**

   Program 5.1: The simplest case—regeneration on the work.
   Program 5.2: The simplest case—regeneration on the wheel.
   Program 5.3: The effect of contact stiffness.
   Program 5.4: Flexible wheel.
   Program 5.5: Inclined mode.
   Program 5.6: Torsion of the work and its drive.
   Program 5.7: Double regeneration—'work chatter' against work speed.
   Program 5.8: Double regeneration—'wheel chatter' against work speed.
   Program 5.9: Double regeneration—'wheel chatter' against wheel speed.
   Program 5.10: Double regeneration—'work chatter' against wheel speed.

Appendix A: Basic vibration theory

Program A.1: Undamped vibration of a spring/mass system without damping.
Program A.2: Damped vibration of a spring/mass system with viscous damping.
Program A.3: Critically damped vibration of a spring/mass system with viscous damping.
Program A.4: Over-damped vibration of a spring/mass system with viscous damping.
Program A.5: Sinusoidal forced vibration of a one degree of freedom system.
Program A.6: Definition of phase.
Program A.7: Forced response of a one degree of freedom system.
Program A.8: The connection between rotating vectors, sinusoidal motion and phase.

Appendix B: Two degree of freedom vibration

Appendix C: Systems approach to vibration

Index

B. Stone, *Chatter and Machine Tools*, DOI: 10.1007/978-3-319-05236-6,

MIX
Papier aus verantwortungsvollen Quellen
Paper from responsible sources
FSC® C105338

If you have any concerns about our products, you can contact us on
ProductSafety@springernature.com

In case Publisher is established outside the EU, the EU authorized representative is:
Springer Nature Customer Service Center GmbH
Europaplatz 3, 69115 Heidelberg, Germany

Printed by Libri Plureos GmbH
in Hamburg, Germany